Applied Superconductivity

Volume I

CONTRIBUTORS

C. L. Bertin
L. S. Cosentino
William S. Goree
Victor W. Hesterman
Walter H. Hogan
R. M. Katz
K. Rose
A. R. Sass
Arnold H. Silver
W. C. Stewart
James E. Zimmerman

APPLIED SUPERCONDUCTIVITY

Edited by

VERNON L. NEWHOUSE

School of Electrical Engineering
Purdue University
Lafayette, Indiana

Volume I

ACADEMIC PRESS New York San Francisco London 1975

A Subsidiary of Harcourt Brace Jovanovich, Publishers

ACADEMIC PRESS, INC.
111 Fifth Avenue, New York, New York 10003

United Kingdom Edition published by
ACADEMIC PRESS, INC. (LONDON) LTD.
24/28 Oval Road, London NW1

Library of Congress Cataloging in Publication Data

Newhouse, Vernon L
Applied superconductivity.

Includes bibliographies.
1. Superconductors. I. Title.
TK7872.S8N42 621.39 74-1633
ISBN 0-12-517701-1

PRINTED IN THE UNITED STATES OF AMERICA

Contents

4. Radiation Detectors

K. Rose, C. L. Bertin, and R. M. Katz

5. Refrigerators and Cryostats for Superconducting Devices

Walter H. Hogan

List of Contributors

Numbers in parentheses indicate the pages on which the authors' contributions begin.

C. L. Bertin (267), IBM System Development Division, Manassas, Virginia

L. S. Cosentino (209), RCA Laboratories, Princeton, New Jersey

William S. Goree (113), Superconducting Technology, Inc., Mountain View, California

Victor W. Hesterman (113), Superconducting Technology, Inc., Mountain View, California

Walter H. Hogan (309), Cryogenic Technology, Inc., Waltham, Massachusetts

R. M. Katz (267), MITRE Corporation, Westgate Research Park, McLean, Virginia

K. Rose (267), Electrophysics and Electronic Engineering Division, Rensselaer Polytechnic Institute, Troy, New York

A. R. Sass (209),* Quasar Electronics Corporation, Franklin Park, Illinois

Arnold H. Silver (1), Electronics Research Laboratory, The Aerospace Corporation, El Segundo, California

W. C. Stewart (209), RCA Laboratories, Princeton, New Jersey

James E. Zimmerman (1), Institute for Basic Standards, National Bureau of Standards, Boulder, Colorado

* Formerly of IBM System Products Division, East Fishkill, New York.

Preface

Attempts to exploit the fascinating properties of superconductors started shortly after the original discovery of the effect by Onnes when he tried to produce dissipation-free high-field electromagnets. As is well known, these attempts failed owing to the relatively low critical field of the superconductors known at that time. Although later efforts to use superconductors were more successful, e.g., the use of magnetic-field-controlled switches by Casimir–Jonkers and de Haas in the early 1930s, applied superconductivity can only be said to have come of age in the 1960s when high-field superconducting magnets began to be used widely, even for experiments at room temperature. This very important application provides a good example of the impact of basic research on applied science, since the high critical field of Nb_3Sn, which made these magnets possible, was discovered in the course of fundamental researches aimed at elucidating the origins of superconductivity itself. An example of how applied research leads to improved fundamental understanding is exemplified by the fact that research on high-field superconductors led to the rediscovery of Abrikosov's work on class-II superconductors, which might otherwise have continued to be ignored for many years.

Since superconductors exhibit zero resistance at low frequencies, they are already important in the production of large magnetic fields and show promise in the production and transport of large quantities of electric power. Since they operate close to zero temperature where Johnson noise becomes small, they either already are or promise soon to become the most sensitive detectors of magnetic fields and of radiation at all frequencies. Furthermore since superconductive circuits can easily store single-flux quanta, they promise to become the most compact and, therefore, the fastest means of handling and storing information.

The current interest in pure superconductivity is proved by the award, in the last few years, of separate Nobel prizes for the theory of superconductivity as well as for superconductive and semiconductor tunneling. The speed of progress in the field of applied superconductivity is exemplified by the fact that high-field superconductors were unknown in 1962 when John Bremer published "Superconductive Devices," the initial work in this field, and that the Josephson effect, which is the basis of most of the radiation detection and magnetic-field measurement devices mentioned above, had not yet been discovered at the time of publication of this editor's book, "Applied Superconductivity," in 1964!

Since the subject of applied superconductivity has now grown to an extent where it can no longer be covered exhaustively by a single author, this treatise is divided into chapters on the various areas, each written by one or more authorities on the subject in question. The work is divided into two volumes, the first of which deals with electronic applications and radiation detection and contains a chapter on liquid helium refrigeration.

The second volume discusses magnets, electromechanical applications, accelerators, microwave and R. F. devices, and ends with a chapter on future prospects in applied superconductivity.

A corollary to being an authority on a subject is the many demands made upon one's time. Thus a deadline must often be secondary to other responsibilities. For reasons of this sort, the original versions of some chapters in this treatise were completed before others, and we were never able to obtain a chapter on high-power rotating machinery.

Each chapter in these two volumes can be read independently, and most assume very little or no background in the physics of superconductivity. The topics treated do not require the use of advanced quantum mechanics; thus the books should be accessible to students or research workers in any branch of engineering or physics. They are intended to serve both as a source of reference material to existing techniques and as a guide to future research. For those wishing to extend their background in the physics of superconductivity, some recent books on the subject, selected from a larger list kindly compiled by Arthur J. Bond, are given in the following Bibliography.

Bibliography

The books listed on the following page are those written recently from the experimentalists' point of view and can be recommended for those wishing to expand their background in the physics of superconductivity.

De Gennes, P. G., (1966). "Superconductivity of Metals and Alloys." Benjamin, New York.

Kulik, I. O. and Yanson, I. K. (1972). "Josephson Effect on Superconducting Tunneling Structures." Halsted, New York.

Kuper, C. G. (1968). "An Introduction to the Theory of Superconductivity." Oxford Univ. Press, London and New York.

Lynton, E. A. (1972). "Superconductivity." Halsted, New York.

Parks, R. D. (1972). "Superconductivity," Vols. 1, and 2. Dekker, New York.

Saint-James, D. (1969). "Type II Superconductivity." Pergamon, Oxford.

Savitsky, *et al.* (1973). "Superconducting Materials." Plenum, New York.

Solymar, L. (1972). "Superconductivity Tunneling and Applications." Wiley, New York.

Williams, J. E. C. (1970). "Superconductivity and Its Applications." Pion, New York.

Contents of Volume II

Chapter 1

Josephson Weak-Link Devices

ARNOLD H. SILVER

Electronics Research Laboratory
The Aerospace Corporation
El Segundo, California

JAMES E. ZIMMERMAN

Institute for Basic Standards
National Bureau of Standards
Boulder, Colorado

I. Introduction

Superconductivity affords us a unique opportunity to observe and utilize the fundamental quantum-mechanical nature of matter, for in superconductors this quantization is on a truly macroscopic scale and is synonymous with the electronic coherence found otherwise only in atomic systems. Josephson weak-link devices are superconducting elements which exhibit an intrinsic quantum electronic behavior. Devices based on such quantum effects exhibit a periodic parametric behavior with respect to very low applied fields, and therefore are highly nonlinear and sensitive at low power levels.

Probably the most significant discoveries precipitating the development of these devices were those of flux quantization and the Josephson effect. Following London's ideas (1950) of more than one decade earlier, Deaver and Fairbank (1961) and Doll and Näbauer (1961) experimentally discovered that the magnetic flux threading a superconducting circuit is quantized in integral multiples of $h/2e$. This unit is equal to 2.07×10^{-15} W (2.07×10^{-7} G cm^2). Josephson (1962) predicted that there should be a quantum-phase-dependent zero-voltage tunneling current between two superconductors separated by a very thin insulating barrier and that in the presence of a nonzero voltage an alternating tunneling current will flow. Attempts to verify Josephson's predictions coupled with renewed interest in the quantum properties of superconductors have spurred the development of these devices.

In this chapter we review a number of devices and device possibilities based on the coherent precession of the quantum phase across weakly superconducting connections such as thin-film tunneling junctions, point contacts, and thin-film bridges. Particular emphasis is placed on low-inductance radio-frequency-biased loop devices since these are particularly amenable to detailed analysis and have certain attractive attributes as practical devices. A weak link shunted by a sufficiently low-inductance loop becomes voltage biased to the extent that the theoretical calculations of the device response are relatively straightforward and in direct quantitative agreement with experiment. Furthermore, in contrast with high-impedance current-biased weak links, these devices are uncluttered by nonsuperconducting or "non-Josephson" effects, which themselves are poorly defined and exceedingly complex. Several reviews in recent years have discussed the Josephson effect (Anderson, 1964). We take as our point of primary emphasis multiply-connected circuits incorporating one, or more, weak links and direct our attention toward device characterization and experimental situations.

Section II discusses the quantum electronics of superconductors, emphasizing the close relation between the Josephson effects and fluxoid quantization. A simple mechanical analog of weak-link phenomena is described and invoked to understand the nature of certain quantum transitions.

In Section III we characterize various device topologies and discuss some of the experimental results. Some of the device construction techniques are reviewed in Section IV, and an array of applications, attempted and proposed, are discussed in Section V.

II. Quantum Electronics of Weak-Link Devices

A. Theoretical Background

The theoretical model which is useful in understanding and postulating weak-link devices can be presented in phenomenological and analog forms. We take as our point of departure an elementary description of the superconducting state in terms of a single complex order parameter Ψ as in Ginzburg–Landau theory. This order parameter can be considered an effective wave function and satisfies an equation similar to the Schrödinger wave equation. The complex function

$$\Psi = |\Psi| e^{i\phi} \tag{1}$$

can be generally dependent on both space and time in both its amplitude and phase. Our major concern will be with phase-dependent effects. Thus we restrict the variations of Ψ to that introduced by ϕ. In standard form then

$$\hbar\phi = \int \mathbf{p} \cdot dl - 2e \int \mu \, dt \tag{2}$$

where $\mathbf{p}$ and μ are the canonical momentum and chemical potential of the electron pairs of the superconductor, respectively.

The current density associated with the order parameter is given from the standard form by

$$\mathbf{j} = (eh/m)|\Psi|^2(\nabla\phi - (2e/\hbar)\mathbf{A}) \tag{3}$$

where $\mathbf{A}$ is the magnetic vector potential and we have introduced the double mass and charge of the electron pairs. For a London superconductor $\nabla\phi = 0$ and we have

$$\mathbf{j} = -(2e^2/m)|\Psi|^2\mathbf{A} \tag{4}$$

which is London's equation for diamagnetism if we associate $2|\Psi|^2$ with n, the density of superelectrons in the two-fluid theory. Thus we note in passing that London theory (1950) corresponds to the real-order-parameter limit of Ginzburg–Landau theory. We interpret $(\hbar/2m)[\nabla\phi - (2e/\hbar)\mathbf{A}]$ as the velocity of the electron pairs. Differentiation of Eq. (3) with time gives

$$\frac{d\mathbf{j}}{dt} = -\frac{2e^2}{m}|\Psi|^2\left[\nabla V + \frac{\partial\mathbf{A}}{\partial t}\right] \tag{5}$$

where we have taken μ equal to the electric potential V. Equation (5) is London's second equation relating the electric field to the time derivative of the current,

$$\frac{d\mathbf{j}}{dt} = \frac{ne^2}{m}\mathbf{E} \tag{6}$$

Thus, starting from a complex order parameter and a quantum-mechanical formalism one can derive London's classical electrodynamic equations.

The conservation, and more specifically the quantization, of magnetic flux in superconductors is readily evident from this approach. The uniqueness of Ψ imposes a condition on ϕ such that for all $(\mathbf{x},t)$

$$\oint \mathbf{p}\cdot dl - 2e\oint \mu\, dt = kh \tag{7}$$

where k is any integer. The stationary behavior then reduces to

$$\oint 2m\mathbf{v}\cdot dl + \oint 2e\mathbf{A}\cdot dl = kh \tag{8}$$

In conventional notation the electric current density j and magnetic flux Φ are

$$\mathbf{j} = ne\mathbf{v} \tag{9}$$

and

$$\Phi = \oint \mathbf{A} \cdot dl \tag{10}$$

which gives

$$\oint (m/ne^2)\mathbf{j} \cdot dl + \Phi = k\Phi_0 \tag{11}$$

where

$$\Phi_0 = h/2e = 2.07 \times 10^{-15}\ \text{Wb} = (2.07 \times 10^{-7}\ \text{G cm}^2) \tag{12}$$

is called the flux quantum. Equation (11) reduces to London's fluxoid conservation relation in the special case $k = 0$. Specifically this predicts quantization of the magnetic flux whenever one has a sufficiently thick superconductor that $j = 0$ along an entire circuit. This condition is satisfied for thicknesses much larger than a penetration depth λ.

B. Linear Weak-Link Approximation

Devices of interest in this chapter involve systems where $\oint \mathbf{j} \cdot dl \neq 0$. In particular, we are interested in systems where j approaches its maximum supercurrent value j_c and where $\oint (m/ne^2)\mathbf{j} \cdot dl$ becomes $\gtrless \Phi_0/4$. Several obvious ways to generate such systems are:

1. utilize surface effects on bulk superconductors;
2. use bulk Type-II superconductors in the vortex flow state;
3. constrict the region of supercurrent flow small compared to a penetration length;
4. produce a region of very low superelectron density and hence make the penetration length larger than the region of current flow.

Those systems which fall in the first two classes are not of interest here. Classes three and four are grouped together under the name Josephson weak-link devices. Generally speaking, class 3 is a superconducting metallic bridge of small lateral cross section while class 4 is a tunneling section between two superconductors, called a Josephson tunneling junction. Many actual devices are really hybrids of these two models.

If we rewrite Eq. (11) as

$$(2\pi/\Phi_0) \oint (m/ne^2)\mathbf{j} \cdot dl + (2\pi\Phi/\Phi_0) = 2\pi k \tag{13}$$

we note that the first term represents the phase angle associated with the current and the second term represents the angle generated by the magnetic flux. Further, restricting the path to $\mathbf{j} = 0$ everywhere except at the weak link we may say

$$\theta + 2\pi\Phi/\Phi_0 = 2\pi k \tag{14}$$

where

$$\theta = (2\pi/\Phi_0) \int_{\text{w.l.}} (m/ne^2)\mathbf{j} \cdot dl \tag{15}$$

As a simplifying assumption, let the weak link have an effective length l_0 and cross section σ small enough that j is uniform. If the total current is denoted by i, the change in phase across the weak link is

$$\theta = (2\pi/\Phi_0)(m/ne^2)(l_0/\sigma)i, \qquad i \leq i_c \tag{16}$$

When this weak link is incorporated in a bulk superconducting ring of inductance L, we have

$$2\pi\gamma Li/\Phi_0 + 2\pi\Phi/\Phi_0 = 2\pi k \tag{17}$$

where

$$\gamma = (m/ne^2)(l_0/\sigma L) \tag{18}$$

is a characteristic parameter of the system.

Equation (16) is a linear current–phase relation, applicable only for systems of uniform j, hence for $\sigma \lesssim \lambda^2$. A reasonable calculation for small superconductors (dimensions comparable to or less than a penetration length) has shown that the maximum supercurrent i_c is $G\Delta$, where Δ is the energy gap in volts and G is the normal-state conductance of the junction (Zimmerman and Silver, 1966a). With reasonable values for the variables involved, the quantum phase across such links can be shown to be of the order of 2π. In fact, it has been experimentally demonstrated that such phase changes are generally limited to $\pi/2$ and do not exceed π (Zimmerman and Silver, 1966b).

Intriguing and novel characteristics result from the nature of the quantum states and transitions between states for these devices (Silver and Zimmerman, 1967b). This is illustrated by the superconducting ring with a weak link described by Eq. (17). In the presence of an applied magnetic field the ring will intercept a portion of the applied flux Φ_x. The total magnetic flux in the ring differs from Φ_x by Li, the flux generated by the circulating current in the ring inductance L,

$$\Phi = \Phi_x + Li \tag{19}$$

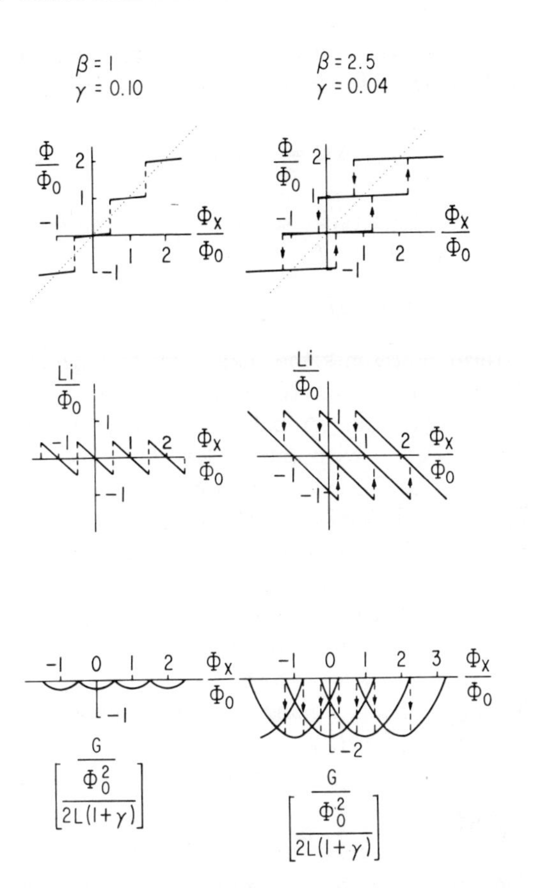

FIG. 1. Theoretical solutions of the quantum states of a weakly connected superconducting ring as a function of the applied magnetic field Φ_x for a linear current–phase-shift relation. β is defined in Eq. (22), γ by Eq. (18), and G is the Gibbs free energy given in Eq. (24).

Solutions of Eqs. (17) and (19) for Φ and Li are

$$Li = -(\Phi_x - k\Phi_0)/(1 + \gamma), \qquad i \leq i_c \tag{20}$$

$$\Phi = (k\Phi_0 + \gamma\Phi_x)/(1 + \gamma) \tag{21}$$

where experimentally γ is usually a small number. Figure 1 shows graphs of these solutions for several values of i_c. We recognize that the important parameter in the problem has been reduced to the ratio $(1 + \gamma)Li_c/\Phi_0$, essentially the ratio of the maximum screened (or trapped) flux to the flux quantum. If we define the dimensionless parameter β as

$$\beta \equiv 2(1 + \gamma)Li_c/\Phi_0, \tag{22}$$

then $\beta = 1$ is the crossover point between overlapping (multiple-valued) and single-valued solutions for $\Phi(\Phi_x)$. The magnetic Gibbs energy

$$\Delta\mathcal{H} = -\int i \, d\Phi_x \tag{23}$$

is also shown with zero arbitrarily taken for $i = i_c$. Integration of Eq. (23) along a stationary state, Eq. (20), gives

$$\mathcal{H} = -\tfrac{1}{2}L(1+\gamma)i_c^2 + [(\Phi_x - k\Phi_0)^2/2L(1+\gamma)] \tag{24}$$

where the first term represents the maximum free energy available, and the second includes both the purely magnetic energy and the kinetic term associated with the weak link. Probable transitions $\Delta k = \pm 1$ at constant Φ_x are indicated in Fig. 1 with $\Delta\Phi = \pm\Phi_0/(1+\gamma)$; the corresponding change in free energy can be calculated for $\beta > 1$ and is indicated by the dashed lines. Hysteretic behavior is also indicated for this region and one can calculate the work done per cycle of the applied field.

Although this linear approximation can be shown to properly describe many features of weak-link devices, a more complete theory is needed to describe the nature of the connection between and the transitions among states. Further, γ is taken as a constant of the motion, neglecting current- (or field-) induced depairing, among other effects. We shall clarify some of these points in the next sections.

C. Josephson Theory

A theory, due originally to Josephson (1962, 1964, 1965), of phase-dependent supercurrents in tunneling junctions has been shown to be essentially correct for all weak links (Fulton, 1970 and others). In simplest form we write for the current density in a weak link

$$j = j_c \sin\theta \tag{25}$$

where θ, identical to that given by Eq. (15), is the optical path length for the electron pairs crossing the junction. In general, θ varies with space and time over the area of a tunneling junction. This gives rise to a total supercurrent of the form

$$i = \int_\sigma j_c(\mathbf{r}) \cdot \sin\theta \, d\sigma \tag{26}$$

where $j_c(r)$ indicates that the tunneling supercurrent may not be spatially uniform. Since θ is the change in the phase of the order parameter, corrected for gauge invariance, we have

$$\theta(x,t) = \hbar^{-1} \int_{\text{w.l.}} (\mathbf{p} - 2e\mathbf{A}) \cdot dl \tag{27}$$

Thus Eq. (26) is a two-dimensional Fourier transform of $j_c(r)$ resulting in a total current $i(\Phi)$ which is a function of the magnetic flux contained in the surface σ. For a wide junction the result is a Fraunhofer diffraction pattern with period Φ_0; for two junctions in parallel connected by superconducting elements the diffraction pattern is superimposed on a two-slit interference pattern (Jaklevic *et al.*, 1965a; Rowell, 1963).

Our interest here is principally with small junctions or weak links such that θ is essentially constant over the area σ. Hence the appropriate expression for the junction becomes

$$i = i_c \sin \theta \tag{28}$$

and γ for a Josephson junction in a superconducting ring is

$$\gamma = (\Phi_0/2\pi Li) \arcsin(i/i_c) \tag{29}$$

We recognize that a Josephson junction is a highly nonlinear element; e.g., γ is a function of the current i. The equation of state for such a weakly connected ring is now given as

$$\arcsin(i/i_c) + 2\pi\Phi/\Phi_0 = 2\pi k \tag{30}$$

which combined with Eq. (19) in the presence of an applied magnetic flux Φ_x gives

$$\Phi + Li_c \sin[(2\pi/\Phi_0)(\Phi - k\Phi_0)] = \Phi_x \tag{31}$$

and

$$Li = -Li_c \sin[(2\pi/\Phi_0)(Li + \Phi_x - k\Phi_0)] \tag{32}$$

The Gibbs free energy calculated from Eq. (23) for any state k and adjusted to zero for $i = i_c$ gives

$$\begin{aligned}\mathcal{H} = &-(i_c\Phi_0/2\pi)\cos[(2\pi/\Phi_0)(\Phi - k\Phi_0)] \\ &+ \tfrac{1}{2}(Li_c^2)\sin^2[(2\pi/\Phi_0)(\Phi - k\Phi_0)] \\ &- (\Phi_0^2/4\pi L) - \tfrac{1}{2}Li_c^2, \qquad i_c > \Phi_0/2\pi L\end{aligned} \tag{33a}$$

$$\begin{aligned}\mathcal{H} = &-(i_c\Phi_0/2\pi)\cos[(2\pi/\Phi_0)(\Phi - k\Phi_0)] \\ &+ \tfrac{1}{2}(Li_c^2)\sin^2[(2\pi/\Phi_0)(\Phi - k\Phi_0)] \\ &- (i_c\Phi_0/2\pi), \qquad i_c \leq \Phi_0/2\pi L\end{aligned} \tag{33b}$$

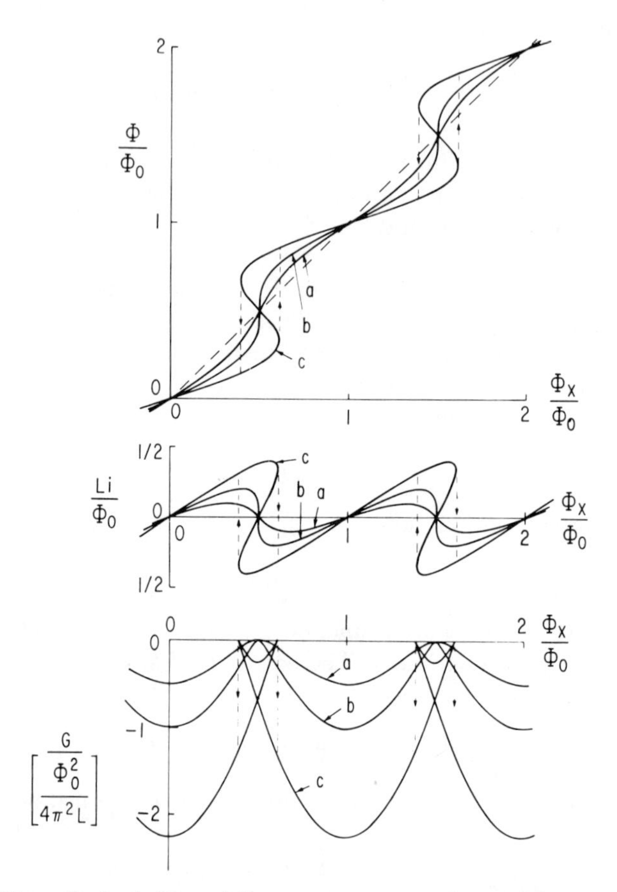

FIG. 2. Theoretical solutions of the quantum states of a weakly connected superconducting ring as a function of the applied magnetic field Φ_x for a sinusoidal current–phase-shift relation $i = i_c \sin \theta$. The curves labeled a, b, c refer to β values equal to $\frac{1}{2}$, 1, 2 respectively, with $\beta = 2\pi L i_c/\Phi_0$. [After Silver and Zimmerman, 1967b.]

The first term represents the phase-dependent binding energy of the junction and the second term the magnetic energy stored in the ring. The ratio $2\pi L i_c/\Phi_0$ for Josephson junctions corresponds to β in the linear approximation, Eq. (22). In general, we will define the symbol β such that $\beta = 1$ is the boundary between single- and multiple-valued solutions for i and Φ as functions of Φ_x. Thus for Josephson tunnel junctions $\beta = 2\pi L i_c/\Phi_0$. Figure 2 presents graphical solutions to these equations as functions of the applied flux for three values of β including both multivalued and single-valued solutions. For $\beta \leq 1$ one can expect a continuous evolution of the system with Φ_x; when β exceeds unity no such smooth evolution is evident, hysteresis is apparent and the model as presented to this point provides no real insight into this process.

One important point can be made at this time. It is readily evident that a weakly-connected superconducting ring has very unusual magnetic properties, attributable to the nature of the weak link. One way of describing this is in terms of a parametric inductance. The effective inductance can be defined in terms of the stationary states by

$$L_{\text{eff}} = -\frac{d\Phi_x}{di} \tag{34}$$

and can be calculated from Eq. (32). This calculation gives

$$L_{\text{eff}} = L + (\Phi_0/2\pi i_c) \sec[(2\pi/\Phi_0)(\Phi - k\Phi_0)] \tag{35}$$

where we identify the inductance of the Josephson junction as

$$L_J = (\Phi_0/2\pi i_c) \sec[(2\pi/\Phi_0)(\Phi - k\Phi_0)] \tag{36}$$

This result was first derived by Josephson (1964) from consideration of voltages across junctions.

D. Generalized Weak Links

Unless otherwise noted, we will in this article assume that in weak-link quantum devices the tunnel junction or metal bridge is narrower than the size of an Abrikosov vortex. This is more than a convenient conceptual simplification. There is good evidence, not generally recognized, that in practical weak-link devices the lateral dimensions of the link are smaller than the superconducting penetration depth, so that the link is too narrow to contain a vortex. Discussions in the literature frequently invoke pictures of a thin-film link in a "vortex-flow" state in which a discrete vortex forms at one edge of the link, moves across and vanishes at the opposite edge; or a pair of oppositely directed vortices form at opposite edges of the link, move in and annihilate each other at the center. For a point contact of circular cross section, a picture of a toroidal vortex nucleating around the contact perimeter and collapsing on itself in the center has been suggested. These pictures are academically attractive because they relate directly to the widely recognized vortex state of Type-II superconductors, but they are misleading and irrelevant when applied to most of the weakly superconducting devices which are the subject of this article. A simple calculation makes this clear. A typical superconducting loop device might have an inductance $L \sim 10^{-9}$ H, so that $\Phi_0/L \sim 10^{-6}$ A. This is the value of circulating current required to sustain one flux quantum within the loop, and the experimental evidence indicates that these devices work best as coherent quantum systems only if the critical current i_c of the weak link is not too much greater than (10 or 100 times) Φ_0/L. On the other hand, the total circulating current required to sustain a flux quantum in a vortex is $\sim 10^{-3}$ A (for a vortex 1 μ in diameter in a thin film 0.1 μ thick). A weak link large enough to sustain such a value of internal circulation should almost certainly exhibit a critical current value i_c of comparable magnitude, and in fact, this has been found to be the case (Hunt, 1966).

It appears, therefore, that for weak links with values of i_c in the range 10^{-4}–10^{-6} A required for low-inductance loop devices, a picture of vortices flowing across the weak link is not appropriate. For such small-area weak links the current density is essentially uniform across the link, and so the total current is the logical variable to use in describing the operation of the device. A more realistic description is that the complex order parameter goes through zero across the entire weak link whenever a flux quantum enters or leaves the loop. It should be emphasized that this picture is not fundamentally different from the picture of a vortex crossing a wide link. The core of a vortex has a zero order parameter in any case, so that a narrow link during flux flow may be regarded as momentarily having the core or a portion of the core of a vortex centered upon it.

Superconducting small-area (see above) weak links are characterized generally by two parallel conducting mechanisms, one a supercurrent i_s which varies periodically with the gauge-invariant phase shift θ across the weak link, and the other a normal current $i_n = GV$ characterized by a conductance G which may be a function of the voltage V across the link. The prototype weak link is the tunnel junction (Josephson junction) for which the supercurrent is given by Eq. (28). This is essentially one limiting case while the linear relation, Eq. (16), is the other likely extreme. The true expression for any type of weak link cannot be grossly different from the Josephson relation, and must in any case preserve the essential periodicity and symmetry.

In tunnel junctions there is also a third current-carrying mechanism in parallel with the other two, namely, the displacement current through the insulating medium, characterized by the junction capacitance. With point contacts and metal bridges the inherent capacitance may be almost negligible but the shunt capacitance of the structure in the immediate neighborhood may still play an important role at high frequencies. The capacitance, whether internal or external, will also play a profound role in determining the so-called "dc" I–V characteristic (see Section III).

The time derivative of θ is proportional to the chemical potential difference across the weak link, Eq. (2), which contains terms in the electrostatic potential, gravitational potential, temperature, and vector potential. For most practical purposes we need only consider the term in the electric potential V, so that

$$\frac{d\theta}{dt} = \frac{2eV}{\hbar} = \frac{2\pi V}{\Phi_0} \tag{37}$$

This fundamental relation is commonly referred to as the "Josephson frequency."

In an ideal tunnel junction well below T_c, $i_n = GV$ is a strong function of V, with G very small for $V < 2\Delta$ and rising sharply to a constant value of the order of $i_c/2\Delta$ for $V > 2\Delta$. In a metal bridge G is not a strong function of V and may be nearly independent of V. The expected response $i(t)$ of a tunnel junction to a voltage $V(t)$ can be written down immediately

$$i = i_c \sin\theta + GV + C\dot{V}$$

$$i = i_c \sin\left((2\pi/\Phi_0)\int V\,dt\right) + GV + C\dot{V} \tag{38}$$

where C is the capacitance of the junction. If V is a constant V_0, the current i oscillates in time at the Josephson frequency V_0/Φ_0; if V varies sinusoidally in time, i can be expanded in a harmonic series whose Fourier coefficients are Bessel functions of $V/\Phi_0 f$. On the other hand, if we current bias the junction by fixing i at a constant level, then, neglecting the term GV and $C\dot{V}$, we have either the trivial case $V = 0$ for $i < i_c$, or no solution for $i > i_c$. To derive the actual response of a junction to current bias (high-impedance source) requires explicitly taking into account any reactive or dissipative elements, including the inherent conductance $G(V)$ and capacitance C, which may be coupled to it. Such additional elements greatly complicate the analysis and in only the very simplest cases can analytical solutions be obtained (McCumber, 1968; Stewart, 1968). Because of this complexity, a great deal of effort, not always successful, has been devoted to the theoretical interpretation of experiments on current-biased junctions and metal bridges, and in particular on their dc I–V characteristics. Such efforts have resulted in much fundamental enlightenment and a number of practical applications, but at the same time have given recognition to many strange or unexplained effects. We suggest that these unexplained effects, and the difficulty in interpreting others, should not obscure the fundamental simplicity of the Josephson effect. As a particular example statements have been made, based on experiments on dc I–V characteristics, to the effect that tunnel junctions exhibit "Josephson behavior" while metal bridges do not. These statements are misleading. Josephson behavior is implied by the dependence of the supercurrent on the quantum phase, and this dependence is the same, except for possible subtle differences noted above, for all weak links.

In point of fact, most dynamical experiments on weak links, voltage biased by small shunt inductance and resistance elements, avoid the complexity of external circuits and internal normal conductance mechanisms. They have been shown to be specified in all important details by the Josephson relations and the fundamentals of flux quantization.

E. Mechanical Analogs of Weak-Link Circuits

The fact that the Josephson coupling energy $-(\Phi_0 i_c/2\pi)\cos\theta$ of a tunnel junction has a direct mechanical analog in the gravitational potential energy $(mgl\cos\theta)$ of a simple pendulum of mass m and length l enables one to construct an elegantly simple mechanical model of any configuration of weak links coupled to ordinary circuit elements like resistors and resonant circuits. Even without actually building the model, it is possible to use intuition about the mechanical analog to envision the operation of the tunnel-junction circuit in considerable detail. Although Shin and Schwartz (1966), McCumber (1968), Stewart (1968), and others have mentioned the pendulum analog, we believe that its tutorial value has been underemphasized and suggest that anyone seeking an understanding of Josephson effects begin by studying the analog.

The Hamiltonian functions for a small-area tunnel junction of critical current i_c and capacitance C, and for a pendulum of mass m and length l are, respectively,

$$\begin{aligned}\mathcal{H}_{\text{w.l.}} &= -(\Phi_0 i_c/2\pi)\cos\theta + \tfrac{1}{2}CV^2 \\ &= -(\Phi_0 i_c/2\pi)\cos\theta + (\Phi_0^2 C/8\pi^2)\dot{\theta}^2 \end{aligned} \tag{39}$$

and

$$\mathcal{H}_{\text{pend}} = -mgl\cos\theta + \tfrac{1}{2}ml^2\dot{\theta}^2 \tag{40}$$

Torque is the analog of current, so that the total current, Eq. (38), expressed in terms of the phase θ:

$$i = i_c\sin\theta + (\Phi_0 C/2\pi)\ddot{\theta} + (G\Phi_0/2\pi)\dot{\theta} \tag{41}$$

is the equivalent of

$$T = mgl\sin\theta + ml^2\ddot{\theta} + F\dot{\theta} \tag{42}$$

Angular velocity $\dot{\theta}$ obviously is the analog of junction voltage V. Since $\int V\,dt$ has dimensions of flux, the angular displacement θ of the pendulum will be the analog of flux crossing from one side of the junction to the other, a concept which has more obvious meaning in the case of closed-loop devices. One complete revolution of the pendulum is the analog of one flux quantum crossing the junction, and so on. The loss term $F\dot{\theta}$ for the pendulum is the analog of the normal or quasiparticle current GV of the junction. If the coefficient F is velocity independent, as in the case of ideal fluid friction, then F is the direct analog of a voltage-independent conductance G. On the other hand, the conductance of a good tunnel junction is very low for voltages less than twice the energy gap 2Δ, and then rises abruptly. An approximate mechanical analog would be a preloaded centrifugal brake which imposes no drag until a predetermined angular velocity is reached, above which the drag rises steeply.

Recently Scott (1969) has described a mechanical analog of a long rectangular tunnel junction, consisting of a closely-spaced series of pendulums attached to a horizontal torsion bar. All of the essential properties of the extended tunnel junction; e.g., Josephson penetration length, vortex-flow state, diffraction effect, plasma resonance, etc., can be demonstrated with such an analog. We will describe later the mechanical analogs of some particular devices.

Shin and Schwartz (1966) suggested the analog of a weak link with a shunt capacitance C in a loop of inductance L. To Eqs. (39) and (40) one adds the terms

$$(\Phi_0^2/8\pi^2 L)(\theta - 2\pi\Phi_x/\Phi_0)^2 \qquad \text{and} \qquad \tfrac{1}{2}K(\theta - \theta_x)^2$$

respectively, where K is the spring constant of a horizontal torsion bar, one end of which is connected to a rigid pendulum and the other end is attached to a rigid support whose angular displacement θ_x can be varied. It is easy to visualize in terms of this analog how reversible transitions occur if $mgl < 2\pi K$ $(\beta < 1)$, and irreversible transitions with the proper selection rule occur if $mgl > 2\pi K$ $(\beta > 1)$. This will be discussed in Section III.B.9.

III. Device Characterization

A. Simply-Connected Small-Area Devices

1. *Time-averaged characteristics (dc I–V curves)*

A representative dc I–V characteristic of a small-area tunnel junction between identical superconductors with no applied magnetic field is shown in the Fig. 3a. Figure 3b shows a more informative plot of a hypothetical

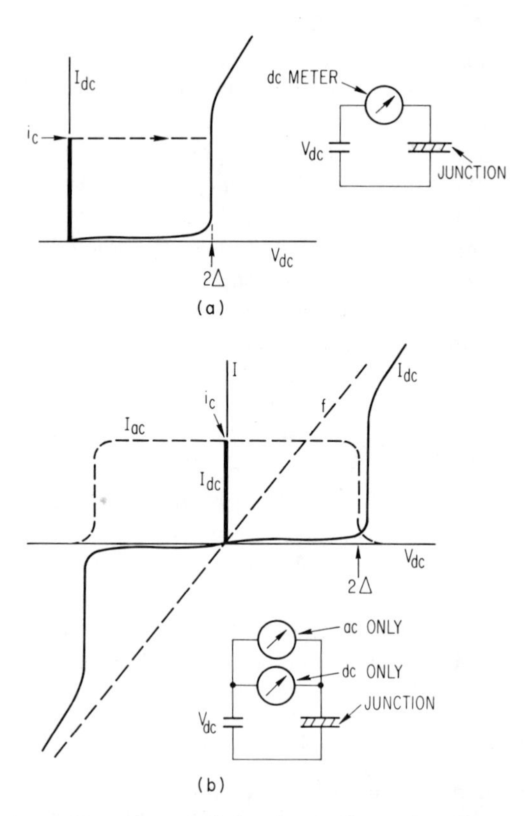

FIG. 3. Current–voltage characteristic of a small-area Josephson tunneling junction. (a) Idealized experimental curve of the dc I–V curves. (b) Hypothetical I–V characteristic showing the ac and dc currents as functions of the dc voltage as measured in the ideal circuit shown.

measurement by two ammeters in parallel, one of which passes only alternating current and measures its amplitude I_{ac} and frequency f, and the other passes and measures only the direct current I_{dc}. No such measurement of I_{ac} has been made, of course, over such a broad frequency range, and the curve may not be correct in detail. In particular, there is experimental evidence that I_{ac} may be appreciable well beyond 2Δ (McDonald *et al.*, 1969). A peak in the neighborhood of 2Δ has been postulated (Reidel, 1964).

Figure 3 is representative of a tunnel junction well below the transition temperature T_c. As T approaches T_c, the normal conductance G of the junction increases and at the same time thermal fluctuations begin to trigger the junction from the superconducting to the normal state. The result is that the I–V characteristic becomes an average of the two states,

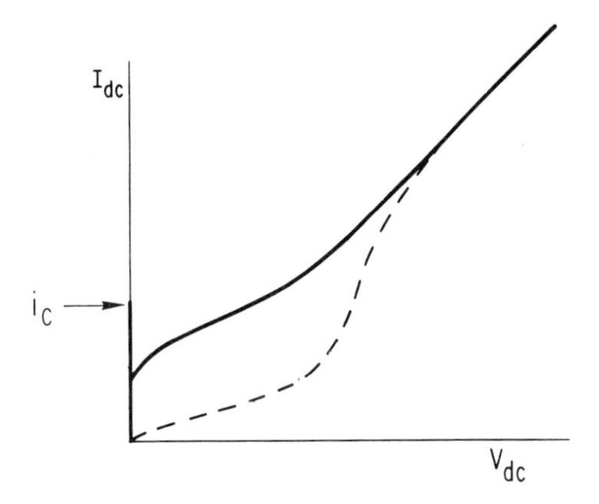

FIG. 4. Idealized dc current–voltage characteristic of a small-area Josephson tunneling junction near T_c under the influence of thermal fluctuations. The dashed line is the normal or single-particle tunneling conductance $G(V)$.

and close to T_c looks approximately as shown in Fig. 4. The parameter which characterizes the thermal averaging process is the ratio of Josephson coupling energy to thermal energy, $\Phi_0 i_c/k_B T$. The zero-voltage supercurrent is suppressed when this parameter is of order unity (Anderson and Goldman, 1969).

In the experimental measurement of tunnel-junction I–V characteristics, the usual practice is to current bias the junction and measure the dc voltage. Under this condition, it is an accident of nature that a result is obtained, which is easily interpreted as a zero-voltage supercurrent and a small normal current at finite voltages less than 2Δ (Fig. 3a). At finite voltages $<2\Delta$ the ac supercurrent (Fig. 3b) is shunted through the large junction capacitance so that essentially no ac voltage is developed. In the absence of shunt capacitance, the ac voltage accompanying the oscillating supercurrent would distort the waveform and lead to a dc term in the supercurrent at finite voltage, so that the finite-voltage branch of the curve could not be interpreted as a pure normal current. This should become clear by considering the case of a point contact or metal bridge having negligible inherent shunt capacitance.

We assume that a weak link in the form of a point contact or a metal bridge embodies only two parallel current-carrying mechanisms: (1) a supercurrent obeying the Josephson relation $i_s = i_c \sin\theta$, and (2) a normal current characterized by a voltage-independent conductance G:

$$i = i_c \sin\theta + GV = i_c \sin\left((2\pi/\Phi_0) \int V\, dt\right) + GV \qquad (43)$$

If we could measure the dc I–V characteristics using a zero-impedance

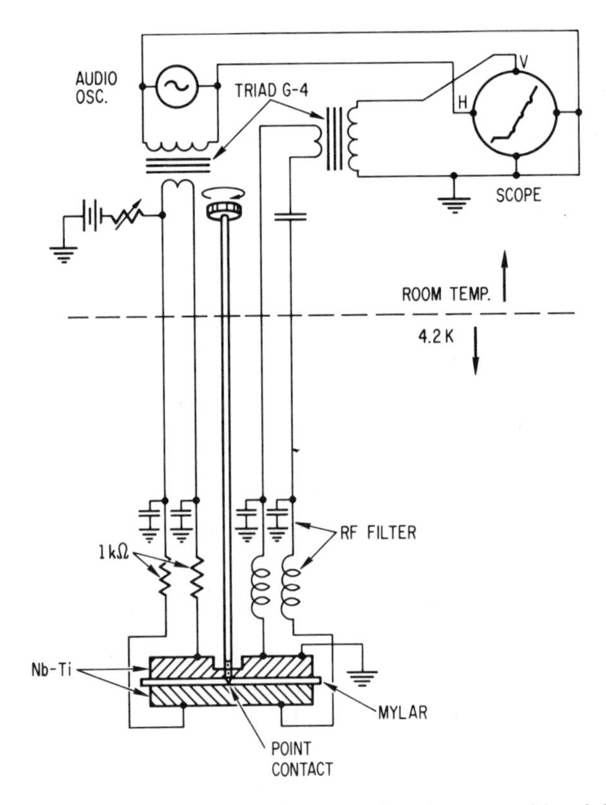

FIG. 6a. Experimental arrangement for measuring the current-biased dc I–V characteristics of a superconducting point contact bypassed at radio frequencies by a superconducting circular parallel plate capacitor with 650 pF.

approximately 50–50 atomic composition, and the operating temperature was 4.2K.

Another consequence of this simple model is that the ac component of the supercurrent persists for bias currents I_{dc} well above i_c (and $V_{dc} < 2\Delta$). Therefore, if external circuit elements, e.g., one or more resonant cavities, are coupled to the weak link, these can absorb power from the ac supercurrent and perturb the dc characteristic in the region well above i_c and at voltages related to the characteristic frequencies of the elements by $V_{dc} = (m/n)\Phi_0 f$, where m and n are integers. The literature is replete with experimental observations of such effects in the "normal" region of the dc characteristic, i.e., in the region above i_c. This portion of the curve has been called the "resistive-superconductive" region (deWaele *et al.*, 1967) in reference to the fact that an oscillating supercurrent flows in parallel with the normal current.

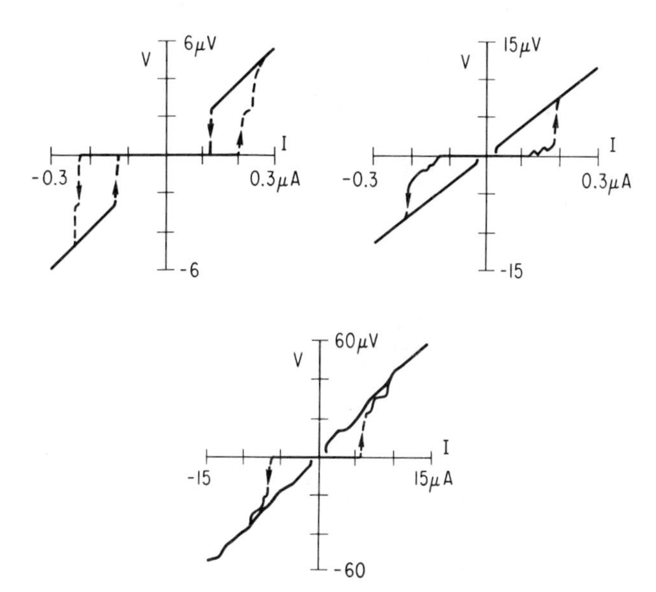

FIG. 6b. Experimental dc characteristics of the bypassed point contact of Fig. 6a in the low-voltage region showing behavior similar to that of Fig. 5a.

The configuration shown in Fig. 6a, when observed at larger voltages, no longer has this simple characteristic (Zimmerman, 1971a). The parallel disk geometry defines a set of microwave resonances having Bessel-function spatial variation of the electric and magnetic fields. The only modes which will couple to the oscillating supercurrent in the point contact at the center of the cavity are the zero-order modes whose electric fields vary spatially as $J_0(2\pi\rho/\lambda_n)$, where ρ is the distance from the center and λ_n is the wave length of the mode having n (radial) nodes. The theoretical positions of the resonances are indicated in Fig. 7 using the Josephson relation $V_{\mathrm{dc}}(0,n) = \Phi_0 f_{0,n}$, where $f_{0,n}$ is the frequency of the nth mode, and agree with the experimental results as shown. Self-induced steps in the dc I–V characteristics were first observed for the self-resonant modes of thin-film Josephson tunnel junctions (Eck *et al.*, 1964a, b; Coon and Fiske, 1965; Dimitrenko *et al.*, 1965), and for a point contact in a microwave cavity (Dayem and Grimes, 1966). The shape of the step characteristic as shown in Fig. 7 is not understood.

A very useful property of superconducting weak links is their response to applied high-frequency signals. It was first shown by Shapiro (1963) and Shapiro *et al.* (1964) that the dc characteristic of a current-biased Josephson junction exposed to a microwave field exhibited I–V steps obeying the Josephson frequency relation between the step voltages V_n and harmonics of the microwave frequency f, that is, $V_n = n\Phi_0 f$.

These radiation-induced steps are closely related to the self-induced steps shown above which result from the absorption of power from the Josephson oscillation at particular frequencies. Figure 8 shows a particularly good example (Grimes and Shapiro, 1968) of radiation-induced steps in the dc characteristic of a point contact at various microwave power

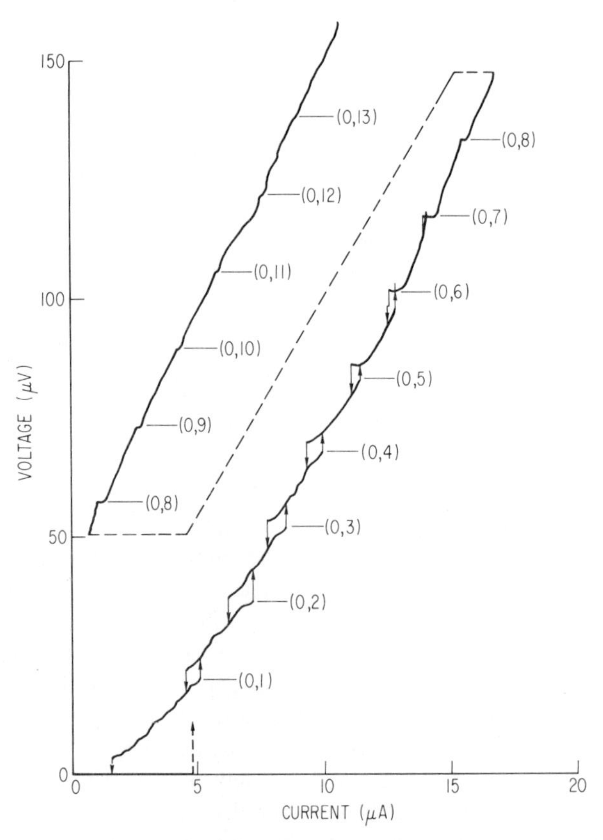

FIG. 7. Higher voltage region dc characteristics of the configuration of Fig. 6a showing the step effect of the $(0,n)$ radial modes of the circular disk.

levels. The point contact, placed across the narrow dimension of the waveguide, was current-biased and had essentially no shunt capacitance. The characteristic with no microwave power is similar to that shown in Fig. 5b, while the step amplitudes ΔI_n vary approximately as Bessel functions of the applied signal amplitude v_1, i.e.,

$$\Delta I_n \propto J_n(2\pi v_1/\Phi_0 f)$$

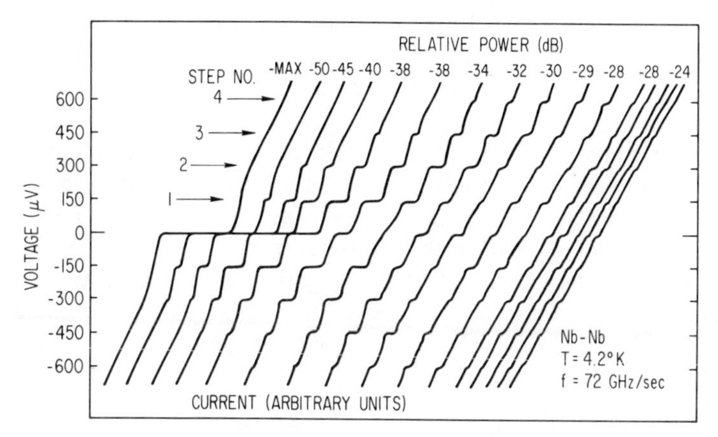

FIG. 8. Microwave-induced steps in the dc characteristic of weak links as a function of the microwave power level. [After Grimes and Shapiro, 1968.]

where f is the signal frequency and n is the number of the step as indicated in the figure.

When the capacitively bypassed point contact of Fig. 6 is subjected to an applied microwave signal at one of the mode frequencies $f_{0,n}$ a very striking constant-voltage step pattern can be produced. Over a certain range of signal amplitude the steps at $V_{\mathrm{dc}} = \pm\Phi_0 f_{0,n}$ become so broad as to cross the zero-current axis as shown in Fig. 9. This device can be used as zero-resistance constant-voltage power source at zero dc bias. The polarity of

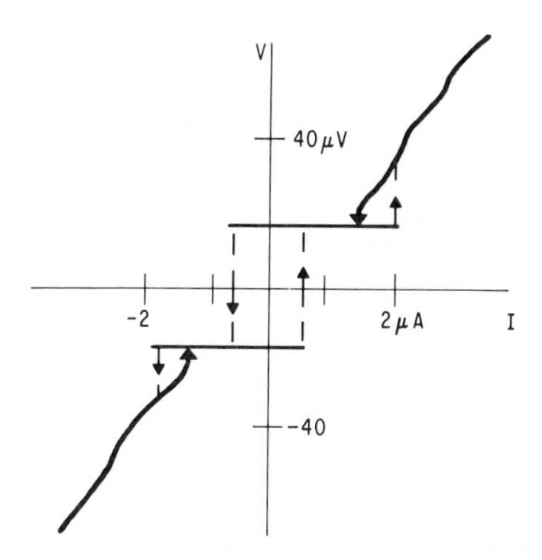

FIG. 9. Microwave-induced step for the capacitively bypassed point contact of Fig. 6a when the microwave frequency matches one of the resonant mode frequencies $f_{0,n}$.

the source can be switched by appropriate manipulation of I_{dc}. An effect of this type was noted by Shapiro (1963) in his original paper on radiation-induced step structure of thin-film tunnel junctions when a signal was applied at a resonant frequency of the junction.

The upper limits of frequency and voltage at which step structure can be induced is generally thought to be determined by the energy gap. In several experiments, however, structure has been seen quite far above the energy gap. McDonald *et al.* (1969), have seen microwave-induced steps at voltages greater than 15 mV. This is several times the energy gap in the material used (Nb point contacts), and they suggest that the upper limit may be determined by the ratio of Fermi velocity to mean free path, rather than by the energy gap.

2. *The mechanical analog of current-biased tunnel junctions*

A simple pendulum of mass m hung from a stiff support of length l with the other end of the support clamped to a horizontal shaft with frictionless bearings, is a very useful analog of a Josephson junction. As discussed in Section II.E, the energy of the pendulum is

$$\mathcal{H} = -mgl\cos\theta + \tfrac{1}{2}ml^2\dot{\theta}^2 \tag{46}$$

where θ is the angular displacement from vertical; that of the tunnel junction is

$$\mathcal{H} = -(1/2\pi)\Phi_0 i_c \cos\theta + \tfrac{1}{2}C(\Phi_0/2\pi)^2\dot{\theta}^2 \tag{47}$$

Recalling that the torque T applied to the horizontal shaft is the analog of a bias current i, for $T < mgl$ (analogous to $i < i_c$), the pendulum assumes a stable deflection $\theta < \pi/2$; for $T > mgl$ the pendulum undergoes a monotonically increasing angular velocity until a dynamical steady state is reached where the applied torque is balanced by air friction (normal conductance of the tunnel junction). The mean steady-state angular velocity $\dot{\theta}$ is the analog of the dc voltage across the tunnel junction. The time for the pendulum to reach the steady state is proportional to its rotational inertia ml^2, which is the analog of the capacitance C of the tunnel junction. Once the pendulum has started to rotate, because the critical torque has been exceeded, the steady-state angular velocity will vary linearly with the applied torque as long as the friction is proportional to velocity; in other words, the relation between applied torque and mean angular velocity is just like that indicated in Fig. 5a for capacitively bypassed weak links with constant conductance.

B. Weakly-Connected Loop Devices

1. *Method of analysis*

In this section we describe in detail how a low-inductance loop device interacts with a resonant circuit to which it is inductively coupled and which is driven by a constant-current rf bias, as indicated in Fig. 10. The idea underlying the description is that the dynamic behavior of the device can be derived from a quantum-periodic magnetic response function, which is either continuous and reversible, or discontinuous and hysteretic, depending upon the critical current of the contact i_c and the inductance L. The parameter which identifies this behavior is β, defined in Sections II.B and II.C. Values of $\beta > 1$ define discontinuous behavior; if β is not *much* greater than unity the selection rule $\Delta k = \pm 1$ applies. For $\beta \gg 1$, higher order transitions $|\Delta k| > 1$ are observed (Silver and Zimmerman, 1967b; Zimmerman and Silver, 1967).

Two general forms of the response function have been derived, Eqs. (20) and (21) and Fig. 1, and Eqs. (30) and (31) and Fig. 2. These two functions will be utilized interchangeably as convenience dictates throughout this chapter since the predicted properties are sufficiently similar.

In the following pages we consider first some simple ideal cases for which analytic solutions are obtainable. Next we discuss several modes of operation of actual practical devices, and finally we describe the mechanical analogs of these devices from which a clear picture of the physical nature of the transitions between quantum states is obtained.

2. *Response of an isolated, low-β, weakly-connected loop to an applied rf field*

Consider a ring containing a weak Josephson tunnel junction for which $i = i_c \sin\theta$ and $i_c \ll \Phi_0/2\pi L$, i.e., $\beta \ll 1$. We assume the voltages and frequencies are low enough that the junction conductance and capacitance can be neglected. The restriction on β insures that the magnetic response will be continuous and reversible, and that an analytic solution can be obtained. Suppose we apply to the ring a field

$$\Phi_x = \Phi_x^0 + \Phi_x^1 \sin\omega t \tag{48}$$

where Φ_x^0, Φ_x^1, and ω are adjustable parameters. Substituting Eq. (48) into Eq. (31), the solution (correct to first order in Li_c/Φ_0) can be written

$$\begin{aligned}\Phi = {} & \Phi_x^1 \sin\omega t + \Phi_x^0 - 2Li_c \sin(\theta_0) \\ & \times \{\tfrac{1}{2}J_0(\theta_1) + J_2(\theta_1)\cos 2\omega t + \cdots\} - 2Li_c\cos(\theta_0) \\ & \times \{J_1(\theta_1)\sin\omega t + J_3(\theta_1)\sin 3\omega t + \cdots\},\end{aligned} \tag{49}$$

where

$$\theta_0 = 2\pi\Phi_x{}^0/\Phi_0, \qquad \theta_1 = 2\pi\Phi_x{}^1/\Phi_0 \tag{50}$$

Differentiating Eq. (49) with time gives the voltage $V(t)$ which one measures across a one-turn coil tightly coupled to the loop. This analytic solution is instructive in that it exhibits many qualitative features of the response of real devices to applied dc and rf fields (Silver and Zimmerman, 1967b). In particular, the periodic variation of the amplitude of the Fourier components of V with applied dc field $\Phi_x{}^0$ and the Bessel-function-like variation with applied rf field amplitude $\Phi_x{}^1$ are well documented experimentally. As harmonic generators, such devices are unique in that the amplitude of any particular harmonic can be "peaked up" by appropriate adjustment of the applied field; furthermore, the peak amplitudes of the harmonics drop off quite slowly with increasing order of the harmonic.

The above example is not representative of practical devices for several reasons: first, because the restriction $\beta \ll 1$ means that the maximum available power output of the device (at ω or any harmonic) is correspondingly small; second, because coupling the device to a vacuum tube or transistor rf amplifier perturbs the response and must be included in the calculation; and third, because thermal fluctuations are not usually negligible compared to the magnetic free energy of the device and so should not be ignored.

3. *Response of a low-β weakly-connected loop and resonant circuit to an applied rf field*

An analytical expression for the response of the same single-junction loop coupled to a resonant circuit can also be written down. The circuit is that of Fig. 10 where we again assume $i = i_c \sin\theta$, $\beta \ll 1$, and neglect the

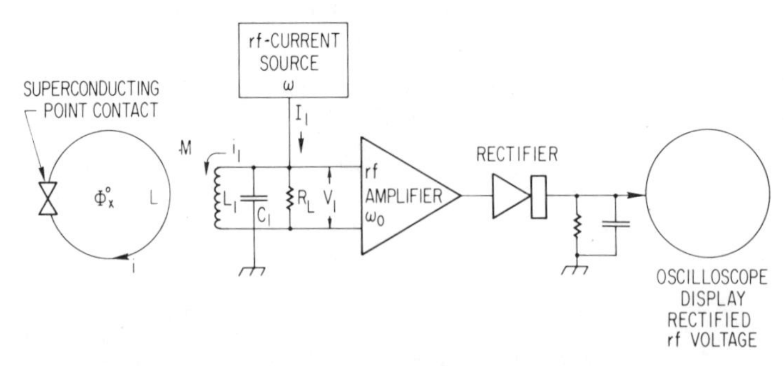

FIG. 10. Schematic representation of point-contact loop device coupled to a resonant circuit driven by an rf-current source I_1 at frequency ω.

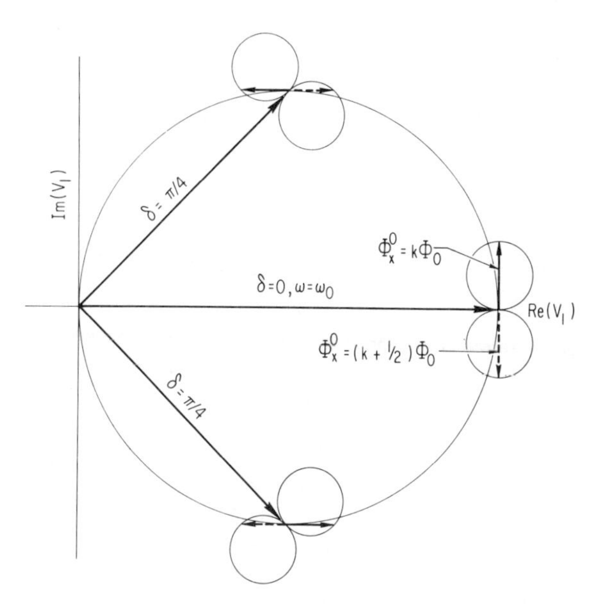

FIG. 11a. Complex response of the rf voltage V_1 in Fig. 10 for $\beta < 1$ and a variable applied field $\Phi_x{}^0$.

junction conductance and capacitance. Without going through the details of the derivation, the voltage across the resonant circuit at the fundamental frequency, correct to the first order in Li_c/Φ_0, is

$$V_1 = |Z|I_1 \sin(\omega t - \delta) + 2\frac{M}{L_1} i_c|Z|J_1$$

$$\times \left[\frac{2\pi}{\Phi_0}\left(\frac{L}{L_1}\right)^{1/2} \frac{|Z|I_1}{\omega}\right] \cos\theta_0 \sin\left(\omega t - 2\delta - \frac{\pi}{2}\right) \tag{51}$$

where

$$Z(i\omega) = (1/R_L + i\omega C_1 + 1/i\omega L_1)^{-1} = |Z|e^{i\delta} \tag{52}$$

and I_1 is the amplitude of the rf bias current at ω. Obviously the first term is the response of the unperturbed resonant circuit and the second is the signal due to field modulation of the weakly-connected loop impedance. A vector diagram of the two terms in V_1 at three frequencies and for $\theta_0/2\pi = k$, $k + \frac{1}{2}$ is given in Fig. 11a. On resonance ($\omega^2 = 1/L_1C_1$) the two terms ("carrier" and "signal") are in quadrature, so to first order the total amplitude is not perturbed by the signal. Off resonance the signal has a component in phase with the carrier so that V_1 is amplitude-modulated

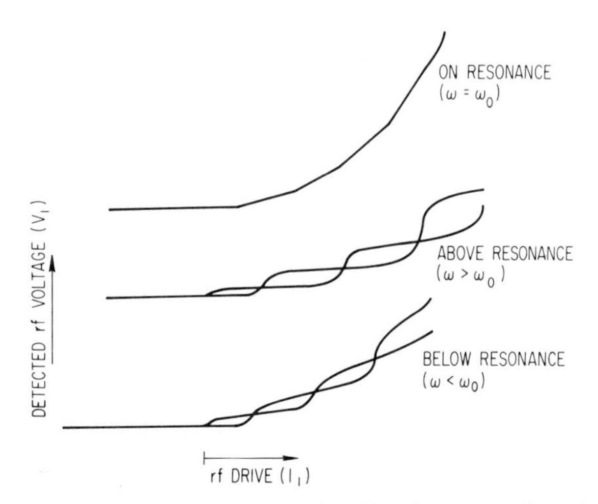

FIG. 11b. Experimental rf V–I characteristics of a point-contact loop device as shown schematically in Fig. 10. The three curves represent changes in the source frequency relative to the tuned circuit. The rf-current source I_1 is amplitude modulated in each case and the two superimposed curves represent extreme variations due to changes in Φ_x^0.

as Φ_x^0 varies from integral to half-integral multiples of Φ_0. The absolute maximum amplitude modulation, however, comes when $|Z| \sin \delta$ is maximized, assuming that I_1 is always adjusted to maximize the Bessel function, i.e., $J_1(x)|_{\max} \cong 0.58$. These points occur at the half-power points, $\delta = \pm\pi/4$. Thus, if V_1 is amplitude-detected, as discussed in the following pages, the bias oscillator must be set above or below resonance in order to see the signal (Fig. 11b). Clearly a better detection scheme would be to phase shift the carrier to quadrature at the detector and set the bias oscillator on resonance. However, the point is somewhat academic since practical devices have greater critical currents ($\beta \sim 1$) so that this first-order analysis is inadequate. As we shall see, however, the bias-oscillator frequency in most practical cases ($\beta \gtrsim 1$) is set above the unperturbed resonance frequency $(L_1C_1)^{-1/2}$.

4. *Operation of practical devices* ($\beta \sim 1$)

We now consider the operation of practical devices. Instead of the Josephson relation $i = i_c \sin \theta$ we will now use the simpler periodic linear relation given by Eqs. (20) and (21). Although the formulation is not analytically precise, we are able to handle in a fairly quantitative way the important practical case where β is not small compared to 1, as was assumed in the preceding paragraphs. For β larger, but not much larger, than unity, transitions between the quantum states of Fig. 1 will obey the selection rule $\Delta k = \pm 1$, and it is this regime of operation which is of greatest practical interest. The case $\beta \gg 1$, for which higher order transitions ($|\Delta k| > 1$) take place, will be discussed briefly in the following section and in connection with the mechanical analog of loop devices.

Consider again a superconducting ring of inductance L, incorporating a weak link whose critical current is i_c, inductively coupled to a tuned circuit in the input of an rf amplifier (Fig. 10). The losses in the tuned circuit and the input resistance of the amplifier are lumped together and are represented by a resistance R_L across the tuned circuit. An rf bias or "pump" oscillator is weakly coupled to the tuned circuit through either a very small capacitor or a very small mutual inductance, the requirement being that the oscillator itself should not load the tuned circuit. The rf voltage V_1 is amplified, detected by a diode rectifier, and displayed on an oscilloscope or chart recorder.

Consider first the case where β is just small enough that the magnetic response of the ring is reversible, but not so small that the signal is difficult to observe, and suppose the rf bias level is adjusted so that the ac flux amplitude Φ_x^1 applied to the ring is of the order of $\Phi_0/4$. Now it is evident from Fig. 1 that if the dc component Φ_x^0 of the applied field is an integral multiple of Φ_0, then the ring acts like a shorted turn in the field of the coil, since $d\Phi/d\Phi_x \approx 0$. Under this condition the resonance frequency of the system lies above that of the tuned circuit alone. Similarly, if Φ_x^0 is an odd half-integral multiple of Φ_0, then $d\Phi/d\Phi_x \gg 1$, the ring acts like a paramagnetic body in the field of the coil, and the resonance frequency of the system lies below that of the tuned circuit alone. This parametric effect on the inductance, Eq. (34), can be observed by setting the bias frequency ω either above or below the resonance frequency $\omega_0 = (L_1C_1)^{-1/2}$ of the tuned circuit and then varying the applied dc field (Fig. 11b). The periodic variation of inductance with dc field causes V_1 to amplitude modulate on the side of the resonance curve. The modulation amplitude goes through zero (to first order) and reverses phase as ω is shifted from one side of ω_0 to the other. Further, the resonance frequency shift with Φ_x^0 has been directly observed by Silver and Zimmerman (1967b).

Modulation of the rf amplitude by a dc field was indicated in the analytic solution for the ideal isolated loop (Section III.B.2). The tuned circuit is not, in principle, necessary for observing these effects. Its real function is to serve as an impedance transformer between the device and the rf amplifier.

Quantum periodicity can also be seen (in the real system) if the dc field is held constant and the rf bias level is varied. This is seen as a stepwise increase in the oscillation level of the tuned circuit as the bias-oscillator level is increased uniformly. Here also the amplitude of the steps goes through zero and reverses phase as ω is shifted from one side of ω_0 to the other. A qualitative analysis of this will be left as an exercise for the reader. Note that in the analytic solution, Eq. (49), for the simple system there is a similar stepwise increase in the fundamental response, embodied in the terms

$$[\Phi_x^1 - 2Li_c \cos(\theta_0) J_1(\theta_1)] \sin \omega t \tag{53}$$

These characteristics of a real system are demonstrated in Fig. 11b, which shows the diode detector output V_1 as a function of bias oscillator amplitude I_1 with dc field and frequency as parameters. These patterns were traced from oscilloscope photographs, and do not show superimposed random fluctuations which are apparent in the original pictures. The increasing pattern amplitude with increasing rf bias level is a consequence of the square-law characteristic of the diode detector. This type of pattern can be realized with any of the single-contact loop devices described in Section IV if the critical current is set at a low level. A general discussion of such devices with experimental data has been given by Silver and Zimmerman (1967b).

Consider now $\beta > 1$ and the rf bias frequency ω equal to the low-level resonance frequency of system. At low levels (such that i never reaches i_c) the device looks like a shorted turn in the field of the coil. Therefore ω is greater than the resonance frequency of the tank circuit in the absence of the device, or with the contact open, or at high bias levels. The rf voltage V_1 across the tank circuit is a linear function of the rf bias current I_1 as long as the peak current in the superconducting loop is always less than i_c. For the case where the dc applied field is zero (or $k\Phi_0$), the critical current is reached at the applied flux level (taking $\gamma \sim 0$)

$$\Phi_c = Li_c \tag{54}$$

The rf flux amplitude Φ_x^1 applied to the device is

$$\Phi_x^1 = Mi_1 = V_1 M/\omega L_1; \qquad M^2 \ll LL_1 \tag{55}$$

where i_1 is the current in the tank coil whose inductance is L_1 and M is the mutual inductance. Hence, when $\Phi_x^1 = \Phi_c$ we have

$$V_1 M/\omega L_1 = Li_c \tag{56}$$

and denoting these respective values of V_1 and I_1 by V_{1c} and I_{1c}, we can write

$$V_{1c} = \omega L_1 L i_c / M = I_{1c} R_L \tag{57}$$

At this bias level (point A in Fig. 12) a transition to one of the two adjacent

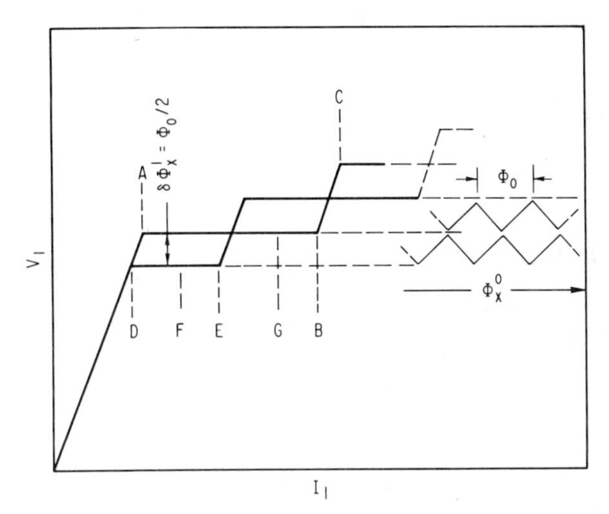

FIG. 12. Expected behavior of the rf voltage V_1 as a function of the rf-current drive I_1 and the applied magnetic field Φ_x^0. The inset at the right of the graph shows the variation of V_1 with Φ_x^0 for fixed I_1.

states and back will take place. Thus the tank-circuit energy is abruptly reduced by the area of one hysteresis loop, that is (Silver and Zimmerman, 1967b),

$$\Delta E = 2\Phi_0 i_c - \Phi_0^2/L = (\beta - 1)\Phi_0^2/L \tag{58}$$

This is equivalent to shock-exciting the tank circuit out of phase with the driven oscillation. No further transitions can take place until the shock excitation dies out, i.e., until the oscillation level again builds up to the critical level. Consequently, the system undergoes a low-frequency, low-amplitude, sawtooth modulation in time. As the bias current I_1 is further increased above I_{1c}, the buildup of the oscillation level V_1 is more rapid and the sawtooth modulation frequency increases. However, *the modulation amplitude and the average oscillation level* $\bar{V}_1$ *remain fixed*, the former being proportional to ΔE and the latter being essentially equal to V_{1c} minus half the modulation amplitude. Both ΔE and V_{1c} are fixed parameters of the system. Thus $\bar{V}_1$ is limited at this level until two hysteresis loops, one above and one below the dc field level Φ_x^0 are traversed on every rf cycle. At this point (Fig. 12, point B), $\bar{V}_1$ again increases until the second pair of hysteresis loops is encountered, at which point (Fig. 12, point C) $\bar{V}_1$ is again limited by the mechanism described above. The total response may be described as a linear rise in $\bar{V}_1$ interrupted by an equally-spaced series of plateaus (Zimmerman *et al.*, 1970; Simmonds and Parker, 1971).

This model predicts that if an average-reading detector is used there should be an overshoot (not shown in Fig. 12; see below) at the leading edge of each plateau, the height of which is half the peak-to-peak modulation amplitude. The sawtooth modulation amplitude is given by

$$\frac{\Delta V_1}{V_{1c}} \cong \frac{1}{2}\frac{\Delta E}{E} = \frac{2\Phi_0 i_c - (\Phi_0^2/L)}{C_1(\omega L_1 L i_c/M)^2} \tag{59}$$

where E is the tank-circuit energy. Therefore, the voltage modulation is

$$\Delta V_1 \cong [2\Phi_0 i_c - (\Phi_0^2/L)]M/\omega L_1 L i_c C_1 \tag{60}$$

which becomes particularly simple if i_c is considerably greater than Φ_0/L

$$\Delta V_1 \cong \Phi_0 \omega M/L \qquad \text{for} \quad \beta \gg 1 \tag{61}$$

The amplitude of the overshoot is therefore

$$\Delta_s \cong \tfrac{1}{2}\Phi_0 \omega M/L \tag{62}$$

For the case where $\Phi_x{}^0 = (k + \frac{1}{2})\Phi_0$ the first plateau in voltage can be shown by similar reasoning to occur at about

$$V_{1c} = (L i_c - \tfrac{1}{2}\Phi_0)\omega L_1/M \tag{63}$$

In this case limiting is effected by one hysteresis loop rather than two, so the first plateau (Fig. 12, points D–E) is half as long in I_1 as succeeding ones.

Finally, it is easy to show by extension of these arguments that for particular rf bias levels (Fig. 12, point F, for example) $\bar{V}_1$ increases linearly in the regions $(k - \frac{1}{2}) < \Phi_x{}^0/\Phi_0 < k$, and decreases linearly in the regions $k < \Phi_x{}^0/\Phi_0 < (k + \frac{1}{2})$, i.e., the response as a function of dc field is a triangular wave. Furthermore, this triangular wave reverses phase as the rf current I_1 is increased so as to encompass the next adjacent pair of hysteresis loops. Such phase reversals occur when the rf flux amplitude at the superconducting loop increases by $\frac{1}{2}\Phi_0$ (Fig. 12, point G).

The vertical separation of the plateaus in $\bar{V}_1$, that is, the peak-to-peak variation of the response versus $\Phi_x{}^0$, is given by the difference of Eqs. (63) and (57),

$$\delta\bar{V}_1 = \tfrac{1}{2}\omega\Phi_0 L_1/M \tag{64}$$

The height of the overshoot at the leading edge of each plateau is given by Eq. (62). We have assumed in Eq. (55) that $M \ll (LL_1)^{1/2}$, from which it follows that $\Delta_s \ll \delta\bar{V}_1$. Thus the overshoots are very small and are not shown in Fig. 12. In fact, they have not been seen experimentally, probably for at least two reasons: (1) We generally use a diode peak detector so that what we measure corresponds more closely to the peak value of V_1 than

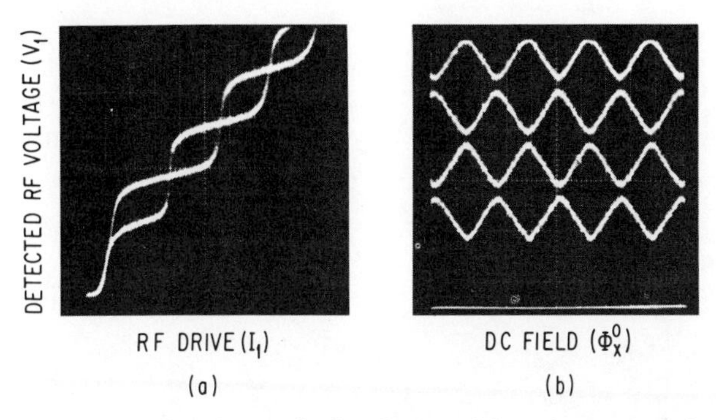

FIG. 13. Experimentally observed rf voltage with a loop-coupled point-contact device and a varying magnetic field using the circuit shown schematically in Fig. 10. (a) rf V–I curves shown with the magnetic field varied as indicated in Fig. 12. (b) The rectified voltage V_1 as a function of Φ_x^0 for selected values of the rf-drive current I_1. The voltage scale is the same for both (a) and (b).

to $\bar{V}_1$; and (2) high-frequency fluctuations superimposed on the applied rf flux causes premature triggering of quantum transitions, so that the edges of each plateau are actually rounded off. The difference between $\bar{V}_1$ and the peak value of V_1 is in any case very small.

The above description of events is valid only if the tank circuit has a fairly high Q, and the mutual inductance M is large enough that the inherent energy loss per cycle $2\pi V_1^2/\omega R_L$ is considerably smaller than the area of a hysteresis loop.

Experimental evidence obtained under these conditions support this detailed description. Figure 13 is typical of patterns obtained with devices of the types shown in Section IV, with a tank circuit $Q \sim 100$, $M/(LL_1)^{1/2} \sim 0.2$, $L \sim 4 \times 10^{-10}$ H, an overall system bandwidth of 10^5 Hz, a tank-circuit capacitance $C_1 \sim 2 \times 10^{-10}$ F, and $\omega/2\pi = 30$ MHz. The tank coil, a ten-turn coil of No. 2 copper about 1.5-mm o.d. was inserted directly in one hole of the device. The peak-to-peak amplitude of the patterns of Fig. 13 is about 14 μV referred to the preamplifier input.

5. *Higher-order transitions between quantum states and anomalous effects at $\beta \gg 1$*

In the foregoing description we assumed that transitions between states (for $\beta > 1$) were instantaneous and occurred only between adjacent states, $\Delta k = \pm 1$. As a matter of experimental fact, Nb point-contact loop devices with larger critical currents ($\beta \gtrsim 5$) usually obey a higher-order selection rule $\Delta k \geq \pm 2$; with a very large critical current Δk may be of the order of 10^3 or more. The fundamental nature of the transitions can be quite

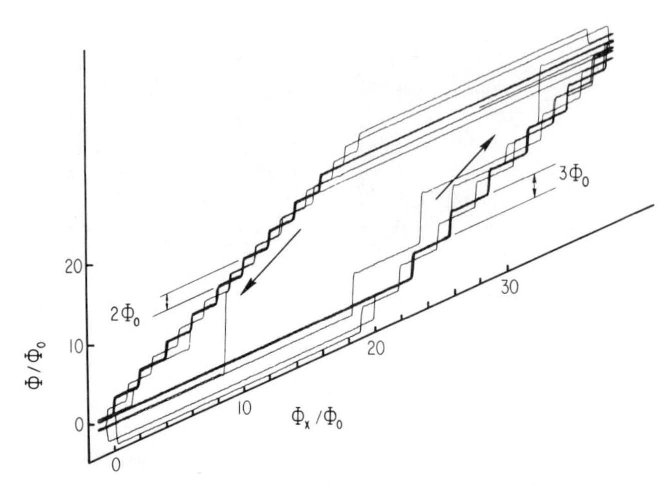

FIG. 14. Experimental magnetization curves of a loop-coupled point-contact device with large β showing high-order quantum transitions and asymmetry in field scan. Values of Δk as large as 6 were observed although the predominant values are 3 (increasing) and 2 (decreasing). The arrows indicate the direction of scan and the scale zeros are arbitrary. [After Silver and Zimmerman, 1967b.]

accurately described for a tunnel junction in a superconducting or partly resistive loop. This will be discussed in Section III.B.9 in connection with the mechanical analog of these devices.

Certain anomalous effects are frequently seen at large critical current. For example, the selection rule may be asymmetric, that is $|\Delta k| = N^+$ for field increasing and $|\Delta k| = N^-$ for field decreasing, where $N^+ \neq N^-$. Such behavior is shown in Fig. 14 where $N^+ = 3$ and $N^- = 2$. Another type of asymmetry is that the dc field patterns (Fig. 13) may be sawtooth instead of triangular, or skew symmetric but not mirror symmetric. One can show that skew symmetry in the dc field patterns implies a nonsymmetric current–phase shift relation, that is $i(\theta) \neq -i(-\theta)$. These effects are almost certainly related to microscopic geometric asymmetry of the weak link, but no detailed model has been advanced.

6. *Analysis of resistive-loop devices as coherent oscillators*

The partly-resistive weakly-connected loop was originally conceived (Zimmerman *et al.*, 1966) as a device for generating Josephson oscillations at an arbitrary frequency determined by the dc voltage V_0 across the resistance. At the same time it was realized that the Fourier spectrum of the oscillation should be essentially that of a superconducting loop device of the same geometry under the influence of a steadily increasing applied

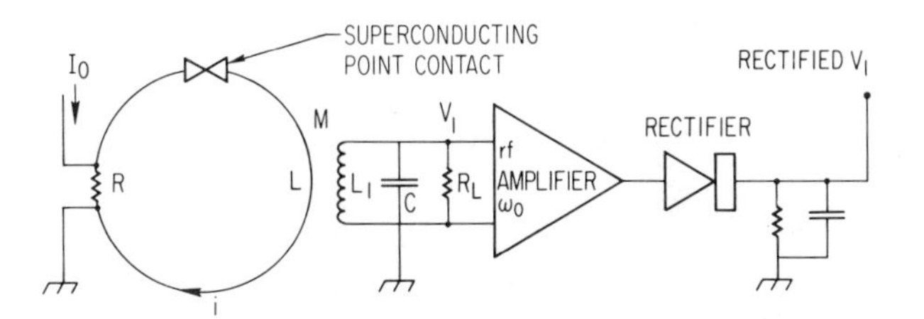

FIG. 15. Schematic representation of the resistive loop oscillator circuit.

flux such that $d\Phi_x/dt = V_0$. In other words, the behavior of the resistive loop device was predictable in detail from the experimental knowledge of the superconducting loop devices that had been studied previously.

The solution has already been given, Eq. (44), for the response of a tunnel junction, internally shunted by a constant conductance, to a constant current. A slightly more complicated problem can also be solved analytically, namely, a small-area Josephson junction externally shunted by a resistance R in series with a small inductance L and connected to a bias current I_0 across the resistance as shown in Fig. 15. We assume I_0R is small enough that the shunt capacitance can be ignored at the oscillation frequency and that the junction conductance G is negligible. Under these conditions

$$I_0R - iR - L\frac{di}{dt} - V = 0 \tag{65}$$

where i and V are, respectively, the instantaneous junction current and voltage. The solution of this equation with Eqs. (28) and (37) is (Zimmerman *et al.*, 1966)

$$(i - \alpha i_c)/|i_c - \alpha i| = \sin\{\omega_0(1 - \alpha^2)^{1/2}[t + (L/R)\log(1 - \alpha i/i_c)]\} \tag{66}$$

where

$$\alpha = i_c/I_0 = Ri_c/V_0 < 1 \tag{67}$$

and

$$\omega_0/2\pi = RI_0/\Phi_0 = V_0/\Phi_0 \tag{68}$$

This function is plotted in Fig. 16 for several values of α and β. For both α and β very small i varies sinusoidally with time. When $\alpha = 0$ the solution is identical to that of a superconducting ring with a steadily increasing applied field, $\Phi_x = V_0t$, as shown in Fig. 2. Increasing values of α cause the positive half-cycle to lengthen and the negative half-cycle to shorten, while the frequency is reduced by the factor $(1 - \alpha^2)^{1/2}$. This is qualitatively easily understandable since the voltage drop iR adds to the applied emf I_0R on the negative half-cycle and subtracts on the positive half-cycle. In-

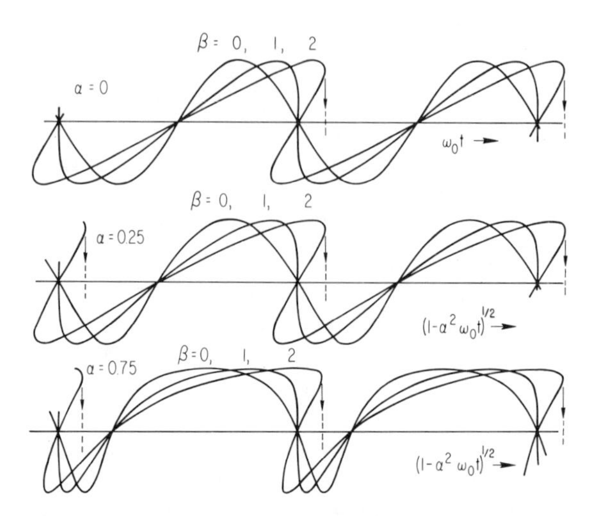

FIG. 16. Time evolution of the weak-link supercurrent in a resistive loop device. Selected values of $\alpha = i_c/I$ and $\beta = 2\pi L i_c/\Phi_0$ are shown to illustrate the effect of resistance and critical current.

creasing β causes the waveform to skew to the right, as also shown in Fig. 2. For $Li_c > \Phi_0/2\pi$ $(\beta > 1)$, discontinuous transitions between successive branches of the curve take place. The description of these transitions is analytically complicated, though straightforward in principle (see Section III.B.9). It is apparent that the oscillation becomes rich in harmonics as either α approaches unity or as β becomes greater than unity. It should be realized that "large" values of R and of L are nevertheless very small by the usual standards. For example, if $i_c = 10^{-4}$ A and $\omega_0/2\pi = 10^8$ sec^{-1}, then $\alpha \approx 1$ requires $R \approx 0.002\ \Omega$. Similarly $\beta \approx 1$ requires $L \sim 10^{-10}$ H.

An additional term in the total loop emf consists of Johnson noise generated in the resistance. If the loop is dc biased to oscillate at a high frequency, then the noise voltage added to the bias emf $V_0 = RI_0$ frequency modulates the Josephson oscillation. The resulting line shape calculated by Burgess (1967) is Lorentzian with the line width given by

$$\delta f = 4\pi k_B T R/\Phi_0^2 = [4.05 \times 10^7\ \Omega^{-1}\ \mathrm{sec}^{-1}\ \mathrm{deg}^{-1}] RT \qquad (69)$$

Experimental confirmation of this was reported by Silver *et al.* (1967) for a resistive loop oscillating at 30 MHz.

7. *Operation of practical oscillators*

The output, at radio frequencies, of a resistive loop oscillator can be detected with the circuit shown in Fig. 15. This is the same as Fig. 10, with the omission of the rf bias and the addition of a constant bias current I_0 through the resistance R. If $\alpha \ll 1$, that is, if the detector frequency ω_0 is much greater than $2\pi i_c R/\Phi_0$, and also if $\beta > 1$, then the current oscillation in the loop is essentially a sawtooth in time (Fig. 16) accompanied by voltage pulses of area somewhat less than Φ_0 across the link. In the same spirit as adopted for superconducting rings in Section III.B.4 we ignore the precise analytic solution Eq. (66). We again assume a linear current–phase relation and that the quantum transitions take place in a time short compared to $1/\omega = \Phi_0/2\pi V_0$. The Fourier spectrum of the current sawtooth wave is a series of terms at $n\omega$: $n = 0, 1, 2, \ldots$, whose coefficients vary as $1/n$ and are independent of ω,

$$i \cong i_c - \Phi_0/2L - (\Phi_0/2\pi L) \sum (1/n) \sin n\omega t \tag{70}$$

while the voltage expansion coefficients are proportional to ω and independent of n in this approximation (Silver and Zimmerman, 1967b)

$$V \cong \Phi_0\omega/2\pi + (\Phi_0\omega/\pi) \sum \cos n\omega t \tag{71}$$

At a frequency of $\omega/2\pi = 30$ MHz, the voltage amplitudes are about 0.1 μV. In the circuit of Fig. 15 the impedance transformation effected by the resonant input circuit steps up the voltage level at ω to the order of 10 μV at the preamplifier input. In order to observe the harmonic amplitudes using a fixed frequency detector at ω_0, it is necessary to set the dc bias level such that $V_0 = \Phi_0\omega_0/2\pi n = \Phi_0\omega/2\pi$ in order that the nth harmonic of the oscillation at ω fall at the detector frequency ω_0. With this experimental technique the harmonic amplitudes will vary as

$$V_n \cong \Phi_0\omega/\pi = \Phi_0\omega_0/\pi n \tag{72}$$

Thus the maximum available power in any harmonic is the product of current and voltage coefficients and varies as $1/n^2$,

$$P_n \cong \Phi_0^2\omega_0/2\pi^2 n^2 L \tag{73}$$

Practical signal-to-noise ratios are such that harmonics beyond the third or fourth harmonic are rarely seen. Typical results are shown in Fig. 17 for coherent oscillations at 28.7 MHz and 9.45 GHz using a device for which $R = 25\ \mu\Omega$. In Fig. 17a the second and third harmonics, appearing at $V_0 = \Phi_0\omega_0/4\pi$ and $\Phi_0\omega_0/8\pi$ are observable. A small signal at $\Phi_0\omega_0/\pi$, as a consequence of $|\Delta k| = 2$ transitions, is also shown.

With somewhat greater critical currents such that higher-order transitions (Section III.B.5) take place, the fundamental frequency of the oscillation becomes (Zimmerman and Silver, 1967)

$$\omega = 2\pi V_0/N\Phi_0 \tag{74}$$

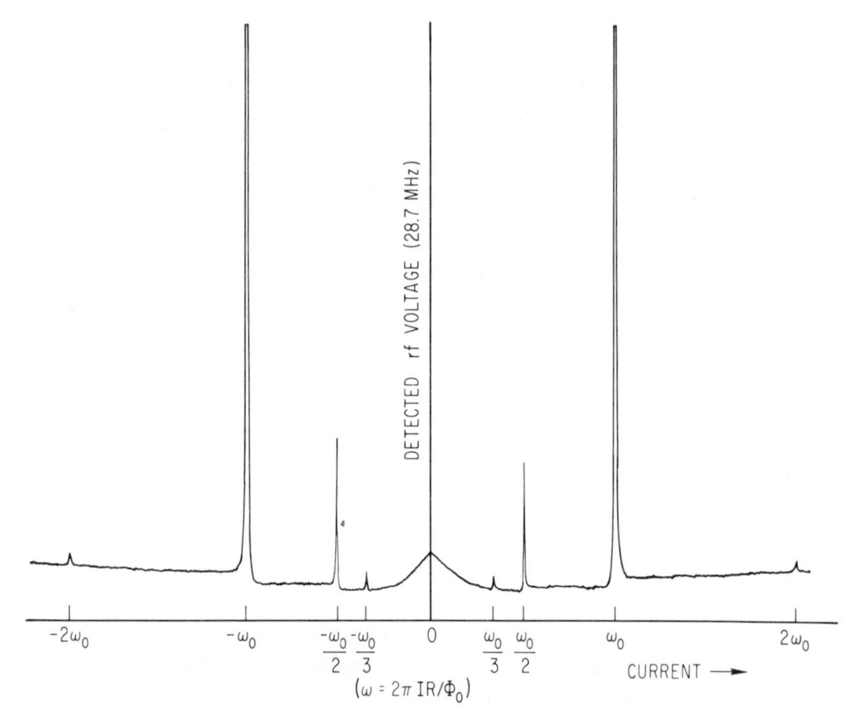

FIG. 17a. Experimentally measured signal at 28.7 MHz for a resistive loop oscillating device where R ≈ 25 μΩ. The fundamental signal at $\omega_0/2\pi$ = 28.7 MHz is off scale; smaller signals due to harmonic generation occur at $\omega_0/2$, $\omega_0/3, \cdots$. A small signal at $2\omega_0$ is believed to be occasional $|\Delta k| = 2$ transitions.

where $N = 2, 3, 4, \ldots$ is the order of the transition. The available power is proportional to N^2, and is calculated by replacing Φ_0 by $N\Phi_0$ in Eq. (74), so that

$$P_{N,n} \cong N^2\Phi_0^2\omega/2\pi^2n^2L \tag{75}$$

Thus the power output is dramatically enhanced by this mechanism, since values of N of the order of several hundred have been reported (Zimmerman and Silver, 1967), and high-order harmonics are easily resolved. Some striking examples of experimental spectra are shown in Fig. 18. A consequence of the theory is that the power in the Nth harmonic of an N-order oscillation should be equal to the power from the $N = 1$ Josephson oscillation, or $P_{N,N} = P_1$. It must be remembered that the detector frequency is fixed at ω_0; hence the fundamental frequency ω_N of the N-order oscillation must be set so that the Nth harmonic falls at ω_0, or $N\omega_N = \omega_0$. We have experimentally confirmed the equality $P_{N,N} = P_1$ for N as large as 15.

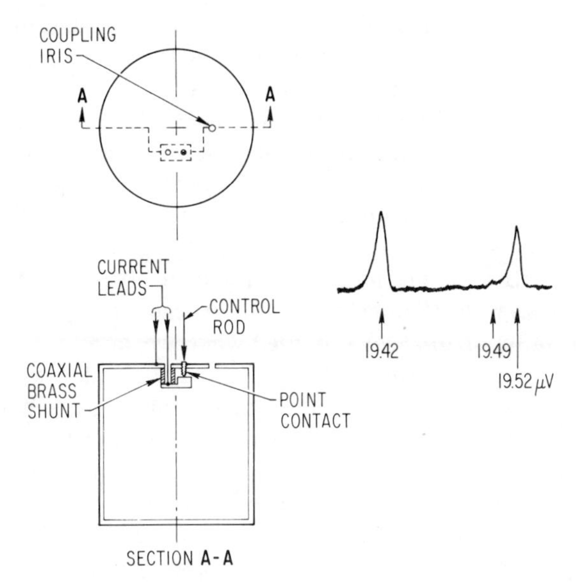

Fig. 17b. Cavity configuration and signal measured at 9.45 GHz using the resistive device of Fig. 17a. The multiple response occurs because of nearly degenerate modes of this cavity.

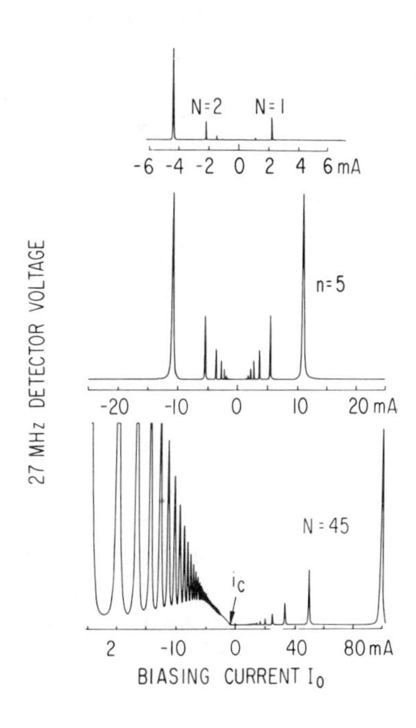

Fig. 18. Observed spectra at 27 MHz of high-order quantum transitions of a resistive loop weak-link device connected as shown in Fig. 15. The transition number N varies from 1 to 45 and R is approximately 25 $\mu\Omega$. The upper curve shows asymmetry upon reversing the bias current I_0, and the lower curve is expanded for negative current values to demonstrate the observed structure.

At the time of the original observation higher-order quantum transitions were regarded as anomalous since the result is a large fundamental Fourier term at a subharmonic of the Josephson frequency. However, an understanding of the physical mechanism involved (see Section III.B.9) makes it apparent that such subharmonic oscillation is a necessary consequence of high critical current and low damping in the weak link, and should therefore be more prone to occur with good (low-conductance) tunnel junctions than with point contacts or metal-bridge devices.

Experiments with tunnel junctions in a loop of very large inductance were reported and clearly understood by Vernon and Pedersen (1968). With β of the order of 10^6, they observed what was aptly termed relaxation oscillations. Frequency of the oscillation varied from a few kHz to 500 kHz for voltages up to about 0.5 mV. In the language of the preceding paragraph, this would correspond to quantum transitions N of order of 10^6; that is, the frequency of the relaxation oscillation at 0.5 mV was 10^6 times smaller than the Josephson frequency V_0/Φ_0. The fact that N was of the same order of magnitude as β is consistent with the observation that the system relaxed to the ground state (zero current in the junction) as soon as the critical current was slightly exceeded, an observation which was invoked by Vernon and Pedersen in the phenomenological interpretation of their results. The relationship between these observations and those of Zimmerman and Silver (1967) had not been explicitly recognized.

8. *Behavior of resistive-loop devices with rf bias*

The behavior of weakly-connected partly-resistive loop devices (Fig. 15) can be described in a manner similar to that for superconducting loops if we include an additional term $\theta_0(t)$ in the quantum phase proportional to the time integral of the voltage V_R across the resistive section. Thus instead of the Eq. (14) we have

$$\theta_0(t) + \theta + 2\pi\Phi/\Phi_0 = 2\pi k \tag{76}$$

with

$$\theta_0 = (2\pi/\Phi_0) \int V_R \, dt \tag{77}$$

where V_R is the sum of at least three terms: V_N the Johnson noise emf, V_T the thermal emf, and $-i_R R$, where i_R is the total current in the resistance. All of these terms approach zero as R approaches zero; so the dynamic behavior of the partly-resistive loop in the limit always approaches that of the superconducting loop.

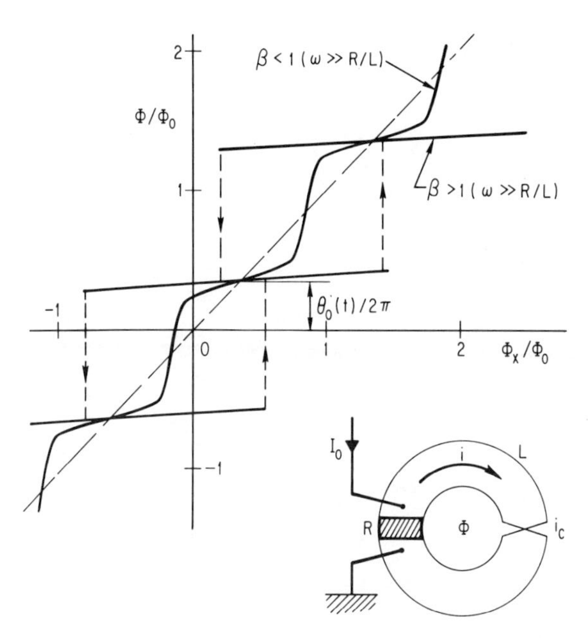

FIG. 19. Quasi-stationary states of resistive loop weak-link devices with idealized forms of $i(\theta)$ for $\beta < 1$ and $\beta > 1$ chosen as sinusoidal and linear functions, respectively. The frequency ω given by $\omega = d\theta_0/dt$ measures the rate at which these "states" progress along the $\Phi = \Phi_x$ direction.

We can describe quasi-stationary quantum states for these resistive loop devices by replacing $2\pi k$ with $[2\pi k - \theta_0(t)]$ in Eqs. (20) and (21) and Eqs. (31) and (32) which define the states of weakly connected superconducting rings (Zimmerman and Silver, 1968a, b). Thus Fig. 19 is equivalent to Figs. 1 and 2 where $\theta_0(t)$ is a continuous function of time which drives the quantum states diagonally along the $\Phi = \Phi_x$, $i = 0$ line. If a time varying magnetic field is applied with $2\pi d(\Phi_x/\Phi_0)/dt \gg d\theta_0/dt$ and $d\Phi_x/dt > Ri_c$, then quantum transitions with the same selection rules for Δk will occur as described earlier for completely superconducting rings. On the other hand, at constant Φ_x the behavior is also understood from Fig. 19. We note that time differentiation of Eq. (76) gives a result of the form of Eq. (65). The oscillator discussed above for $\alpha < 1$ is described by the time evolution of $\theta_0(t)$; the fundamental frequency is just $\langle d\theta_0/dt \rangle$. If $\alpha > 1$, θ_0 is a constant except for fluctuating terms which will be discussed below.

We consider now the operation of a resistive loop device driven with an applied rf magnetic field in the same manner as discussed in Section III.B.4

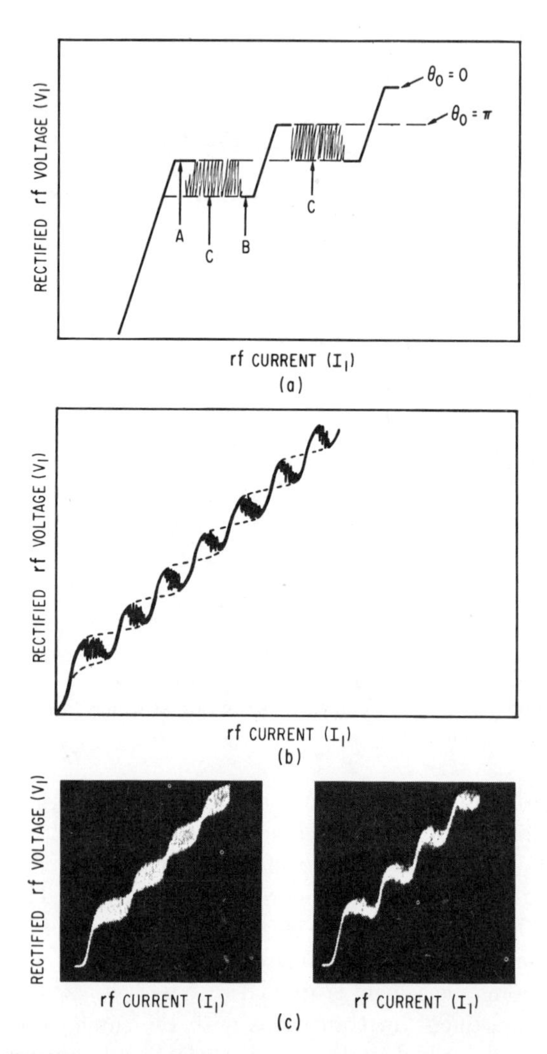

FIG. 20. Radio-frequency response of a resistive loop device operated in a circuit similar to Fig. 10. (a) Expected rf V–I characteristic. (b) Experimental curves showing the instability of the two solutions. (c) Oscilloscope photographs of the rf V–I curves demonstrating oscillation near the zeros of $J_0(\theta_1)$. On the left a dc voltage is superimposed on the resistance, while on the right one sees only the thermally induced instability.

and shown in Fig. 10 for a superconducting loop. The response of an rf-biased resistive loop device with $\omega \gg R/L$ is shown in Fig. 20. This particular device incorporated a resistance of 18 $\mu\Omega$ in the structure, but was otherwise almost identical to the superconducting device with which Fig. 13

was obtained. The response of the latter for the two cases $\theta_0 = 0$ and $\theta_0 = \pi$ [that is, $\Phi_x{}^0 = k\Phi_0$ and $\Phi_x{}^0 = (k + \frac{1}{2})\Phi_0$] is sketched in dashed lines. Thus the average value of θ_0 for the resistive loop alternates between zero and π as I_1 is increased, with intermediate regions of large fluctuations. These phase fluctuations are due to the Johnson noise voltage V_N in the 18 $\mu\Omega$ resistance, and they are largest at the rf bias levels where the time average of the Josephson coupling energy over an rf cycle is essentially independent of θ_0, that is, where $J_0(\theta_1) = 0$. At these bias levels (for example, point C in Fig. 20) the variation of θ_0 in time, under the influence of V_N, is similar to the Brownian motion of a free particle in one dimension (Harding and Zimmerman, 1970). In particular, θ_0 undergoes a random walk such that

$$\langle|\theta_0(t) - \theta_0(0)|^2\rangle = (2\pi)^2 2k_B TRt/\Phi_0{}^2 \tag{78}$$

At other bias levels (points A and B in Fig. 20), or with no rf bias, the Josephson coupling energy averaged over an rf cycle varies as $\cos\theta_0$, and so acts as a constraint to reduce the amplitude of the phase fluctuations. If $\beta \gg 1$ and the size of the loop is such that $k_B T \ll \frac{1}{2}\Phi_0{}^2/L$, for $J_0(\theta_1) = 0$, θ_0 will be constrained to the neighborhood of zero or π. On the other hand, with zero bias, θ_0 will be constrained to the neighborhood of zero if $k_B T \ll \Phi_0 i_c/2\pi$. The latter case, which was discussed in detail by Burgess (1967), is analogous to a particle trapped in a minimum of a periodic potential, where thermal energy is less than the height of the barriers between minima.

If a steady emf V_0 is inserted in series with the resistive loop (e.g., by addition of a current bias $I_0 = V_0/R$), with the rf bias level set at point A in Fig. 20, then θ_0 will steadily increase in time, according to

$$\theta_0 = (2\pi/\Phi_0) \int (V_0 + V_N)\, dt = 2\pi V_0 t/\Phi_0 + \text{fluctuations} \tag{79}$$

Equation (79) is precise without regard to the junction current $i = i_c \sin\theta$ because when $J_0(\theta_1) = 0$ both the average Josephson coupling energy and the dc current [Eqs. (19) and (49)] through the junction are zero. Hence none of the current I_0 flows through the weak link and the dc voltage across R is $I_0 R$. The rf voltage V_1 will therefore be amplitude modulated by an amount given previously for the superconducting loop Eq. (65)

$$\delta\bar{V}_1 = \tfrac{1}{2}\omega\Phi_0 L_1/M$$

and the frequency of this modulation will be given by

$$f = \frac{1}{2\pi}\frac{d\theta}{dt} = \frac{V_0 + V_N(t)}{\Phi_0} \tag{80}$$

The rectified output is centered at $f_0 = V_0/\Phi_0$ and is frequency modulated by the noise $V_N(t)$. To the extent that our description of the operation is accurate, the power spectrum of the detected signal is identical to that of a dc-biased resistive loop given in Section III.B.6. The spectrum centered at f_0 is Lorentzian in shape, with the full width at half-power given by

$$\delta f = 4\pi k_B TR/\Phi_0^2 \tag{81}$$

With very low resistance devices ($R \sim 10^{-10}\ \Omega$) it is possible by this mechanism to measure Josephson oscillations at frequencies as low as 10^{-2} Hz, corresponding to a voltage $V_0 \sim 10^{-17}$ V. The line-width formula was confirmed to within a few percent by Silver *et al.* (1967) for values of R from 10^{-5} to $10^{-10}\ \Omega$ and T from 1.4 to 8K. The range of line widths was from about 10^{-2} to 10^4 Hz.

Note that in describing the response of the rf-biased device to a series emf V_0 we implicitly ignore any iR drop and assume that the emf appears directly across the weak link. The justification for this was already stated above, namely, that at the operational rf bias levels, the coupling energy and the circulating current average to zero over an rf cycle, independent of θ_0 to first order. Thus to this approximation there is no circulating current in response to a dc or quasi-static series emf. An analytic verification for this assumption in the low critical current limit, and experimental verification at higher critical current, was given by Harding and Zimmerman (1970). A corollary of this assumption, applicable to superconducting loop devices, is that at operational rf bias levels (Fig. 12, points F or G, for example) no screening current flows in response to an arbitrary applied flux Φ_x^0. The loop therefore appears to be open circuited as far as slowly varying applied fields are concerned.

An emf V_0 can be introduced into the resistive loop in at least two ways. One way is by the thermoelectric effect. A temperature difference of a degree or two across the device produces a thermal emf in the resistance which, depending on the material, may be as high as a microvolt or more, corresponding to a Josephson frequency of ~500 MHz. In many experiments by the authors a small temperature gradient across the device was usually present because of unintentional heat leaks in the cryostat, giving rise to Josephson oscillations at frequencies of a few hertz to a few kilohertz. The other way is to current bias the resistive section from an external source I_0, which is equivalent to a series emf of magnitude I_0R.

9. *Mechanical analog of single-junction loop devices*

The mechanical analog of a weak link with a shunt capacitance C in a loop of inductance L was introduced in Section II.E. The Hamiltonian for

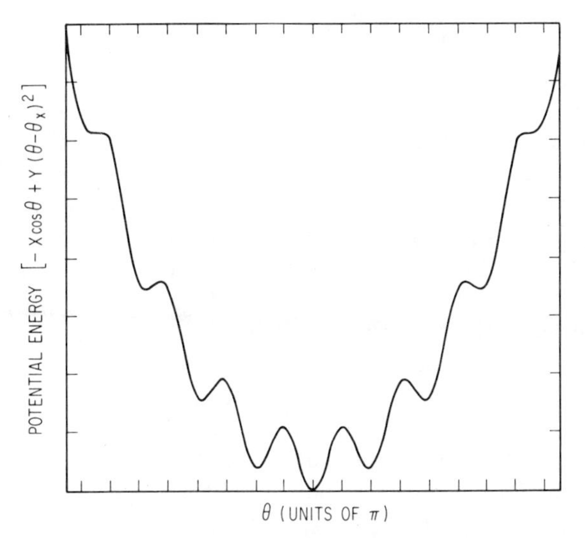

FIG. 21. Potential energy of a single weak-link loop device or mechanical analog as a function of θ. X represents $-\Phi_0 i_c/2\pi$ or $-mgl$; Y represents $\Phi_0^2/8\pi^2 L$ or K/2 for the two systems, respectively. The curve is plotted for θ_x (or Φ_x) = 0 and β = 25. The shape and positions of the potential minima are influenced by both θ_x and β.

the superconducting system is

$$\mathcal{H} = -\frac{\Phi_0 i_c}{2\pi}\cos\theta + \tfrac{1}{2}Li^2 + \tfrac{1}{2}CV^2$$

$$= -\frac{\Phi_0 i_c}{2\pi}\cos\theta + \frac{\Phi_0^2}{8\pi^2 L}\left(\theta - 2\pi\frac{\Phi_x}{\Phi_0}\right)^2 + \frac{\Phi_0^2 C}{8\pi^2}\dot{\theta}^2 \tag{82}$$

An exact mechanical analog is a rigid pendulum of length l and mass m attached to a horizontal torsion bar of torque constant K, the other end of which is attached to a rigid support whose angular displacement θ_x can be varied. For this system we have

$$\mathcal{H} = -mgl\cos\theta + \tfrac{1}{2}K(\theta - \theta_x)^2 + \tfrac{1}{2}ml^2\dot{\theta}^2 \tag{83}$$

The sum of the first two terms of the Hamiltonian, which constitutes the potential energy of the system, is plotted in Fig. 21 for $\theta_x = 0$ and for $\beta = 2\pi L i_c/\Phi_0 = mgl/K = 25$. For this value of θ_x there are nine stationary states for the system, two of which are marginally stable, while for $\theta_x = \pi$ there are eight stationary states (not shown).

In terms of the analog it is quite easy to visualize how irreversible transitions between states come about, and to visualize to some extent what actually happens during the transition. As we slowly increase θ_x (the analog of the applied flux Φ_x) by rotating the end of the torsion bar, the pendulum angle θ also will rotate, but more slowly than θ_x as potential energy is stored in the torsion bar. When the pendulum reaches some angle θ_c between $\pi/2$ and π, it will spontaneously go over the top, and thereafter, in the absence of damping, will execute nonsinusoidal oscillations, as described by Shin and Schwartz (1966). On the other hand, with large damping, the pendulum will simply flop over and come to rest in a new position θ_2 where $\theta_c < \theta_2 < \theta_c + 2\pi$. This is the analog of a transition of the superconducting system to the adjacent fluxoid state. In terms of the potential function, Fig. 21, the system has been tipped out of the highest stationary state and slides down the potential curve to the bottom of the next minimum.

With less damping, the rotational inertia (capacitance) of the system may carry it through one or more minima of the potential function, so that the final angle θ_2 is given by $\theta_c + 2\pi(N - 1) < \theta_2 < \theta_c + 2\pi N$, with $N = 2, 3, 4, \ldots$. This is the analog of the so-called "higher-order" quantum transitions (Sections III.B.5 and III.B.7).

The pendulum analog makes it obvious that higher-order transitions will certainly take place if the loop inductance is increased sufficiently; in other words, the order N of the transition is a function of the total system and not just of the weak link itself. It is also apparent that high-order transitions (and subharmonic radiation) will be much more likely to occur with a tunnel junction operating well below critical temperature (large C, low G) than with a metal bridge operating close to the transition temperature (small C, large G), for the same value of β. Thus the Josephson tunnel junction is more prone to "non-Josephson" behavior than the metal bridge. It has been experimentally confirmed that thin-film metal-bridge loop devices operating near T_c are well behaved (i.e., Josephson-like), in some cases with i_c as large as $10^3\Phi_0/L$. Point-contact devices, on the other hand, operating well below the critical temperature T_c usually undergo higher-order transitions if i_c is greater than 5–10 times Φ_0/L. Point contacts probably are also metal bridges in many cases. However, they simulate tunnel junctions in two respects: first because there may be some shunt capacitance built into the structure, and second, because the conductance of a contact well below T_c is relatively small, as compared to that of a metal bridge of the same critical current operating close to T_c. In general, the values of C and G for point contacts will be between those for metal bridges and for tunnel junctions.

Figure 22 is a photograph of two mechanical analogs which differ by

FIG. 22. Photographs of two mechanical analog models.

several orders of magnitude in cost of materials, but are practically equivalent in tutorial value. The more expensive model features variable eddy-current damping (by moving the magnet relative to the aluminum disk), variable-speed motor drive on the right end of the torsion bar, adjustable pendulum length, and adaptability to a number of different analogs. As shown, it is set up to represent a single-junction loop device with a voltage-independent junction conductance, under the application of a steadily-increasing applied field or of a zero-resistance emf in series with the loop. The economy model represents a symmetric, single-weak-link, double-hole device described in Section IV. The plastic card attached to the paper clip provides damping through air friction, and the application of magnetic flux to one or the other of two holes in the device is simulated by turning one or the other of the cranks at the ends of the rubber band.

The analog of a resistive loop device can be constructed by inserting a viscous clutch in the torsion bar, which permits an additional nonconservative phase slippage between θ and θ_x.

C. Multiple Weak-Link Devices

1. *Quantum interferometer*

Experiments on superconducting quantum interference preceded the superconducting quantum electronic device developments discussed above (Jaklevic *et al.*, 1964a; Zimmerman and Silver, 1964; Lambe *et al.*, 1964). The configuration of two superconducting weak links connected in parallel in a superconducting circuit was widely advertised as the prototype of long-range quantum interference in superconductors, analogous to the optical double-slit interference effect. While this early analogy was conceptually attractive, it was of little value in the quantitative understanding and development of practical devices. In fact, the Josephson theory applied to these interference devices obscured for some time the fundamental simplicity of the quantum electronics of superconductors. For these reasons we have postponed discussion of such devices beyond that of the single-weak-link devices above.

Following the prediction of phase-dependent superconductive tunneling, the magnetic-field-induced diffraction pattern of a single junction was reported by Rowell (1963). However, the sensitivity (magnetic field required for one period of the diffraction pattern) was limited to fields of the order of 1 G or greater. This limit is determined by the size of a vortex in a Josephson junction and is a practical limit on useful junction apertures. In order to confirm the origin of very small magnetic field periodicity discovered in the microwave (X-band) impedance of thin-film superconducting

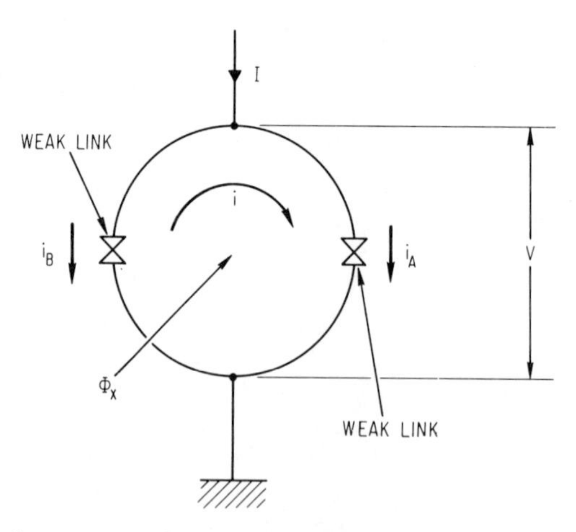

Fig. 23. Interferometer configuration of two weak links in a superconducting ring, biased with a current source I.

bridges (Lambe *et al.*, 1964) experiments were designed and performed on Josephson junction interferometers (Jaklevic *et al.*, 1964a, b, 1965a, b). We discuss now the nature and observation of this effect.

The interferometric behavior is related to the periodic behavior of the screening current in superconducting rings and manifests itself as a periodic variation of the critical current as a function of the magnetic flux applied through the loop. For a ring containing two identical junctions as shown in Fig. 23, the defining equation is from Eq. (14)

$$\theta_A - \theta_B + 2\pi(\Phi/\Phi_0) = 2\pi k, \qquad k = 0, \pm 1, \pm 2, \ldots \tag{84}$$

and

$$\tfrac{1}{2}(i_A - i_B) = (\Phi - \Phi_x)/L \tag{85}$$

where θ_A and θ_B are the phase shifts across junctions A and B, respectively, the positive direction defined from the lower superconductor to the upper. The total current through the external circuit is

$$I = i_A + i_B \tag{86}$$

and the inductance of each half-ring is assumed equal to $L/2$. If we define a circulating current in the ring then

$$i = \tfrac{1}{2}(i_A - i_B) \tag{87}$$

These equations coupled with the appropriate current–phase-shift relation for a weak link, e.g., the Josephson relation $i = i_c \sin\theta$, completely specify the zero-voltage response of the device.

Jaklevic *et al.* (1964a, b, 1965a, b) demonstrated elegantly the nature of the maximum zero-voltage current for two Josephson tunneling junctions in a quantum interferometer. Figure 24 shows the maximum supercurrent as a function of applied magnetic field with a periodicity given by Φ_0. Shortly thereafter Zimmerman and Silver (1964, 1966a) showed similar results utilizing small-area metal contacts. Coupled with the microwave results on thin films (Lambe *et al.*, 1964), the insensitivity of the quantum interferometer to the specific function $i(\theta)$ was in reality established, and specifically alluded to by Zimmerman and Silver (1966a, b). For identical weak links with $i_{A,B} = i_c \sin\theta_{A,B}$ one has

$$I_c = 2i_c|\cos(\pi\Phi/\Phi_0)| \tag{88}$$

This apparent prediction of the modulation of I_{max} from zero to $2i_c$ with period Φ_0 is not observed; rather modulation is generally restricted to less than Φ_0/L.

An excellent analysis, supported by precise experimental data, has been given by deWaele and deBruyn Ouboter (1969). They showed that for

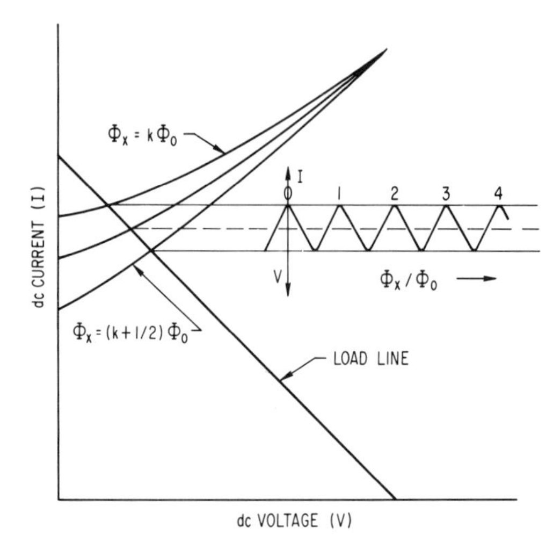

FIG. 27. Idealized dc I–V response of a current-biased double-point-contact interferometer showing the effect of an applied magnetic field on the measured current or voltage.

in agreement with the exact theory of deBruyn Ouboter and deWaele (1969).

Where practical devices are concerned, the zero-voltage response of the double-link system is only of peripheral importance. What we need to know is the magnitude of output power that is available from the device in response to the input Φ_x. This requires a knowledge of the dc I–V characteristic. The simplest case to discuss is that of weak links with constant conductance and negligible shunt capacitance. The I–V characteristic for a single link of this type was given in Fig. 5b. For a double link the characteristic is similar but varies with applied flux as shown qualitatively in Fig. 27, where we have assumed that the critical current of each link is of the order of or greater than Φ_0/L (Zimmerman and Silver, 1966a). The maximum available power from the device is of the order

$$P \approx 0.1(\Phi_0 i_c/LG) \tag{92}$$

which is achieved with a bias current $I \sim 2i_c$ and a load line as shown in Fig. 27, Φ_x varying from $k\Phi_0$ to $(k + \frac{1}{2})\Phi_0$. Making the reasonable assumption that G and i_c are both proportional to the cross section of the link, it follows that P is approximately independent of i_c for links in the size range where $i_c > \Phi_0/L$. On the other hand, the ratio i_c/G decreases rapidly with temperature and vanishes at the critical temperature T_c so that quantum interference is inherently difficult to observe in devices operating close to

T_c. Probably for this reason, the phenomenon has only recently been experimentally observed (Notaries, private communication) in thin-film double-bridge devices, though it was seen several years ago in tunnel-junction and point-contact configurations operating at lower reduced temperatures. In view of these results a comment regarding the original discovery of quantum interference in thin-film double-bridge devices (Lambe *et al.*, 1964) is in order. Although these microwave experiments were at that time regarded as double-bridge interference, it is apparent from the extensive rf studies of single-point-contact devices that in those earliest experiments only one bridge functioned as a weak link. Therefore, the microwave observations fall in the class of phenomena discussed in Section III.B.

Thin-film tunnel junction devices have relatively large inherent shunt capacitance, and many point-contact devices have considerable shunt capacitance built into the structure. This leads to an I–V characteristic with two branches, a zero-voltage supercurrent branch and a finite-voltage normal conductance branch as shown in Figs. 3 and 6. Only the supercurrent branch varies periodically with applied flux Φ_x. In order to use these devices as flux sensors it is necessary to repeatedly sample the critical current I_c, namely, the current at which the device switches from zero to finite voltage. This is most conveniently done by applying an alternating current sweep $I_{ac} > I_c$ and using a phase-locked detector to measure the average voltage across the device as a function of Φ_x (Jaklevic *et al.*, 1965b). When this is done it can be shown qualitatively that the maximum available power, for Φ_x varying between $k\Phi_0$ and $(k + \frac{1}{2})\Phi_0$, is the same order of magnitude as given by Eq. (92). Another way of operating these devices has been suggested, but has not been used in practice so far as we know. This is to apply a high-frequency bias current $I_{rf} > I_c$ to the device. The high-frequency current averages the two branches of the dc characteristic into a single-valued curve, similar to the effect of thermal noise on the characteristics of very weak junctions (see Fig. 4). The device can then be dc biased and operated as explained above for devices without shunt capacitance.

The importance of thermodynamic considerations in the dc quantum interferometer is demonstrated by an interesting effect observed with Josephson tunneling junctions but not reported in similar experiments with point contacts. The observation of a multiplicity of critical current values for a given applied field Φ_x has been reported by Fiske (1964), Jalkevic *et al.* (1965b), Goldman *et al.* (1965), and others, in tunneling junction devices where there are a multiplicity of allowed quantum states. Such an effect is easily understood in terms of the quantum states with no external current, $I = 0$. At the critical-current value I_c the device switches from the zero voltage to the finite voltage branch (see Fig. 3) of the I–V characteristic. When I is now returned to zero, the voltage reduces to zero and the double-junction loop switches to the superconducting or zero-voltage state, it may occupy one of several metastable current carrying states, $i \neq 0$, corresponding to different quantum numbers k in Eq. (84). For $\beta > 1$ there will be more than one state, the number depending on β, Φ_x, and the current–phase relation. The lowest energy state is that one with the smallest circulating current i, independent of direction; the distribution over the various metastable states (with $I = 0$) is determined by the function

$$p(k) \propto \exp[-E(k)/k_{\mathrm{B}}T] \tag{93}$$

where $E(k)$ is the energy of the kth state (Goldman *et al.*, 1965). Since the initial ($I = 0$) value of i determines I_c as discussed above, there will be a different critical current for each initial metastable state. Smaller observed values of I_c correspond to higher energy metastable states.

When a particular critical current value is reached by increasing the applied current I there are actually two possibilities for the subsequent evolution of the system. One is that it switches to the finite-voltage branch as stated above. The other possibility is that the system makes a single (or perhaps higher order) quantum transition to a lower metastable state, such that the supercurrent flow is maintained until the circulating current drops to the ground-state value. By this latter mechanism only the largest critical-current value will be observed in the I–V characteristic, corresponding to the smallest allowed circulating current. It was pointed out previously (Section III.B.9) that the larger conductance and lower capacitance of point contacts, compared to tunnel junctions, favor single quantum transitions. By the above argument point-contact devices should be less likely to exhibit multiple critical-current values. This expectation is borne out by the fact that multiple critical currents have not yet been reported for point-contact devices.

The previous result that $\Delta I_c \lesssim \Phi_0/L$, Eq. (91), is no longer strictly required when $\beta \gg 1$ and multiple values of I_c can be observed. If one were to perform the experiment in such a way that only one metastable state were examined over its entire range of validity, the critical current would vary from $2i_c$ to zero and the naive analogy with optical interference would be applicable.

A very useful physical picture of why the usual superconducting quantum interferometer response does not modulate in the manner of optical interference can be had with reference to Fig. 28 which shows the magnetic behavior of the interferometer with $I = 0$. The experimental path as Φ_x is scanned and I varies between zero and $2i_c$ is shown by dashed lines. We see immediately that a region of Φ-space near $(k + \frac{1}{2})\Phi_0$ is *never* traversed along the path defined by the quantum state. This is analogous to omitting

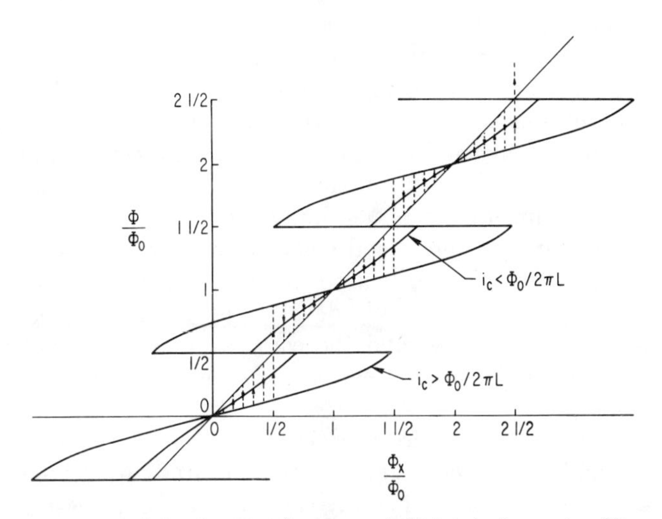

FIG. 28. Magnetic behavior of a double-weak-link interferometer. The solid curves represent the stable magnetic states for two general cases, $i_c \lessgtr \Phi_0/2\pi L$. The dashed lines represent the behavior as the external current oscillates beyond the total critical current with Φ_x increased at a rate small compared with the variation in total current.

a region near $(k + \frac{1}{2})2\pi$ in optical path length in the optical interferometer. For any $i_c > 0$ this inaccessible region will exist; consequently, a double-junction ring with identical critical currents cannot traverse the region near $\Phi = \Phi_x = (k + \frac{1}{2})\Phi_0$ reversibly. This result is also consistent with our previous discussion of single-junction rings when reversibility was ensured for $\beta < 1$. The load for one of the junctions is the loop inductance in series with the second junction. Hence one junction sees an effective inductance equal to the loop inductance L and the Josephson inductance L_J of the second junction, given earlier by Eq. (36), restated here as

$$L_J = (\Phi_0/2\pi i_c) \sec\theta \tag{94}$$

Since each identical junction reaches $\theta = \pi/2$, $i = i_c$ at the same applied field, the junction inductance diverges and the effective β at i_c is therefore always much greater than unity.

IV. Device Construction

A. Thin-Film Tunnel Junctions with Insulating Barriers

Tunnel junctions are constructed by laying down a thin film of superconductor on any suitable substrate, overlaying this with a very thin

($\sim 10^{-7}$ cm) film of insulating material, and finally adding a second film of superconductor. Probably the major technological problem in making these junctions is that of producing a uniform barrier layer of the correct thickness, free of shorts, with long-term reliability under conditions of room-temperature storage and repeated cooling to cryogenic temperatures. For tunnel junctions made with soft superconductors such as tin and lead, the lifetime of the junction depends dramatically on how it is prepared. Originally, most barriers were metal oxides formed by room-temperature exposure of the first metal layer to air or to pure oxygen for a few minutes. Junctions made in this way usually develop shorts when stored at room temperature for several days, and the early experimenters had to keep turning out new junctions on a regular basis, or else store them at low temperatures.

Metal-oxide barriers have also been made by glow-discharge oxidation of the first metal layer. A recent, and continuing, study by Schroen (1968) shows that junctions made in this way can survive, for at least 1 yr, with storage at room temperature with repeated cycling to low temperatures. These junctions were made in arrays, and it was shown that within an

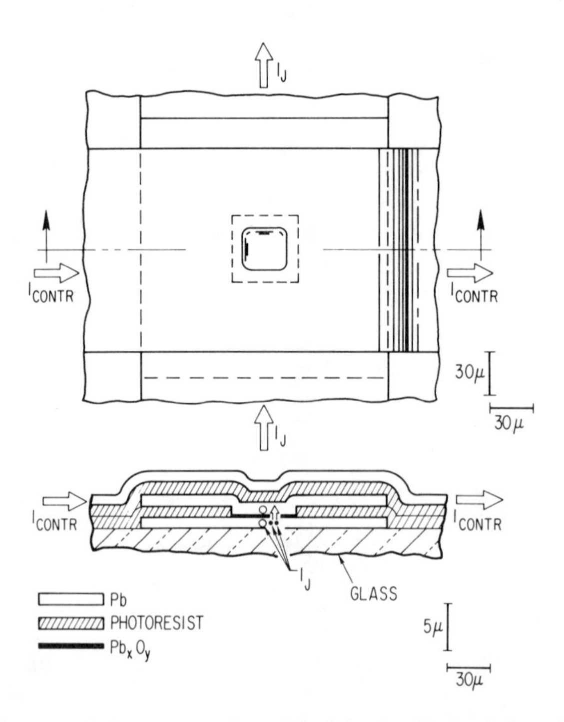

FIG. 29. Diagram of the construction of lead lead-oxide lead Josephson tunneling junctions as performed by Schroen.

array the junctions had nearly the same critical current and switching characteristics between superconducting and normal states. A sketch of one of the junctions is shown in Fig. 29. A thin-film strip of Pb 150 μ wide was vacuum deposited on a clean glass substrate, followed by a 1 μ thick layer of photoresist. An opening the size of the desired junction was made in the photoresist, and the insulating barrier was then formed by glow discharge on the exposed metal surface. A second layer of Pb was evaporated over the photoresist and the barrier to complete the tunnel junction. A third Pb strip over an insulating layer shown in Fig. 29 was used as a control element to apply a magnetic field to the junction and need not concern us at the moment. The insulating barrier was formed by a glow discharge in a low-pressure, 10^{-2} to 10^{-4} Torr, oxygen atmosphere. It is this part of the process, along with the fact that both the sides and the edges of the junction are sealed against atmospheric effects, which is unique and may be responsible for the long-term reliability of the junction. Any model of what happens at the Pb surface during the glow discharge must be regarded as tentative, but it appears that the energetic atomic or molecular species produced by the glow discharge lead to relatively stable Pb_xO_y compounds, as compared to those formed by exposure to a passive oxygen atmosphere at room temperature. Since room-temperature diffusion and annealing take place quite rapidly in soft metals like Pb, the long-term reliability of these junctions is all the more encouraging for being not exactly what one would expect.

Some attention has been given to the construction of thin-film tunnel junctions from hard metals like Nb, or combinations of hard and soft metals. Kamper *et al.* (1969) have made many junctions with Nb and Pb on opposite sides of the barrier. All of these have survived, with "no detectable deterioration," several months storage at room temperature and several hundred cycles of cooling to liquid-nitrogen temperature. Their procedure was to evaporate a Nb film onto a single-crystal sapphire substrate at 400C in a vacuum $\sim 10^{-9}$ Torr. An elevated substrate temperature is essential to produce niobium films with high transition temperatures; that is, lower substrate temperatures result in lower transition temperatures, but the physical basis for this effect has not been clearly established. Nb films approximately 10^{-5} cm thick or greater and made as above had transition temperatures fairly near that of the bulk metal, while thinner films had reduced transition temperatures, e.g., 6.9K at 200 Å, 7.9K at 300 Å, and 8.7K at 600 Å. To make a tunnel junction, the niobium films were cooled to about 80K by circulating liquid nitrogen through a duct in the substrate holder, and then exposed to a glow discharge in pure O_2, or O_2 with 10% N_2, at a pressure of 10^{-2} Torr for several minutes. With the high vacuum restored, the films were warmed momentarily to $\sim$200K, then cooled again to 80K, and then the Pb film was evaporated over the

Nb. Finally, the films were warmed to room temperature, and after being removed from the vacuum system were coated with a thin layer of polyimide plastic to protect the Pb from attack by atmospheric water vapor. The essential process involved in this rather complicated procedure of forming a barrier is to trap on the Nb surface some, but not all, of the reactive products formed by the flow discharge. These products react with the Pb film, which is deposited subsequently, to form the insulating barrier. Thus the barrier may be physically rather similar to those made by Schroen, where the reaction took place simultaneously with the glow discharge.

Nordman (1969) has also fabricated junctions of Nb and Pb, the insulating barrier formed by exposing the Nb film at a temperature of about 100C to oxygen gas at 0.3 Torr for several minutes. These junctions had lifetimes of at least several months, but many of them suffered a considerable increase of conductance with age.

Schwidtal and Finnegan (1969) made Pb junctions with the barrier formed by a glow discharge in an atmosphere of oxygen at 0.02 Torr. These junctions were not particularly stable when stored at room temperature, although the technique of preparation was essentially the same as that of Schroen. On the other hand, the junctions apparently were not protected from attack by atmospheric moisture.

It appears at this juncture that while some workers have developed reliable techniques for fabrication, this technology is not readily transferable.

B. Thin-Film Bridge Devices

Quantum effects in thin-film bridges were first observed by Lambe *et al.* (1964) and by Anderson and Dayem (1964), both using microwave techniques. At the same time it was recognized, at least by implication, that the bridge had to be quite narrow, but narrow compared to what was not then (and perhaps is not now) clearly established. For studies of the dc I–V characteristics of thin-film bridges in a microwave field, the bridges were formed by evaporating through a double V-shaped mask as shown in Fig. 30. The points of the mask were carefully lapped to microscopic sharp-

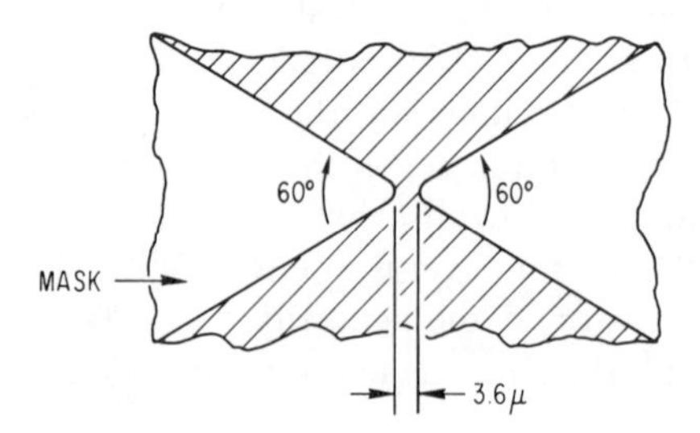

Fig. 30. Geometry of thin-film bridge device. [After Anderson and Dayem, 1964.]

TABLE I

EXPERIMENTAL VALUES OF CRITICAL CURRENT DENSITIES, EXTRAPOLATED TO 0 K, FOR THIN FILMS[a]

Material	Thickness (Å)	Bridge width (μ)	Critical current density (A/cm^2)	Theoretical[b]
Sn	500	1.9	1.76×10^7	1.92×10^7
Sn	295	1.0	1.54×10^7	1.57×10^7
Sn	170	2.9	1.11×10^7	1.28×10^7
Pb	475	1.0	5.26×10^7	7.46×10^7
Al	500	3.0	2.9×10^6	3.16×10^6

[a] From Hunt (1966).

[b] Calculated from H_0/λ_0, the ratio of critical field to London penetration depth at absolute zero.

ness, about 0.1 μ, and were positioned over the substrate so as to make bridges 2 to 6 μ wide. It was important that the bridges were of zero length. No quantum effects were seen in rectangular bridges 4 μ wide and 7 μ long. Lambe *et al.* (1964) observed quantum effects with rectangular bridges 20 μ wide and long. However, these bridges were made by mechanically cutting and removing material from the film; consequently they were rather irregular, and quantum-phase precession probably was localized at some accidental constriction in the bridge.

Nisenoff (Mercereau, 1967) has made numerous thin-film bridge devices from Al and Sn and operated them at 30 MHz. The thin film was deposited as a band on the surface of a glass rod which was rotated during the evaporation, and the bridge was made by mechanically removing material from the band so as to leave a bridge as short as possible and a few microns to a few tens of microns in width.

In all these experiments, except for those on Al, the quantum effects of interest were observed in a restricted temperature range just below the critical temperature. At temperatures well below the critical temperature the critical current of the bridge becomes so large relative to the amplitude of the quantum effects that the latter becomes difficult to observe. The problem in extending the usable temperature range of these devices is then one of making exceedingly narrow bridges in very thin films. So far as rf-biased loop devices of the type developed by Nisenoff are concerned, a pertinent parameter is the critical current relative to Φ_0/L, where L is the loop inductance. Critical-current densities in Sn, Pb, and Al as measured by Hunt (1966) are given in Table I. These measurements were made on bridges narrow enough that the current density could be considered nearly uniform across the bridge.

To illustrate the use of these figures, consider the design of an rf-biased thin-film loop device 1 mm in diam, for which $L \sim 10^{-9}$ H, or $\Phi_0/L \sim 1\ \mu$A. In order to achieve a critical current approximately 10 Φ_0/L with films 100 Å thick the bridge width must be about 400 Å for Al and even narrower for Sn and Pb, which is obviously difficult to achieve with present technology. Nevertheless, Goodkind and Stolfa (1970) made Al devices which function at temperatures of a few milliKelvin, with critical currents approximately 20–100 Φ_0/L. Their bridges, $\sim$1 μ wide, were made by carefully notching the film with a razor blade mounted in a micromanipulator. The films were nominally 100 Å thick, but it was estimated that atmospheric oxidation reduced the actual thickness to $\sim$30 Å. These authors also made Pb devices which operated at 4.2 K, using bridges which were first formed by cutting the film with a razor blade and were then etched down to final dimensions with distilled water. The resistance of the bridge was monitored and the etching process was halted when the resistance became a few ohms.

Thin-film bridges a few microns wide, which can be made by less exacting techniques, will generally function as quantum devices if the temperature is maintained in the neighborhood of T_c where the critical-current density is correspondingly reduced.

No data have been published on the long-term reliability of metal bridges. With soft materials like Sn and Pb in which room-temperature diffusion and annealing take place, it is not obvious that a very narrow bridge would maintain its characteristics over a long period of time. Certainly this is a problem with tunnel junctions made with these materials. On the other hand, it seems that devices made of Nb or Ta films should be highly stable, and deserving of much more attention than they have been given thus far.

C. Point-Contact Devices

So many different configurations of point-contact devices have been used, or are possible, that it would be difficult to review all of them. These configurations have one thing in common; they all work, and except for varying degrees of mechanical instability, they work well. There is no restriction on contact material, provided it is superconducting, nor on operating temperature, provided it is below the transition temperature of the material. Contact materials which have been used fairly extensively include Sn, In, Pb, Ta, V, Nb, Nb–Zr alloys, Nb–Ti alloys, Nb_3Sn, and others, as well as combinations with two different materials on either side of the contact. Usually the materials are in bulk form, however, the technique can be adapted to thin films, as in an experiment where long-range phase coherence was demonstrated in vanadium thin films. Also, we have made single-

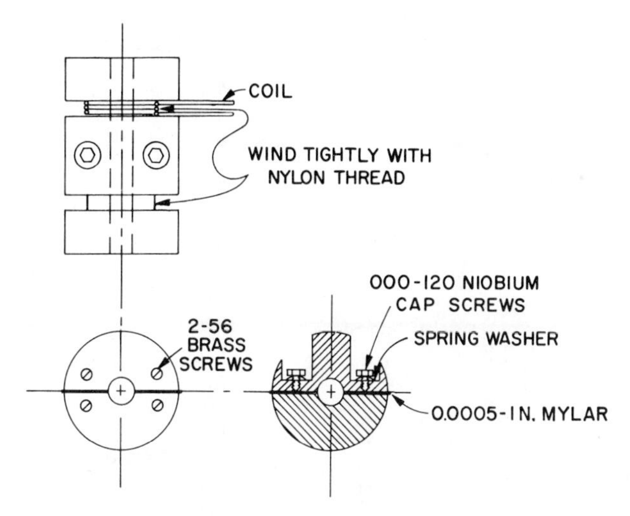

FIG. 31. Drawing of an adjustable point-contact device using niobium pointed screws and niobium rings, not to scale. [After Silver and Zimmerman, 1967b.]

contact loop devices by wrapping 25 μ Nb foil around an insulating rod and applying pressure with a pointed member to make contact in the region of overlap. This type of device has the virtue that the loop area is rather well defined, so it can be used as an absolute magnetometer.

The tendency to mechanical instability of point contacts, noted above, is wholly dependent on the supporting structures and can be eliminated by good mechanical design. A very satisfactory design is shown in Fig. 31. So far as mechanical stability is concerned, three features of the device are essential: the mating surfaces should be lapped to optical flatness and the two halves clamped tightly together; a stiff spring washer in the form of a curved piece of beryllium copper is necessary to eliminate backlash and stabilize the screw against mechanical vibration; and the adjusting rod should slide vertically in a guide and be able to be removed after the contact is adjusted. Devices like this, once adjusted, have been cycled a number of times to room temperature with no significant change of contact characteristics. However, gross changes seem to occur sooner or later, necessitating readjustment. These changes are not necessarily associated with temperature cycling as such, but may be caused by the mechanical effect of moisture condensing during warmup, working into the screw threads, and subsequently freezing. A very common effect with double-contact current-biased devices like that shown is that quite low-power electrical transients in the neighborhood can burn out or grossly alter the contact.

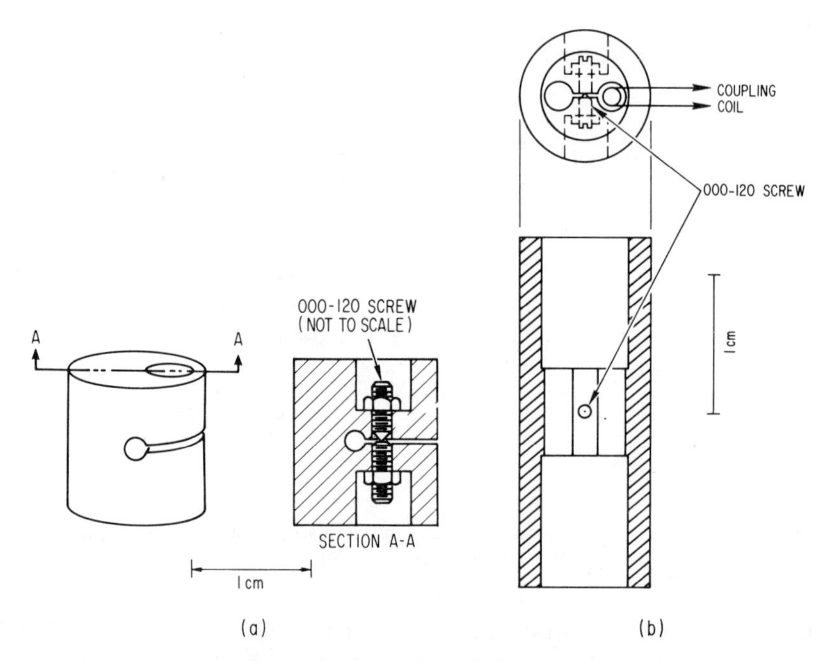

FIG. 32. Point-contact loop devices with fixed point contacts and good long-term stability. (a) A C-shaped block structure with a preset and locked 000-120 Nb screw contact. (b) A symmetric structure with a fixed screw contact and the coupling accomplished with the coil inserted in one of the two holes.

Because of the mechanical delicacy of point contacts, few serious efforts were made to produce reliable fixed-contact devices as compared with the efforts on thin-film bridge and tunnel junction devices. Nevertheless, for most applications such devices are clearly desirable. Consequently, Zimmerman *et al.* (1970) have made rf-biased loop devices with fixed point contacts which have proven to be quite reliable in long-term operation. In fact, the first such device which was built in January 1968 was used repeatedly with no further adjustment, most recently by Cohen *et al.* (1970) in an experimental demonstration of magnetocardiography.

The devices were made entirely of one material (niobium) in order to eliminate differential expansion, and were sufficiently rugged to withstand shock and vibration. In addition, they were mounted and operated in a closed cryostat or probe so that they were never subjected to atmospheric condensation during warmup and cool-down cycles. Two types of stable devices that have been used extensively are shown in Fig. 32a. One is a C-shaped block of material with a point contact between two opposing 000-120 screws bridging the gap. This device with a 1.5 mm hole has a quantum periodicity of about 10^{-5} G applied field. The other is a symmetric, two-hole structure with the contact bridging a narrow slot between the two holes. The latter structure is the more rugged of two, and has some unique magnetic properties as well. By symmetry, it exhibits no response to uniform applied fields, but will respond to an inhomogeneous field which tends to produce a greater local field in one hole than in the other, i.e., a field with a nonzero off-diagonal component of the gradient (e.g., $\partial H_z/\partial x$). As a consequence of the close spacing of the holes and the shielding effect of the long cylindrical body, this device is rather insensitive to laboratory fields, uniform or otherwise. Coupling to the device is effected by means of a coil inserted directly in one or the other of the two holes, as indicated in Fig. 32b (see also Section V). The tip of one screw is carefully machined and honed to make a sharp, 90° point. (Some workers prefer to use a chemical etch to make an extremely sharp point.)

The contact screw is adjusted and locked *at room temperature* to give a resistance of 10 to 100 Ω. Obviously, the contact resistance cannot be measured by dc methods in the closed-loop structure, however it can be inferred from high-frequency measurement of the impedance of a resonant circuit to which the loop is inductively coupled. The circuit used for this purpose is basically the same as that given in Fig. 10 except that everything is at room temperature, and a relatively-high rf bias level and a low-gain amplifier can be used. As the contact is gently closed, the Q of the input circuit will decrease linearly with the decrease of contact resistance. If the contact is screwed down more tightly the resonance frequency will shift upward, but such a contact will have too high a critical current by some orders of magnitude ($\beta \ggg 1$). Since the correlation between room-temperature contact resistance and critical current is rather uncertain (Zimmerman and Silver, 1966a), room-temperature adjustment of a device is essentially a cut-and-try process which may involve several trials.

Construction of resistive devices involves the use of two different materials, and differential expansion is a serious problem. The cross section of a device is given in Fig. 33. With this design, differential expansion between the niobium part of the structure and the resistive part (a dilute copper–germanium alloy) amounts to 2 or 3 μ, so that the contact will open completely upon warming the device from 4.2 K to room temperature. A design which, in principle, compensates for the effects of differential expansion is shown in Fig. 34. By dividing the resistive section into two identical parts, symmetrically arranged, the thermal expansion integral around the loop vanishes to a first approximation. Several such devices were constructed. The resistive material was copper with up to 10 at. % germanium to increase the resistivity. The resistive alloy was bonded to the niobium by induction melting in a pure helium atmosphere. One technique employed was to mill

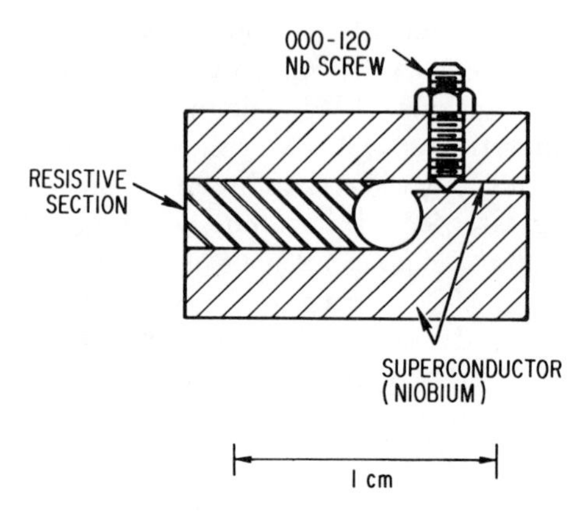

FIG. 33. Construction of a resistive loop device employing a point contact.

a slot in a niobium block, place a slab of the alloy in the slot, and then heat the block to a temperature well above the melting point of the alloy. Very good bubble-free bonds were obtained if the helium atmosphere was sufficiently pure and the metal parts were well cleaned. The device was machined from the composite block. As far as long-term reliability is concerned, this design was only marginally successful. A device with a 0.15 mm thick resistive section was temperature cycled 14 times in 45 days with no

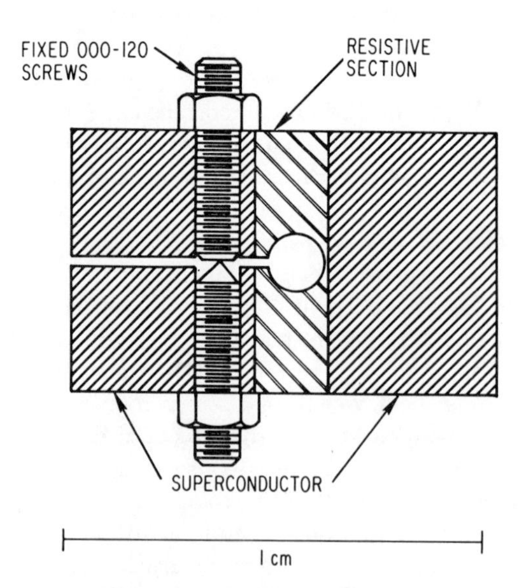

FIG. 34. Construction of a resistive loop device with a fixed point contact stabilized against thermal expansion degradation of the contact.

appreciable change of characteristics, but several devices with 1.5 mm thick resistive sections survived no more than a few temperature cycles, the contact generally getting weaker with each cycle.

The response of these single-point-contact devices to an rf bias field and a quasi-static or dc field, as described in Section III.B is quite reproducible from one device to another (of the same geometry) provided the contacts are preset so that the critical-current values lie within the appropriate range. It is quite difficult to set the critical current exactly to a particular value by room-temperature adjustment, but fortunately, as discussed in the earlier section, the amplitude of the quantum effects [Eq. (44)] does not depend upon the critical current within this limited range, and so there is no need to set the critical current precisely.

The critical current of point contacts frequently varies, periodically or otherwise, with gross changes ($\sim$1 G) of ambient field. This is interpreted as a quantum interference effect (Section III.C) between two or more parallel conducting paths within the microscopic contact area (Zimmerman and Silver, 1964). Such variations present a problem if a single-contact rf-biased device is to be used over a wide range of dc field, since the rf bias level required for a particular small range of dc field may be quite inappropriate for other values of dc field, because of the change of critical current. The effect is minimized by using the sharpest possible point and by taking care not to damage it during adjustment. In the case of the symmetric device (Fig. 32b), the dc field at the contact can be held constant by maintaining symmetry in the coupling coil, as discussed in Section V.

D. The Clarke Device

In this chapter we have concentrated mainly on thin-film and point-contact devices whose geometry and construction can be rather accurately specified. As often happens in practice, devices can be made by exceedingly simple methods which do not fit this description, but which are nevertheless very useful for practical applications. Such a device, invented by Clarke (1966), consists of a length of Nb wire surrounded in a small region by a lump of soft solder, and provided with bias leads as sketched in Fig. 35. The natural oxide layer on the surface of Nb provides an insulating layer between the Nb and the solder, but there is a tendency for superconducting weak links (tunnel junctions or metal bridges) to form at each end of the solder lump. Therefore, the solder lump and Nb wire together constitute a double-weak-link quantum interferometer whose dc characteristics have been described in Section III.C. The aperture of the interferometer is the annular insulating space between the wire and the solder lump. The annular space is totally shielded from external field. However, flux can be induced in the interferometer by passing a current I_x through the Nb wire, such that in its simplest form this device is a current sensor.

Several variations of the basic device are possible for particular applications. For example, if the ends of the Nb wire are bent around and joined together to form a closed superconducting loop, an externally applied magnetic flux Φ_x applied to the loop will induce a persistent current I_x. This modification converts the device into a magnetometer.

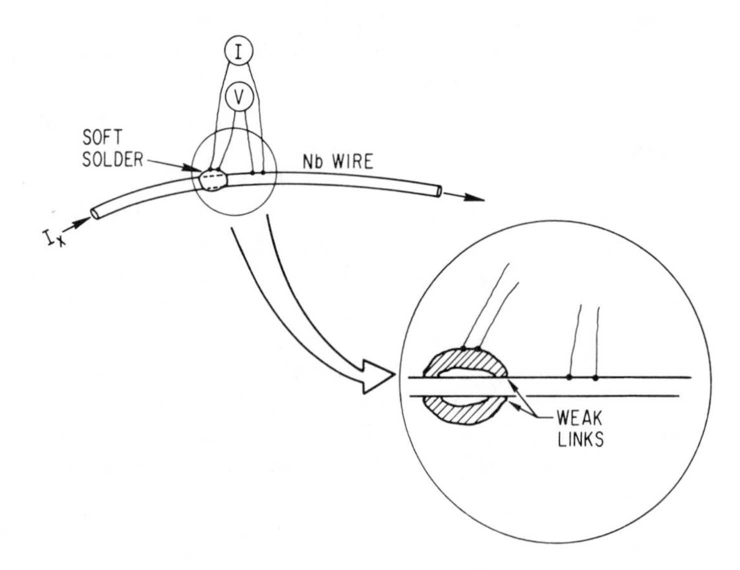

FIG. 35. Solder-superconductor device of Clarke where the weak links are annular regions at the edge of the solder lump. I_x is the external current which generates the flux coupled to the device, Φ_x.

Clarke (private communication) has found that these devices can exhibit good long-term reliability under conditions of thermal and mechanical stress encountered in normal handling, room-temperature storage, etc. It is not expected that the characteristics would be reproducible from one device to another, but this is offset by the ease with which large numbers can be produced, from which devices with suitable characteristics can be selected.

Because of its very small effective volume (see Section V.A), the sensitivity of this device as a magnetometer or galvanometer probably is not as high as can be achieved with the larger point-contact and thin-film loop devices.

V. Applications

A. Magnetometry

1. *General considerations*

Josephson tunnel junctions, point contacts, and superconducting metal bridges, singly or in parallel, dc or rf biased, whether or not incorporated into loop structure, are all potentially useful for magnetic field sensing. As a general rule, devices of large effective volume will be more sensitive than those of small volume. The sensitivity to magnetic field is proportional to the square root of the fraction of field energy to which the device is coupled. Thus, weak-link loop devices like those described in Sections III.B and III.C, with effective volumes of the order of 0.01 cm^3, are likely to be much more sensitive than a single dc-biased tunnel junction with a volume of the order of 10^{-6} cm^3 or less. It should be emphasized that the effective *volume*, not the effective area, is the pertinent parameter in determining the sensitivity. We can operationally define the effective volume of a loop device of inductance L by the equation

$$V_{\mathrm{eff}} B_x^2/2\mu_0 = \Phi_x^2/2L \tag{95}$$

where Φ_x is the flux which links the device as a result of an applied field B_x. The factor $B_x^2/2\mu_0$ is the energy density in the applied field. If we define the effective area as $A_{\mathrm{eff}} = \Phi_x/B_x$, then

$$V_{\mathrm{eff}} = \mu_0 A_{\mathrm{eff}}^2/L \tag{96}$$

Now, since $L \sim \mu_0 a/l$, where a and l are the area and length of the loop aperture, respectively, we have

$$V_{\mathrm{eff}} \sim al \tag{97}$$

That is, the effective volume of a device is the order of its geometric volume, except in the case of peculiar geometries (see below).

With most magnetometers sensitivity can be arbitrarily enhanced simply by increasing the size (that is, the volume) of the sensor. Superconducting quantum devices, however, are in a class by themselves in that the response to field is periodic (period $\Delta B_x = \Phi_0/A_{\mathrm{eff}}$), so that, while sensitivity can be increased by scaling up the device, the apparent dynamic range decreases correspondingly. The limiting size of a device is given by the condition that thermal energy $\frac{1}{2}k_B T$ should be appreciably less than the energy of a flux quantum $\Phi_0^2/2L$, or

$$L \ll \Phi_0^2/k_B T \tag{98}$$

For $T = 4$ K this requires $L \ll 10^{-7}$ H. A 1 cm diam, double-contact interference device with $L \sim 10^{-8}$ H, which was operated several years ago by the authors (Zimmerman and Silver, 1966a), is of the order of the limiting size.

The above discussion notwithstanding, the sensitivity of a superconducting loop device can be arbitrarily enhanced by increasing the effective volume in a way which does not increase the inductance. This can be done by adding slotted disks or flanges to the loop, as indicated in Fig. 36b, or through the use of a flux transformer, as in Fig. 36c.

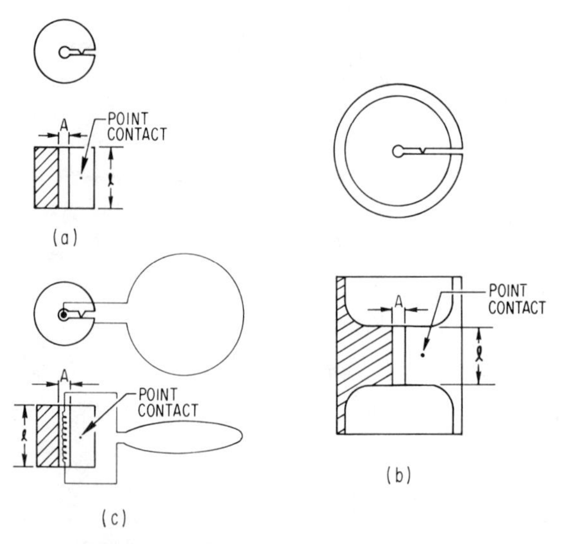

FIG. 36. Superconducting loop devices of enhanced sensitivity in (b) and (c) compared to that of (a), although the aperture near the point contact has the same area A and length l in each case.

Figure 36 presents two variations of the same fundamental idea of using additional superconducting material to distort the applied field in such a way as to enhance the flux in the aperture of the device. These are examples of the "peculiar" geometries referred to above. If we postulate that the device aperture (the central hole) has the same area and length in all three geometries, then it follows that the device inductance of either Fig. 36b or 36c will be less than that of Fig. 36a, because of the presence of additional diamagnetic material. Thus, these configurations enhance the signal-to-noise ratio by two mechanisms, first by reducing the periodicity in field and second, by increasing the amplitude of the response by reducing the inductance (see Eq. (44) with M proportional to $L^{1/2}$).

2. *Sensitivity of loop devices as determined by thermodynamic fluctuations*

We consider three noise sources which contribute to the total fluctuation level of a weakly connected loop device. We will specifically discuss the rf-biased single-weak-link loop device, although the greater part of the discussion will apply to a dc-biased double-link quantum interferometer also. The noise sources are (1) "intrinsic" noise produced by thermal fluctuations of the normal electrons in the weak link, (2) thermal-fluctuation noise from normal-metal parts in the neighborhood of the device, such as a room-temperature metal cylinder enclosing the cryostat, and (3) noise generated in the preamplifier input.

If the conduction mechanism in the weak link consists of two parallel processes [Eq. (43)], one the quantum-phase-dependent supercurrent, and the other a normal current characterized by a voltage-independent conductance G, then the normal current is subject to thermal fluctuations (Burgess, 1967). The fluctuation emf is given by the Nyquist formula

$$\frac{d\langle V_N^2\rangle}{d\omega} = \frac{(2/\pi)k_B T}{G} \tag{99}$$

In the absence of any supercurrent, this produces a flux-fluctuation spectrum in the loop given by

$$\frac{d\langle \Phi_N^2\rangle}{d\omega} = \frac{(2/\pi)L^2 k_B T G}{1 + \omega^2 L^2 G^2} \tag{100}$$

Here we digress to point out again (see Section III.B.8 and Harding and Zimmerman, 1970) that at the operational rf bias levels there is essentially no dc or low-frequency supercurrent in response to dc or low-frequency applied fields. There is, however, an influence on the rf supercurrent in response to these applied fields. It is the modulation of the rf supercurrent which flows in response to the rf bias current which results in the modulation envelopes shown in Sections III.B.5 and III.B.8. Therefore, the low-frequency part of the fluctuation spectrum of Eq. (100) is not shorted out by the supercurrent as it would be in the absence of rf bias (Burgess, 1967). The integral of Eq. (100) gives the total fluctuation of the magnetic flux

$$\langle \Phi_N^2\rangle = L k_B T \tag{101}$$

and the spectrum is plotted in Fig. 37. It is maximum at zero frequency and drops to one-half at $\omega = 1/GL$. The zero-frequency value can be written

$$\frac{1}{\Phi_0^2}\frac{d\langle \Phi_N^2\rangle}{d\omega} = \frac{2Lk_B T}{\pi\Phi_0^2}(LG) \sim 10^{-2}LG \tag{102}$$

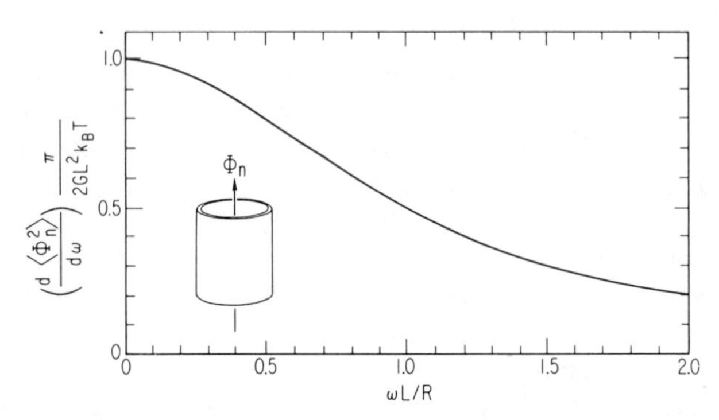

FIG. 37. Intrinsic flux-fluctuation spectrum in a conducting loop in thermal equilibrium at a temperature T. The fluctuation is normalized to unity at zero frequency. G and L are the low-frequency conductance and inductance of the cylinder, respectively.

The numerical coefficient here is calculated for a typical device ($L \sim 10^{-9}$ H) at 4.2 K. The value of G for point-contact devices is generally 10 Ω or greater, so the rms flux fluctuation per unit bandwidth in the neighborhood of zero frequency is

$$\left(\frac{d\langle\Phi_N^2\rangle}{d\omega}\right)^{1/2}\Bigg|_{\text{intrinsic}} \lesssim 10^{-6}\Phi_0 \sec^{1/2} \qquad (103)$$

The flux-fluctuation spectrum produced by a ring or cylinder of normal metal is calculated in exactly the same manner as above, if L and G are replaced by the values appropriate to the normal-metal ring. A cylindrical normal-metal rf shield enclosing the device is an example. Typically the decay time L_c/R_c for such an enclosure is 10^{-4} sec or longer, so the low-frequency ($\omega \ll R_c/L_c$) spectral intensity is

$$\left(\frac{d\langle\Phi_N^2\rangle}{d\omega}\right)^{1/2}\Bigg|_{\text{external}} \sim 10^{-2}\eta\Phi_0 \sec^{1/2} \qquad \text{for} \quad T = 300 \text{ K}$$

$$\sim 10^{-3}\eta\Phi_0 \sec^{1/2} \qquad \text{for} \quad T = 4.2 \text{ K} \qquad (104)$$

where η is the coupling coefficient between the enclosure and the device. Here it is envisioned that the rf shield might enclose the whole cryostat, or might sit inside the helium bath in much closer proximity to the device. In the first case the flux fluctuation is much greater, but on the other hand a smaller fraction of the fluctuation is seen by the device because of the lower coupling coefficient.

The fluctuation mode considered above is just one of an infinity of "decay modes" of the normal-metal cylinder. These modes are solutions of the diffusion equation for the particular geometry of the normal conductor, the nth mode being characterized by a time constant τ_n and embodying an average magnetic energy of $\frac{1}{2}k_B T$. Thus the fluctuation spectrum cannot, in general, be characterized by a single time constant, since all of the modes of the system, both intrinsic and extrinsic, will contribute. Regardless of the spectral distribution, however, the total thermal magnetic energy in the device is just $\frac{1}{2}k_B T$, since the device is a one-degree-of-freedom system (Harding and Zimmerman, 1968). The presence of normal metal (at temperature T) in the neighborhood of the device can only modify the spectral distribution but not the total magnitude of the thermal magnetic energy.† Furthermore, for a cylindrical normal-metal enclosure surrounding the device, the major contribution will be from the lowest decay mode, so that to good approximation the low-frequency fluctuation spectrum will be of the form given in Fig. 37, characterized by the decay time L_c/R_c.

Consider now noise generated in the rf amplifier input. It has been shown (Zimmerman *et al.*, 1970) that the fluctuation level at the output of the electronics is equivalent to a spectral intensity of $1.6 \times 10^{-4}\Phi_0\ \mathrm{sec}^{1/2}$ in the neighborhood of zero frequency, with a device (either of the types shown in Fig. 32) enclosed in a superconducting shield. This is considerably larger than the estimated intrinsic level given in Eq. (103) above ($\sim 10^{-6}\Phi_0\ \mathrm{sec}^{1/2}$), and indicates that the observed noise level can be ascribed primarily to fluctuations at the preamplifier input circuit, that is,

$$\left.\left(\frac{d\langle \Phi_N{}^2\rangle}{d\omega}\right)^{1/2}\right|_{\mathrm{preamp}} \sim 10^{-4}\Phi_0\ \mathrm{sec}^{1/2} \tag{105}$$

If the device is tightly coupled to a rod or cylinder of normal metal at the same temperature, then the spectral intensity may be $10^{-3}\Phi_0\ \mathrm{sec}^{1/2}$ or greater, Eq. (104), depending upon the time constants of the dominant decay modes. This level is greater than either the intrinsic contribution or the preamplifier contribution, and so is easily observable (Vant-Hull *et al.*, 1967). As long as any high-conductivity normal-metal parts are well removed or decoupled from the device aperture, the sensitivity will be determined by input circuit fluctuations. The order-of-magnitude numerical value given above Eq. (105) can be achieved with a cascade preamplifier using any of several types of vacuum tubes or field effect transistors, for example the 6922, 2N4416A, 2N3819, or TIS88.

† Whether the device "sees" the normal-metal enclosure directly or is coupled to it through a flux transformer does not alter this conclusion.

3. *Effect of interference on device operation. rf shielding and dynamic range*

The very broad frequency response and high sensitivity of superconducting quantum devices has been widely remarked. In fact, this property may be a mixed blessing, since the dynamic range of the devices is very small compared to that of many electronic devices. If the total of all extraneous signals, thermal noise as well as interference such as local radio and radar signals and ignition noise, over the frequency range $0\text{–}10^{12}$ Hz, has an rms magnitude $\Phi_0/2$ or greater in the device, then the quantum periodic response is smeared over half a period or more. This means, in effect, that the utility of the device has been wiped out. The full implications of this limitation on dynamic range will be appreciated if it is recalled that helium-temperature thermal noise *alone* gives a fluctuation of the order of $0.1\Phi_0$ [Eq. (101), with $L \sim 10^{-9}$ H and $T \sim 4$ K].

This discussion makes it clear that a quantum device used as a magnetometer must be enclosed in a shield so that the device is not exposed to the full spectrum of terrestrial interference. In the preceding section it was pointed out that a metallic shield tends to increase the thermal fluctuation spectral intensity in the region of zero frequency, that is, in the very region that is of interest in magnetometry. The increase is enhanced if the shield is of high conductivity and is closely fitted around the device.

The core of the matter is that, for most purposes, a shielding enclosure is essential to keep the device within its dynamic range, but the shield can seriously degrade the magnetometer sensitivity by increasing the low-frequency spectral intensity of thermal noise. Thus the design of an optimum shield is more subtle than one might suppose. It involves considerations of size, geometry, terrestrial noise spectrum, magnetometer bandwidth, and application, etc. For present purposes we will note merely that the shield should have as low conductivity as possible consistent with the requirement of keeping interference to an acceptable level, and it should be as large as possible to reduce the device coupling to the thermally excited decay modes of the shield.

4. *Flux-transformer design*

The use of a flux transformer (also referred to as a dc transformer) to enhance the sensitivity of a device to applied field has been mentioned above (Section V.A.1).

A schematic of a flux-transformer coupled device is given in Fig. 38. The flux transformer is topologically a closed loop, shown here as an external coil L_1 coupled to a magnetic field B_x and connected to a coil L_2 which is tightly coupled to the superconducting quantum device. Because of the

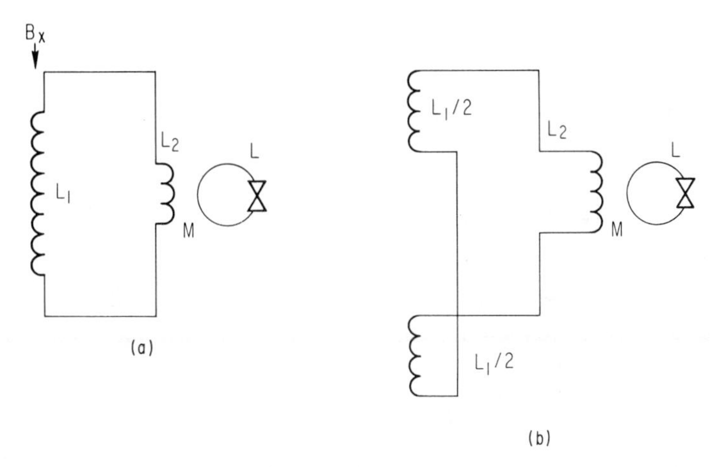

Fig. 38. Schematics of two flux transformers for use with loop devices: (a) field-sensitive transformer; (b) field-gradient-sensitive transformer.

close coupling, the mutual inductance M will approach $(L_2L)^{1/2}$. Two configurations are shown in Fig. 38. On the left, L_1 is a uniform coil, and application of a field B_x causes a current to flow in the topological loop formed by L_1 and L_2. On the right, L_1 is shown as two equal coils in series opposition, so that there is no response to a uniform applied field, but a current will be induced proportional to the gradient of the field in the neighborhood of the two coils. Different transformer configurations can be devised which will couple the device to any component of the field or its derivatives, but these two arrangements seem to be of greatest practical interest. The gradient configuration is particularly useful for susceptibility measurements, where one applies a uniform field to the two coils and measures the unbalance produced by the sample magnetic susceptibility in one of the coils. Such measurements are in progress in several laboratories, using a variety of rf- and dc-biased thin-film bridge and point-contact devices like those described in Section IV.

Flux-transformer design is relatively straightforward. A useful principle is that for a magnetic field applied to a given external coil (or coils) L_1, the flux in the device will be maximum if the coupling coil has the same inductance, that is, if $L_2 = L_1$. Here it is to be understood that L_2 is defined as the *in situ* inductance of the coupling coil, with the weak link open circuited (see Section V.A.2). This requirement for maximizing the response is analogous to impedance matching for maximum power transfer in ordinary circuits. Under this condition the field energy in the device will be one-quarter of the short-circuit free energy of the external coil in the applied field B_x. Thus

$$\tfrac{1}{2}(\Phi^2/L) \leq \tfrac{1}{4}[\tfrac{1}{2}(\Phi_x^2 N_1^2/L_1)] \tag{106}$$

and therefore,

$$\Phi \leq \tfrac{1}{2}(\Phi_x N_1 L^{1/2})/L_1^{1/2} = \tfrac{1}{2}[N_1 B_x A_1/(L_1/L)^{1/2}] \tag{107}$$

$$B_x \geq 2\Phi(L_1/L)^{1/2}/A_1 N_1 \cong 2B(\mathcal{V}/\mathcal{V}_1)^{1/2} \tag{108}$$

$$\Phi_x \cong 2\Phi(A_1 l/l_1 A)^{1/2} \tag{109}$$

where Φ is the flux in the device, Φ_x is the applied flux within the projected area A_1 of the external coil of length l_1 and volume $\mathcal{V}_1$, and N_1 and L_1 are, respectively, the number of turns and the inductance of the external coil. $\mathcal{V}$, A, and l are, respectively, the internal volume, area, and length of the device, with $L \cong \mu_0 A/l$ and $B = \Phi/A$.

If we assume for simplicity that L_1 is a single-layer coil of length l_1 and area A_1, then L_1 is proportional to N_1^2. Therefore, the optimum response $\Phi_{\max}$ is independent of N_1, and is proportional to the square root of the volume of the external coil [substitute $L_1 \propto N_1^2 A_1/l_1$ in Eq. (107)]. We have stated that maximum flux in the device is realized when $L_2 = L_1$. As a matter of fact, this is not quite the condition for maximum signal–to–noise. The signal–to–noise ratio of an rf-biased device is enhanced by the presence of the flux transformer because of the reduction of the effective device inductance. Denoting this by L_{eff}, we have

$$L_{\text{eff}} = L - M^2/(L_1 + L_2) \tag{110}$$

If we replace L in Eq. (101) with L_{eff}, and then maximize the signal–to–noise ratio $\Phi^2/\langle\Phi_N^2\rangle$, we derive the condition

$$L_2/L_1 = (1 - \eta^2)^{-1/2} \tag{111}$$

where $\eta = M(LL_2)^{-1/2}$ is the coupling coefficient. With point-contact devices it is difficult to achieve η greater than about 0.8. For this value of η we have $L_2/L_1 = \tfrac{5}{3}$, which is not grossly different from the condition for maximum flux ($L_2 = L_1$).

Reviewing the main points of the above discussion: (1) maximum flux in the device is realized when $L_2 = L_1$, although maximum signal–to–noise ratio requires L_2 slightly greater than L_1; (2) response is not enhanced by using a multiturn coil for L_1, so that a single-turn coil (or two opposing turns for the gradient configuration) is usually most convenient; (3) sensitivity to field varies as the square root of the volume of the external coil, or as the $\tfrac{3}{2}$ power of its linear dimension. If a single turn is used for the external coil, it is best made from a wide metal band rather than a wire loop of the same diameter, as this will increase the effective volume.

5. *Prospects for ultrasensitive magnetometry*

To achieve and *utilize* the full capability of a superconducting device as a magnetometer requires extraordinary attention to the quality of the environment (including the cryostat) in which it is to be used. An rf-biased loop device with a small flux transformer to enhance the coupling to the external field is certainly capable of sensing field changes of 10^{-10} G or less, but to use this sensitivity in a meaningful way requires stability of the ambient external field to at least one part in 10^{10} of the earth's field. One way of clamping the ambient field is to surround the magnetometer with a superconducting shield, and this undoubtedly is or will become standard practice for high-sensitivity magnetic studies on small samples.

The ambient field can also be reduced and stabilized by means of large, high-permeability magnetic enclosures at room temperature. Cohen *et al.*

FIG. 39. Arrangement for recording of the magnetocardiogram inside a 12 ft magnetically shielded enclosure.

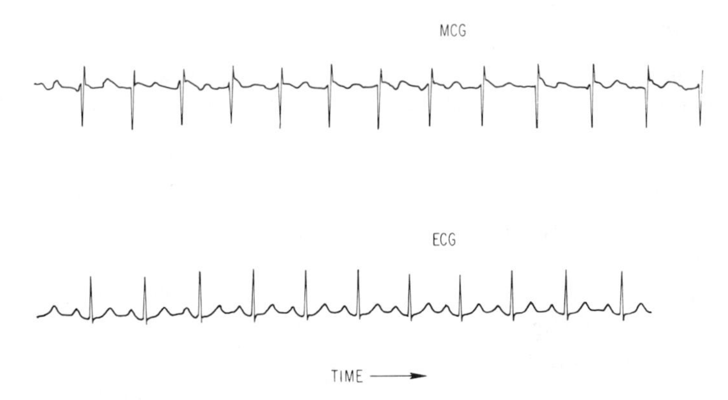

FIG. 40. Comparison of the magnetocardiogram (MCG) and electrocardiogram (ECG) on a single subject.

(1970), working inside an 8-ft-diam enclosure as shown in Fig. 39 achieved an operating sensitivity of 10^{-9} G rms $Hz^{-1/2}$, and recorded the magnetocardiogram of a human subject with a signal–to–noise ratio comparable to that of a good electrocardiogram. A symmetric device of the type shown in Fig. 40 was coupled to the external field by means of a flux transformer whose external coil (L_1 in Fig. 38) was a 2-cm^2 loop of 0.025 cm Nb wire. The coupling coil L_2 was ten turns of the same wire closely fitted into one of the 2-mm holes of the symmetric device. The external coil L_1 was coupled to the human magnetic field by placement of the cryogenic apparatus as shown in Fig. 39. Favorable comparison between the magnetocardiogram (MCG) and an electrocardiogram (ECG) of the same subject is evident in Fig. 40.

Another way of utilizing the very high-sensitivity capability in the terrestrial environment, which may be applicable to magnetic anomaly detection and geophysical prospecting, is to couple the device to the field with a flux transformer in the gradient configuration. This, in principle, eliminates response to the uniform component of the terrestrial field, and to the temporary variation thereof. Since the short-term variations of the earth field are $\sim 10^{-6}$ G $Hz^{-1/2}$, the gradiometer must possess a high degree of symmetry in order that the 10^{-10} G sensitivity be effectively utilized. Gradient sensitivities of 10^{-11} G cm^{-1} $Hz^{-1/2}$, and possibly much lower, will probably be realized.

B. DC AND LOW-FREQUENCY VOLTAGE AND CURRENT SENSING

Much of the preceding discussion on magnetometry is pertinent to the design of high-sensitivity galvanometers (voltmeters, null detectors, etc.).

If a coil of N turns of fine wire is tightly coupled to a superconducting loop device of inductance L, then the minimum detectable current in the coil is

$$\delta i = \eta\delta\Phi/NL, \tag{112}$$

where $\delta\Phi$ is the minimum detectable flux as noted previously and $\eta \lesssim 1$ is the coupling coefficient. The minimum flux detectable with an rf-biased device was given as $\delta\Phi \sim 10^{-4}\ \Phi_0\ \mathrm{Hz}^{-1/2}$. If, for example, $L \sim 10^{-10}$ H and $N = 100$ turns, then the minimum detectable current that can be realized is of the order of 10^{-11} A.

To establish a current of this magnitude in the coil of 100 turns in a time t of 1 sec would require a voltage

$$V = N\delta\Phi/t \sim 10^{-17}\ \mathrm{V} \tag{113}$$

which would be the minimum detectable voltage for this configuration, assuming a noise-free source. A 1-sec time constant would be realized if the source resistance were

$$R = N^2L/t \sim 10^{-6}\ \Omega \tag{114}$$

Johnson noise in the resistance is

$$\langle V_{\mathrm{N}}{}^2\rangle = 4k_{\mathrm{B}}TR/t \sim 10^{-28}\ \mathrm{V}^2 \tag{115}$$

where the numerical value is calculated for $T = 4$ K. Thus the noise voltage is about 1000 times larger than the minimum detectable voltage calculated above for a noise-free (zero-resistance or zero-temperature) source. This corresponds to the fact noted previously that the rms total flux fluctuation in the device amounts to about $\Phi_0/10$. In the above numerical example ($N^2L/R \sim 1$ sec) this flux fluctuation appears in a 1 Hz bandwidth centered at zero frequency, and in this bandwidth it is about 1000 times greater than the inherent noise level of the device and the electronic system ($\sim 10^{-4}\Phi_0$).

As long as the source can be reasonably well matched to the device, i.e., N^2L/R can be made of the order of the desired response time (1 sec in the above example), the sensitivity of a voltmeter will be determined by thermal noise in the source itself (unless the source is at extremely low temperature).

It will appreciated that the number of turns that can be effectively coupled to a low-inductance loop device is limited to a few hundred or thousand at most. Therefore, voltmeters made in this way will be very low-impedance instruments. However, large impedance transformation ratios can be achieved through the use of large superconducting transformers external to the device.

C. Fundamental Electrical Standards

1. *Voltage standards*

Next to magnetometry, the use of the Josephson effect to establish fundamental standards of voltage, current, etc., has perhaps been the most widely studied application. Using the microwave-induced steps (Figs. 8 and 9) in the dc I–V characteristics of point contacts and tunnel junctions, Parker *et al.* (1967, 1969) established the value of $\Phi_0 = h/2e = V_0/f$ in terms of present voltage and frequency standards to within a few parts per million. This monumental work has led to a revision in the accepted value of Planck's constant h, and to a critical reevaluation of the whole system of fundamental constants. It also demonstrated the practicality of using microwave-biased weak links as both primary and secondary voltage standards. Work is now proceeding toward this end in several standards laboratories. Constant-voltage steps in the dc I–V characteristics of point contacts have been produced well above 10 mV (McDonald *et al.*, 1969).

A number of experimental and theoretical studies have dealt with the fundamental reliability of the Josephson frequency relation, or in more practical terms, with how constant the "constant-voltage" steps really are. The major cause for concern is the effect of fluctuations rounding off the corners of the steps as indicated in Fig. 8. Kose and Sullivan (1970) have shown that the voltage at the center of a step remains unperturbed as the level of fluctuations in the system is artificially varied over a considerable range. Both coherent and incoherent low-frequency signals were used in this study to simulate fluctuations. The effect of a small low-frequency signal is to average over a small region of the I–V characteristic, and since each step tends to be symmetrical about its own center, the averaging process gives no change in the dc level when the contact is biased on the center of a step.

In the case of steps that are separated from each other by clearly defined hysteresis loops (Fig. 9), one suspects that the steps are absolutely flat ($dV/dI = 0$). This is to say that the Josephson oscillation is in synchronism with the applied microwave signal, with short- and long-term phase fluctuations much less than 2π. Any fluctuation of order 2π would trigger the system to an adjacent part of the I–V characteristic, especially if the bias current is set near the end of a step. Thus if no such transition is observed during a time interval t_0, it seems highly probable that the average voltage is precise, in principle, to better than one part in ft_0, where f is the frequency of the applied radiation, regardless of where on the step the bias current is set. The implication is that for observation times of the order of seconds and frequencies of the order of 10^{10}, one need not be concerned about the "flatness" of the steps, but this has not been proved experimentally.

The obvious advantage of the microwave-induced steps as voltage standards is that the voltages are given in terms of the fundamental constant Φ_0 and the frequency f, which is measurable with very high precision. The main problem is that the step voltages are inconveniently low for general laboratory use, especially in relation to thermal emf's. A cryogenic voltage divider or potentiometer to compare the step voltage with a much higher voltage source at room temperature would be desirable, and, in fact, is under development in at least one laboratory (D. Sullivan, NBS, Boulder, Colorado, private communication).

2. *Current standards*

Just as a microwave-biased weak link can be used to relate voltage to frequency, so a weakly-connected loop device (Sections III.B and III.C) can be used to relate current to length. If a solenoid of known dimensions is inserted through the aperture of the loop, then the solenoid current I is related to the loop response $\mathcal{R}$ by

$$I = \mathcal{R}\Phi_0 l/\mu_0 NA = \Phi l/\mu_0 NA \tag{116}$$

where l and A are the length and area of the solenoid, N the number of turns, and Φ the flux. $\mathcal{R}$ is defined as the ratio Φ/Φ_0; it is the number of cycles of the interference pattern (Fig. 13 or 24) swept out by the current I.

The basic problem in making absolute measurements of current is to determine the dimensions accurately, particularly the area A, since the diameter of typical devices is only of the order of a few millimeters. Meservey (1968) in a quite detailed proposal has considered the possibility of achieving part–per–million accuracy. He envisioned the use of a 1 cm diam thin-film loop device placed inside the solenoid, in which case the area A in the above equation becomes the device area rather than the solenoid area. A successful thin-film device of this diameter had not yet been produced, and in any case the difficulty of measuring the dimensions to a few parts per million seems rather formidable.

Even if the absolute response of a loop device to a current in a solenoid cannot be (or has not yet been) measured with such high accuracy, it may still be useful as a secondary standard of high relative accuracy. The response $\mathcal{R}$ to a current I should be highly linear over the full range from maximum current (of the order of amperes) to the minimum current detectable ($\sim 10^{-10}$ A, see Section V.B). Reproducibility should be limited mainly by dimensional stability of the device and the solenoid, which should be excellent at cryogenic temperatures.

D. Electronic Systems

This chapter has been an attempt to establish the physical principles and operating characteristics of Josephson weak-link superconducting devices. We have discussed the interface with electronic circuits only as far as deriving a measurement signal is concerned. In (almost) all cases of interest we have indicated how to obtain a voltage periodic in magnetic field, time, or some other parameter. This periodic response is unique among electronic devices and offers some distinct advantages when properly utilized. In this section we shall demonstrate how to effectively use these devices as a magnetometer, voltmeter, and thermometer as examplary cases.

1. *Linear analog magnetometer*

The prototype sensor is a loop-coupled weak-link device coupled to an rf circuit as shown in Fig. 10. Thus the periodic voltage is the rectified rf voltage as shown in Fig. 13. Because a sensitive device has a very small period $\Delta\Phi_x$, it appears from Fig. 13 to have a very small dynamic range. With available state–of–the–art electronic components, i.e., integrated circuits or electronic subsystems, one can develop the required system.

Although the largest responsivity of the device appears at the positions where $|dV_1/d\Phi_x|$ is greatest, it is generally desirable to choose the operating point where $dV_1/d\Phi_x = 0$. The reasoning behind this choice of operating point is that it is insensitive to variations in the bias, whether rf or dc; insensitive to small variations in temperature; invariant with respect to critical current; invariant with respect to external noise; and independent of the quantum phase dependence of i, $i(\theta)$. Thus the invariant positions have zero slope at $\Phi_x = k\Phi_0$ and $(k + \frac{1}{2})\Phi_0$. In order to operate at these zero-slope positions, one modulates the field at the device and then measures the response to the modulating field. This affords not only a stable and reproducible operating point but also *linear* response and expanded dynamic range.

The technique is one of differentiation of the $V_1(\Phi_x)$ response, followed by a servoed field to maintain a *fixed* total magnetic flux at the device. This latter can also be implemented at positions of high response, maximum $|dV_1/d\Phi_x|$, but with the reduced stability indicated above. This general technique was first suggested and used by the authors with a dc interferometer (Zimmerman and Silver, 1966a, b) and with an rf-biased device (Silver and Zimmerman, 1967), and an instrument has been discussed in detail by Forgacs and Warnick (1966). Figure 41 shows the essential components in the block diagram of a linear magnetometer. Using the rf-biased technique of Fig. 10, one develops a periodic $V_1(\Phi_x)$ at the output of the rf

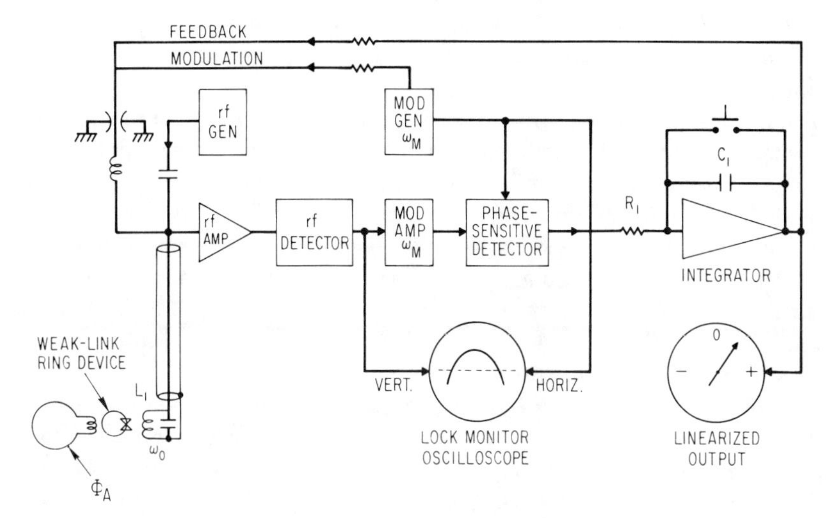

FIG. 41. Schematic diagram of a linear response analog magnetometer using a loop-coupled weak-link device.

detector by adjusting (1) the frequency ω_{rf} as discussed in Section III.B.4, and (2) the rf-bias level to give the lowest pattern of Fig. 13b. Techniques for accomplishing these adjustments are discussed in Section III. The applied flux Φ_x now has several contributions:

$$\Phi_x = \Phi_A + \Phi_M + \Phi_S + \Phi_x^1 \tag{117}$$

where Φ_A is the ambient field to be measured as transformed by the flux transformer, Φ_M is the modulation field, Φ_S is the servo field generated by the integrator, and Φ_x^1 is the rf field. As shown the latter three terms are applied to the device by the tank inductance; other configurations are possible although this is the simplest. We assume Φ_M is harmonic at $\omega_M \ll \omega_{rf}$. If the amplitude of Φ_M is set at approximately $\Phi_0/2$ for this device, the output of the phase-sensitive detector (PSD) will be similar to Fig. 13b but with the phase shifted by $\frac{1}{4}$ period $(\Phi_0/4)$ and with the voltage offset automatically suppressed. Thus zero voltage at the output of the PSD corresponds to the extrema of $V_1(\Phi_x)$ of Fig. 13b and occurs at $\Phi_x = k\Phi_0$ or $\Phi_x = (k + \frac{1}{2})\Phi_0$. If a high-gain integrating amplifier is connected as shown the total system will "lock on" such that

$$\langle \Phi_x \rangle = \Phi_A + \Phi_S = k\Phi_0 \quad \text{or} \quad (k + \tfrac{1}{2})\Phi_0 \tag{118}$$

We neglect Φ_M and Φ_{rf} in evaluating the low-frequency spectrum of Φ_x. Since $\langle \Phi_x \rangle$ is now a constant, which can be set to zero or some other quantized value, Φ_S effectively measures Φ_A. The linear metering system measures Φ_S directly, hence Φ_A. An oscilloscope monitor confirms the continuous "lock" by a stable signal at $2\omega_M$ which is a maximum when that at ω_M is maintained at null.

The key element here is an integrating amplifier which continuously nulls the output of the PSD. The linear range of the magnetometer is determined by the saturation characteristics of the integrator. The range can be further extended, without sacrificing sensitivity by changing the "lock on" position from $\langle\Phi_x\rangle = k\Phi_0$ to $(k + k')\Phi_0$. Specific techniques for accomplishing this are discussed in the following section. Use of a flux transformer, in addition to increasing the sensitivity of the device, also reduces the effect of the magnetometer fields, i.e., Φ_x^1, Φ_M, and Φ_S, on the source volume. This is particularly important in susceptibility measurements.

The response time or overall bandwidth of the linear magnetometer is determined by the integrator. Without this limitation one can sense more rapid variations at the output of the PSD. Here the bandwidth is limited by either the resonant circuit coupled to the device, the rf amplifier bandwidth, the modulation amplifier bandwidth, or some combination of these. Ideally the resonant circuit should determine the bandwidth, with the rf and modulation bandwidths equal to that of the resonant circuit. Hence with an rf frequency in the vicinity of 30 MHz, a bandwidth of several hundred kilohertz can be realized. Sufficiently high modulation frequencies can be utilized by setting the bias frequency ω_{rf} approximately equal to $(\omega_0 - \omega_M)$. In this manner one places the upper signal sideband within the bandpass of the input circuit and rf amplifier.

2. *Digital–analog magnetometer*

The greatest sensitivity of the linear analog magnetometer is achieved when a full scale deflection corresponds to a $\Delta\Phi_A = \pm\Phi_0/2$. Larger values of $\Delta\Phi_A$ can be measured as indicated above by changing the lock-on field from $k\Phi_0$ to $k'\Phi_0$. If $(k' - k)$ is digitally recorded or counted we can maintain a constant analog sensitivity regardless of field change. The primary obstacle to such counting is determining the sign since the $V_1(\Phi_A)$ or $V_{PSD}(\Phi_A)$ functions alone do not have sign information. Forgacs and Warnick (1967) developed a digital–analog magnetometer by using multiple-derivative techniques. We will present a simple extension of the analog magnetometer which possesses the required characteristics.

First consider that the detected rf voltage is represented by

$$V_1 = A_0 + A_1 \cos(2\pi\langle\Phi_x\rangle/\Phi_0) \tag{119}$$

and that the phase-sensitive-detector output is

$$V_{PSD} = A_2 \sin(2\pi\langle\Phi_x\rangle/\Phi_0) \tag{120}$$

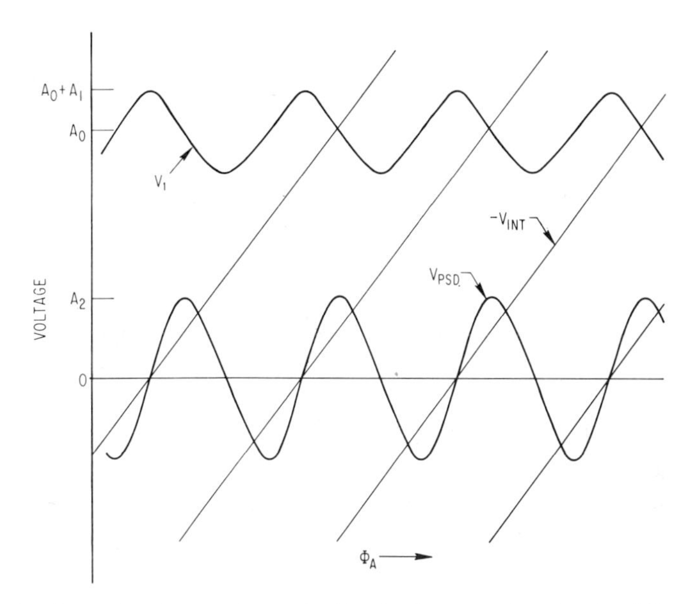

FIG. 42. Typical voltages observed with the linear analog circuit of Fig. 41. The detected rf voltage V_1, phase-sensitive-detector voltage V_{PSD}, and integrator output voltage V_{int} are all shown as functions of the ambient magnetic field Φ_{A}.

Figure 42 shows V_1, V_{PSD}, and the integrator output voltage V_{int} for the circuit of Fig. 41. Since

$$\langle \Phi_x \rangle = \Phi_{\mathrm{A}} + \Phi_{\mathrm{S}} \tag{121}$$

and

$$\Phi_{\mathrm{S}} = F V_{\mathrm{int}} = -(F/R_1 C_1) \int V_{\mathrm{PSD}}\, dt \tag{122}$$

we have for the phase-sensitive-detector output voltage

$$V_{\mathrm{PSD}} = A_2 \sin \left\{ (2\pi/\Phi_0) \left[\Phi_{\mathrm{A}} - (F/R_1 C_1) \int V_{\mathrm{PSD}}\, dt \right] \right\} \tag{123}$$

This equality is satisfied for all time if $V_{\mathrm{PSD}} = 0$ and $V_{\mathrm{int}} = \Phi_{\mathrm{S}}/F$. However, if we replace the integrator with a wide-band dc amplifier by substituting a resistance R_2 for C_1, then

$$\Phi_{\mathrm{S}}' = -(F R_2/R_1) V_{\mathrm{PSD}}' \tag{124}$$

and

$$V_{\mathrm{PSD}}' = A_2 \sin\{ (2\pi/\Phi_0) [\Phi_{\mathrm{A}} - (F R_2/R_1) V_{\mathrm{PSD}}'] \} \tag{125}$$

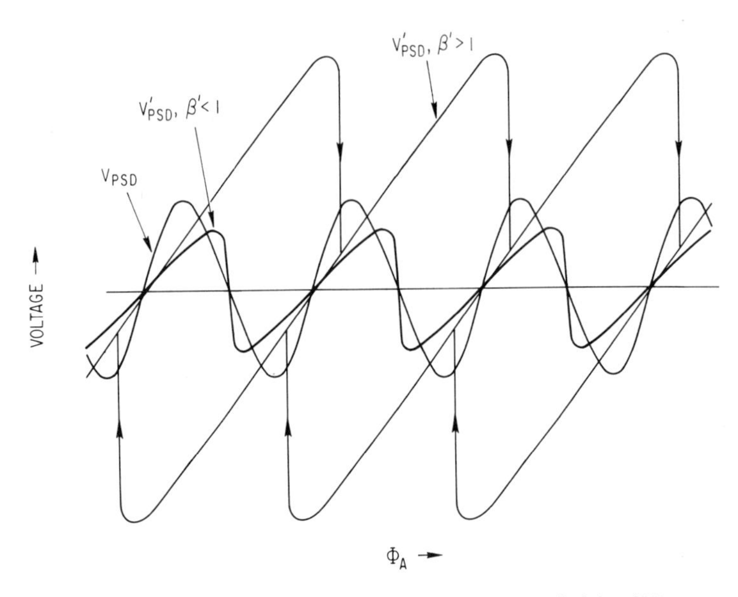

FIG. 43. Several experimental curves of V_{PSD}' versus Φ_A recorded for different values of β' obtained with $\omega_n/2\pi = 167$ Hz and a bandwidth of 8 Hz. The phase-sensitive-detector signal V_{PSD} is shown for reference and corresponding to $\beta' = 0$.

where the primes represent the same variables with the new configuration. This relation is identical in form to Eq. (32) and its solution depends on the adjustment of the various system gains, A_2, F, and R_2/R_1. Depending on the ratio $\beta' = 2\pi A_2 F R_2 / R_1 \Phi_0$, analogous to the ratio β of Section II, V_{PSD}' can be varied from a sinusoidal function of Φ_A/Φ_0 at $\beta' = 0$ to a hysteretic, sawtoothlike function at $\beta' \to \infty$ as shown in Fig. 2 for *Li*. Thus for $\beta' > 1$, V_{PSD}' becomes a signed function which automatically provides the direction of counting from the polarity of the transition pulse. Such an automatic digital counting system can have the full speed of the input circuit and provide rapid counting of flux quanta without loss of memory. Figure 43 shows experimentally recorded functions of V_{PSD}' for various values of β'. Two comments about the actual forms of V_{PSD}': the value of $|k' - k|$ which is observed depends on β' and on the effective damping of the overall system, and the shape of the function depends on $V_1(\Phi_x)$. In general, this function is more triangular-wavelike, resulting in a more nearly linear form of $V_{PSD}'(\Phi_A)$.

Figure 44 shows a composite digital–analog magnetometer in schematic form. Expanding on the linear-analog-magnetometer circuit one adds at the integrator output voltage comparators to sense when the servo field Φ_S corresponds to $\pm\Phi_0$. These levels trigger an appropriately signed reset of

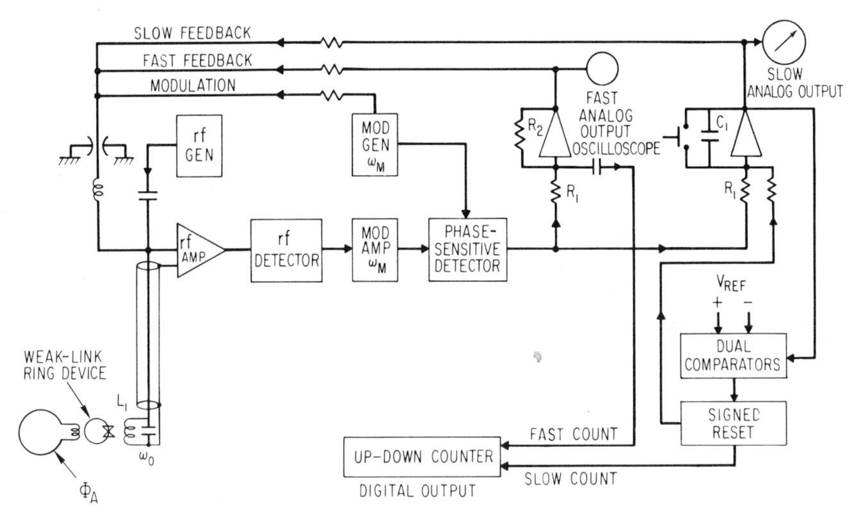

FIG. 44. Diagram of a composite digital–analog magnetometer incorporating fast and slow digital counting and fast and slow analog readout.

the integrator, shifting the lock-on position by $\pm\Phi_0$. Simultaneously the reset pulse is used to actuate an up–down electronic counter. Thus the digital readout is in units of Φ_0, and the SLOW ANALOG OUTPUT measures fractions of Φ_0 when the excursions in Φ_A are less than $\pm\Phi_0$. The digital portion of this system is similar to integrating digital voltmeters. However both the analog and digital readouts are limited in response time because of the integrating amplifier. Serious errors will arise if counts are missed. Therefore one utilizes in parallel the hysteretic nature of a high β' feedback without an integrator. A wide-band amplifier is connected in parallel with the integrator and β' adjusted for a transition when $\Phi_A = \pm\Phi_0$. The transient pulse is appropriately signed and is also counted in the up–down counter. Thus the digital readout accumulates the total integral variation in Φ_A. Further rapid variations in Φ_A, less than $\pm\Phi_0$, can be measured with a FAST ANALOG OUTPUT oscilloscope display at the (amplified) output of the phase-sensitive detector.

The net result is an instrument of high dynamic range, ultrahigh sensitivity, and large bandwidth. Generally the higher the rf bias frequency the greater will be the overall gain bandwidth of the sensor. Thus, if X band were utilized as the biasing frequency (Lambe *et al.*, 1964), the modulation could be approximately 30 MHz and bandwidths of 5–10 MHz could be achieved.

A recent application of the high-speed digital readout is found in low-temperature measurements on gallium (Hanabusa *et al.*, 1970). Magnetic field and temperature dependences of the magnetic susceptibility were observed with the self-resetting digital magnetometer technique and a $\frac{1}{4}$-in.-diam gallium rod inserted in a point-contact loop device. The reported accuracy was limited by the thermal fluctuation noise of the sample itself.

3. *Cryogenic noise thermometer*

Kamper (1967) first proposed the possibility of a Johnson noise thermometer at cryogenic temperatures. The sensitivity to and unique effect of a thermal noise source on the frequency of a Josephson weak-link oscillator makes this an attractive wide-range thermometer. In simple physical terms, the coherence of the quantum oscillations and hence the stability of the frequency of a voltage-biased oscillator (Section III.B) is determined by the thermal fluctuation voltage in the biasing resistance R (see Fig. 15). As noted, the spectrum produced by frequency modulation is Lorentzian with a full width at half power

$$\delta f = 4\pi k_B TR/\Phi_0^2 \tag{126}$$

as given in Eqs. (69) and (71).

We describe here a particular scheme for such a noise thermometer which will measure T by means of averaging the correlation time of the quantum oscillator. While the oscillations can be directly detected most easily at rf or microwave frequencies, several advantages can be gained by using quantum oscillators in the low-frequency regime. In this frequency region the oscillations can be detected by the parametric amplification (through up-conversion) described in Section III.B.8. Lower frequency of oscillation reduces the bias current and associated heating effects, stability requirements for the first local oscillator in the heterodyne receiver, and the effect of instability in the critical current of the weak link. Thus, one would use an rf-biased resistive loop device (Section III.B.8), where the bias frequency must exceed R/L, the cutoff frequency of the resistive loop.

If the line shape $G(f)$ is assumed to be temperature independent, the autocorrelation function is also temperature independent and given by

$$\rho(\tau) = \int_{-\infty}^{\infty} G(f) \cos 2\pi f\tau \, df \tag{127}$$

If $G(f)$ is Lorentzian as predicted, then $\rho(\tau)$ is exponential with a correlation time τ_c given by the reciprocal of Eq. (126). The correlation time can be more directly measured electronically than the line width, and this is accomplished by the method of Fig. 45. The resistive loop device is rf biased to operate near point C in Fig. 20. Thus the rf detector senses any

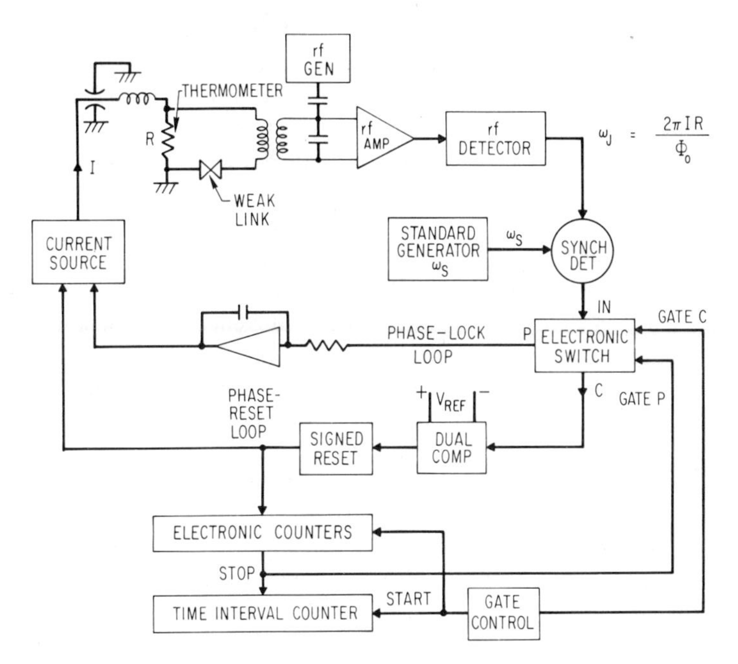

FIG. 45. Diagram of the noise thermometer. The resistance R at the temperature T controls the quantum oscillations of the superconducting weak link. The correlation time, which varies inversely as T, is read from the time-interval counter.

voltage across R as an oscillation frequency ω_J given by

$$\omega_J = 2\pi IR/\Phi_0 \tag{128}$$

where I is the current in R required to produce the voltage. A stable current source generates the current required to match ω_J with a standard reference oscillator ω_S. By means of a synchronous detector and integrator feedback (with the electronic switch in the "P" position) the quantum oscillation ω_J is "phase locked" to the reference oscillator ω_S. Because phase locking occurs when the synchronous detector output is null, a $\pi/2$ phase shift exists between ω_S and ω_J.

When the continuous phase lock is removed (the electronic switch in the "C" position) it is presumed that ω_S is fixed and that ω_J will fluctuate because of the thermal voltages across R. Denoting the difference in phase as $(\pi/2 + \theta_0)$, we find

$$\theta_0(t) = (2\pi/\Phi_0) \int_0^t V_N \, dt \tag{129}$$

where V_N is the instantaneous noise voltage. The output voltage of the synchronous detector varies as $V_S \sin \theta_0$, where $\pm V_S$ is the maximum voltage which occurs when ω_J and ω_S are in phase. Since $V_N(t)$, $\theta_0(t)$, and $\sin \theta_0$ are all stochastic variables, no prediction of $V_S \sin \theta_0(t)$ is possible. However the average over many samples is determined by the correlation function $\rho(\tau)$. If we choose a particular angle θ_0', less than $\pi/2$, such that

$$\theta_0' = \arcsin \epsilon \tag{130}$$

then when $\theta_0 = \pm\theta_0'$, $V_S \sin \theta_0 = \pm\epsilon V_S$. By choosing the reference levels for the dual comparator as $\pm\epsilon V_S$, the signed reset will produce a pulse of proper polarity to readjust the current I and hence the phase of ω_J by $\mp\theta_0'$. The length of this pulse must be very short compared to τ_c, and the net effect is that ω_S and ω_J are restored to the phase-lock condition. If a time interval counter is started at $t = 0$ and stopped when $\theta_0 = \pm\theta_0'$, we have measured a fixed fraction of τ_c. Since the entire system is stochastic, we repeat this measurement χ times, with a derived fractional accuracy of $\chi^{-1/2}$. This is most easily accomplished by requiring χ successive reset pulses from the signed reset to stop the time interval counter. In this manner one measures $\chi\tau'$, where τ' is the time for dephasing θ_0', with a fixed fractional accuracy. As T is lowered and τ_c and τ' increase, the time required for constant precision increases. (This is precisely what would be required in a direct line width measurement. The only time saving in the correlation-time method is in the analysis of the measurement.) The desired temperature varies as

$$T = \Phi_0^2/4\pi k_B R \tau_c \tag{131}$$

After each measurement interval $\chi\tau'$ the phase lock is restored by returning the switch to "P."

The overall reliability of any measuring technique is ultimately determined by the signal–to–noise ratio and the fraction of the available information which is utilized. Compared to direct-counting techniques, the method outlined above processes all of the signal in the synchronous detector and compresses it into the comparator input. More rapid sampling, determined by a smaller ϵV_S, is possible if the S/N ratio at the synchronous detector output is sufficiently large. In general, ϵ will vary inversely as S/N.

Two problems may reasonably be considered here. A systematic offset in I produces a constant frequency difference $(\omega_J - \omega_S)$ and a resultant error in the measurement of the statistical correlation time. This offset manifests itself as a larger number of exclusions $+\theta_0'$ compared to $-\theta_0'$, or the inverse. One can measure the occurrence of such distinct events, χ_+ and χ_-, respectively, and assign acceptable limits on $(\chi_+ - \chi_-)$ compared to $\chi = \chi_+ + \chi_-$. Excess terms $(\chi_+ - \chi_-)$ can be used to correct for the offset.

A second consideration is the practical temperature range of the measuring system. Excessively long correlation times produce high-Q oscillators and put greater emphasis on the reference standard stability. It would therefore be desirable to either reduce ω_S or increase R as T is lowered. Since R is bounded by the requirement that ω_{rf} exceed the low-frequency cutoff of the resistive loop device, it may be necessary to lower ω_S in steps as the temperature range is lowered.

Experimental confirmation of the operating range has recently been extended down to 75 mK by Kamper and Zimmerman (1971). Theoretical limitations on the low-temperature region of validity indicate usefulness to microKelvin temperatures.

E. High-Frequency Radiation Detection

1. *Video detection with current-biased weak links*

The dc response of Josephson tunnel junctions to microwave radiation was first reported by Shapiro (1963), and that of thin-film metal bridges by Anderson and Dayem (1964). Spectral response and sensitivity of point contacts was reported by Grimes *et al.* (1966, 1968), with the response extending up to the neighborhood of the energy gap and beyond. (For Nb, the gap voltage is 2.8 mV, for which the Josephson frequency is 1390 GHz and wavelength is 2.2 mm.) In fact, McDonald *et al.* (1969), found that constant-voltage steps (Section III.A) in the dc characteristic of certain Nb point contacts could be produced as high as 17 mV, six times the energy gap, indicating Josephson currents with a free-space wavelength of 37 μ. Radiation detection in the video mode is accomplished either by measuring the reduction of the supercurrent in the presence of radiation, or by noting the onset of finite-voltage steps. Grimes *et al.* (1968) were able to detect $\sim 10^{-13}$ W using the first method, basically a broad-band detector. The second method provides information about the frequency as well as the intensity of the radiation, through the Josephson frequency relation [Eq. (37)]. This point was emphasized in the work of McDonald *et al.* (1969) whose purpose was to use a point-contact detector as a far-infrared spectrometer.

In most of the above work, the Josephson device was coupled to the radiation field in more or less broad-band fashion. For example, in the work of Grimes *et al.* (1968) the point contact was made between the ends of two wires extending across a circular waveguide ("light pipe"). It was pointed out and demonstrated by Richards and Sterling (1969) that a significant improvement in sensitivity could be realized by coupling a point contact to a resonant cavity, so that a self-induced step (see Section III.A) is produced in the dc characteristic. Incident radiation at or near the cavity frequency (190 GHz) was detected at levels less than 10^{-14} W, with 1 Hz postdetection bandwidth. Richards and Sterling attributed the enhanced sensitivity to "regenerative" narrowing of the cavity mode by mutual interaction of the cavity and the Josephson current, and estimated that the effective width of the response was $\sim$300 MHz, corresponding to an effective Q of 600.

It is expected that the sensitivity of a weak link to radiation should be greater the higher its normal resistance, that is, the lower the normal component of the current. Since the normal current i_n and the supercurrent i_s are parallel conduction mechanisms (as described in Section II.D), a given applied rf voltage V_1 across the link results in a power dissipation $V_1^2G/2$, plus an absorption through the nonlinear interaction of the incident radiation with the supercurrent. The "normal" term $V_1^2G/2$ is in effect wasted and can only degrade the sensitivity. Thus, most thin-film bridge devices should be less sensitive than point-contact or tunneling-junction devices, regardless of impedance matching, because of their higher conductance.

When the weak link is coupled to a resonant cavity, the conductance G influences the parameter which governs the qualitative nature of the response to an applied signal at frequency f_0. In the equivalent circuit of Fig. 46a, the cavity is represented by the lumped constants L_1 and C_1, where C_1 includes the inherent capacitance of the weak-link structure. The parameter

$$\alpha = Ri_c/\Phi_0 f_0 = 2\pi f_J RC_1(f_J/f_0) \tag{132}$$

has already been introduced [Eqs. (44) and (67)]. Here R is the total equivalent shunt resistance of the system at resonance, including the weak-link conductance G, the cavity loses, and the radiation source resistance. The frequency f_J is defined by the weak-link inductance, Eq. (130), and the capacitance C_1 (for an isolated tunnel junction, f_J would be the Josephson plasma resonance frequency), i.e.,

$$f_J = (2\pi)^{-1}(\Phi_0 C_1/2\pi i_c)^{1/2} \tag{133}$$

If the cavity has a high Q the response will be appreciable only when f_0 is close to the cavity frequency f_c. When $\alpha \ll 1$ the bandwidth of the response will be just f_c/Q. However, as α becomes the order of unity or greater, the response peaks up and becomes sharper, as in the case of a regenerative receiver. Richards and Sterling (1969) invoked this mechanism to account for the bandwidth and sensitivity of the observed response. Their experimental arrangement was a cavity 4 mm long and 4 mm in diameter with a point contact between Nb wires positioned axially in the cavity. One wire

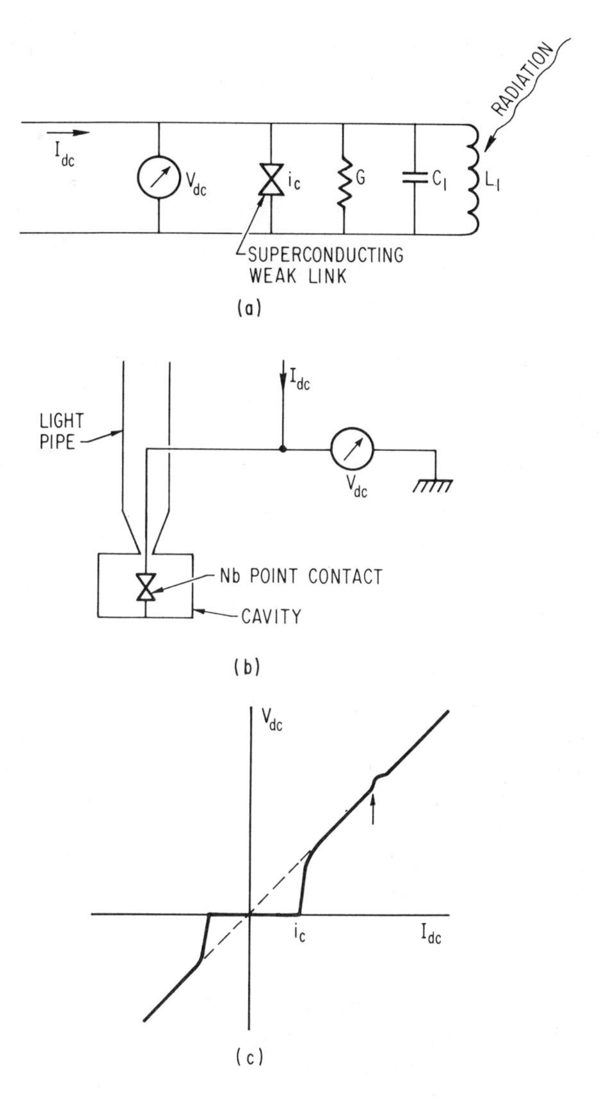

FIG. 46. (a) Equivalent circuit of the video mode detector experiment using point contacts in a microwave cavity. (b) Typical schematic arrangement of a far infrared or millimeter wave detector. (c) Typical V–I curve for a current-biased point contact in a cavity showing the self-induced step (arrow).

extended through a 1.6-mm coupling hole and into a light pipe, by which radiation was fed into the cavity. The presumed arrangement is sketched in Fig. 46b. Maximum sensitivity to radiation at the cavity frequency f_c was realized by setting the bias current I_{dc} at the point of maximum differential resistance, indicated by the arrow in Fig. 46c, rather than on the flat part of the self-induced step. Ulrich (1970) used a similar arrangement as a sensor for millimeter wave astronomy, and reported comparable sensitivity.

Although the use of the Josephson effect as a high-frequency detection mechanism has been widely discussed and attempted, there remains a lack of definitive theoretical and experimental work on the subject. For example, no estimate of the ultimate sensitivity has been made. The minimum detectable power noted above, 10^{-14} W, is many orders of magnitude greater than Johnson noise ($k_B T \sim 10^{-22}$ W Hz^{-1} at 4.0 K) or photon shot noise ($hf \sim 10^{-22}$ W Hz^{-1} at 1 mm wavelength).

2. *Heterodyne detection with weak-link devices*

A rather different mode of operation is to use the Josephson effect device as a heterodyne mixer to convert from millimeter wavelength to a convenient (that is, lower) frequency where conventional infrared amplification can operate. This can be achieved either by feeding the external signal and a local oscillator signal from a conventional source into the device, or by using the Josephson oscillation itself as the local-oscillator signal. These modes of operation have been the subject of a preliminary study by Zimmerman (1970). The ultimate sensitivity was not measured, and the highest frequency used was in the X-band (10-GHz) range. It should be noted that when a parametric device is used in this way to down-convert in frequency, there is a power loss by the ratio of intermediate frequency to signal frequency. Such loss upon down-conversion does not occur in a conventional (nonparametric) mixer.

3. *Noise characteristics*

Ultimate limits on the sensitivity of Josephson devices as radiation detectors will be provided by a proper evaluation of the detection mechanisms and the corresponding noise. We have noted in discussing magnetometers that even in the absence of additional noise (dissipative) sources, there is an intrinsic noise limit associated with the superconducting element itself. Scalapino (1967) and Stephen (1969) have calculated the noise spectrum for the mean squared current of a current-biased Josephson junction.

Kanter and Vernon have recently measured the sensitivity and noise characteristics of Nb–Nb and Nb–Ti point contacts with I–V curves similar to those described theoretically by McCumber (1968), Stewart (1968), and Aslamazov and Larkin (1969). For studying the microwave response the thin niobium wire (75 μ diam) was situated across the narrow dimension of E-band waveguide, the Nb surface being part of the waveguide wall. Operating the point contact in the video detection mode, while illuminating it with the attenuated radiation of a klystron at 90 GHz, a best noise equivalent power (NEP) of 5×10^{-15} W Hz$^{-1/2}$ was obtained. For optimum matching a waveguide short was positioned near the contact. Under this condition the responsitivity was about 20,000 V W^{-1}. Because of the impedance adjustment with the waveguide short, the frequency response near 90 GHz was not flat. Nevertheless an effective bandwidth of 7 GHz was obtained which would have made this a useful detector. The calculated discernible temperature variation of radiation in the 90-GHz region was $\Delta T \sim 0.2$ K when averaging the NEP measured across the frequency region. This value compares well with the best mixer receivers presently in use. It is expected that the principle high-frequency limit of a weak link is determined by the gap frequency, of the order of 10^{12} Hz, in excess of limits attainable with diode mixers.

From the noise-current spectrum to be expected for a Josephson junction, Scalapino (1967) calculated the quasiparticle, and Stephen (1969) the pair contribution of the junction current. Their combined result for the power spectrum is

$$P(\omega) = \langle i(\omega)^2 \rangle = (e/\pi)[I_n(V) \coth(eV/2k_B T) + 2I_s(V) \coth(eV/k_B T)] \tag{134}$$

The expression is applicable for $\omega < eV/\hbar$ and $\omega < k_B T/\hbar$ with V the junction dc potential and $I_n(V)$ and I_s the dc current contributions by quasiparticles and pairs, respectively. Equation (134) is of the shot-noise type with the hyperbolic cotangent factor arising from the sum of the contributions by forward and backward currents. In the limit $eV \gg k_B T$:

$$P(\omega) \rightarrow (1/2\pi)(2eI_n + 4eI_s) \tag{135}$$

which is the shot-noise expression in a more familiar form. Note the double-charge factor from the pair current contribution. For $eV \ll k_B T$:

$$P(\omega) \rightarrow (1/2\pi)4k_B T(I_s + I_n)/V \tag{136}$$

which has the form for Johnson noise with $(I_s + I_n)/V$ the total dc conductance of the junction. Recall that I_s and I_n are the supercurrent and normal-current components of the total current in the weak link.

Of particular interest in Eq. (134) is the fact that the power spectrum is only related to dc current and voltage of the junction, i.e., the power extracted from the battery. Thus, in the spirit of the Callen–Welton theorem, which relates energy dissipation with random driving forces for thermal equilibrium systems, power dissipation due to coherent or incoherent current is expected to produce noise. Since pairs carry double charge thus requiring half the number of carriers for a certain average current, the shot-noise contribution of pairs is twice that of quasiparticles. In addition, as evident from Eq. (136) in the thermal limit ($eV \ll k_B T$), noise is determined by the dc conductance, which becomes very large for supercurrents when $V \to 0$ and thus results in considerable excess noise when compared with normal currents. Since experimental evidence supports this theoretical prediction, the original expectation frequency expressed in the literature, that no noise would result from a coherent current crossing the gap, is not substantiated.

Voltage fluctuations are related to current fluctuations in terms of the dynamic junction impedance $R_D = (dI/dV)^{-1}$ with

$$\langle V^2(\omega) \rangle = \langle i^2(\omega) \rangle [R_D{}^2/(1 + \omega^2 \tau^2)] \tag{137}$$

where τ is the circuit time constant associated with the capacitance and the "normal" resistance of the junction.

Dahm *et al.* (1969) investigated the voltage fluctuations by measuring the line width of the Josephson radiation emitted from tin and lead thin-film junctions. The radiation line width is the result of frequency modulation by the fluctuating junction voltage. These authors found the line width generally in excess of but limited towards the lower side by that calculated with a power spectrum of Eq. (134) for a range of sample currents and temperatures. Individual samples at a fixed temperature, however, showed a greater dependence on current than indicated in Eq. (134). This result was attributed to excessive rf power in the weak link. All the data were obtained for $eV \ll k_B T$ where Eq. (136) is valid.

By directly measuring the voltage fluctuations across point contacts Kanter and Vernon (1970a, b) showed that the lower limit of noise currents is given by Eq. (135) for $eV \gtrless k_B T$ and 6×10^{-7} A $< i_c < 4 \times 10^{-4}$ A. They also confirmed the voltage dependence of the current noise. Figure 47 compares the measured noise with theory, Eq. (134). These authors verified the temperature dependence only to the extent that the shot-noise limit is temperature independent for $eV > k_B T$. It appears that in order not to exceed the shot-noise limit requires $k_B T < eV$.

The overall effect of thermal fluctuations is to distort and smooth out the dc I–V curves of weak links (Ivanchenko and Zil'berman, 1968a, b; Ambegaokar and Halperin, 1969; Kurkijarvi and Ambegaokar, 1970). In the theoretical treatment of these authors the time dependence of the phase difference across the weak link is treated analogously to that of a spatial coordinate of a particle in a gas. The resulting phase slippage produces a finite average voltage across the weak link.

Essential for the derivation of the I–V characteristics influenced by a thermal environment is the fact that the thermal driving force changes rapidly in a time short compared with the natural period of the weak link

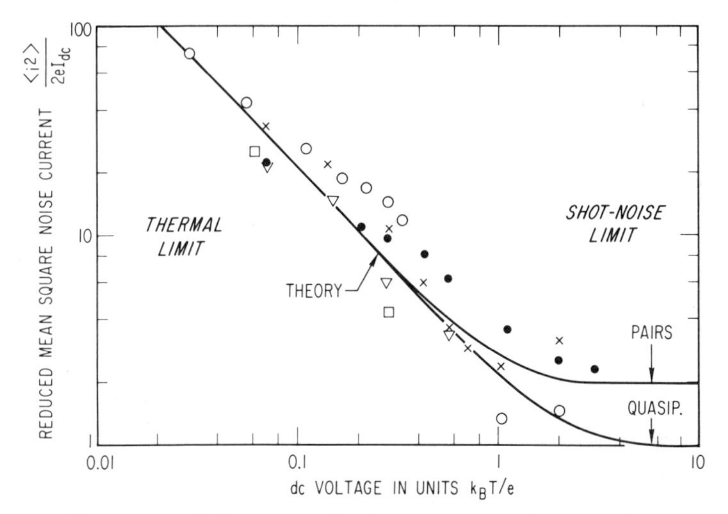

FIG. 47. Comparison of the measured noise currents as a function of dc voltage with theory for point-contact devices [after Kanter and Vernon, 1970a] showing good agreement with the thermal noise limit.

as well as its response time, i.e., the frequency spectrum of the driving force is assumed to be white. The theory does not apply for slowly varying driving forces, i.e., low-frequency noise currents with a maximum noise frequency $\nu_N \ll 2eV/h$ and $\nu_N \ll 1/\tau$. In this case, if the average voltage is measured with a voltmeter of time constant $\tau_V \gg 1/\nu_N$, the measurement results in an average over the instantaneous junction voltage given by the *static* noise-free I–V characteristic. For a Gaussian noise amplitude distribution, the noise-modified I–V characteristics have been determined by Kanter and Vernon (1970c) and shown to be similar but not equal to the thermal noise results. For small voltages the two curves for "fast" and "slow" noise currents are very similar. In fact, both coincide in the limit $V \rightarrow 0$ since for finite ν_N eventually $\nu_N > 2eV/h$.

Anderson and Goldman (1969) and Simmonds and Parker (1971) also measured the low-voltage region on thin-film tunneling junctions and metallic thin-film weak links, respectively. However, in both cases the critical current was lowered by operating arbitrarily close to the transition temperature. Agreement of the resulting I–V characteristics with theoretical predictions for the thermal case was accomplished by adjusting the critical current and effective bath temperature or the critical current. The use by Anderson and Goldman of an effective bath temperature of 10K compared with an experimental value of 3.8K suggests noise currents of other than thermal origin. Simmonds and Parker found deviations from theory for thermal currents for large i_c which may at least in part have its origin in "slow" noise currents. Neither of these two experiments can thus be considered proof of the validity of thermal theory.

F. Other Applications of a Partly-Resistive Loop Device

The resistive loop device discussed in Sections III.6–III.8 and in Section IV.C has several potential applications, none of which has at this time been developed to any extent. We will briefly summarize some of these possibilities in the following paragraphs.

It can function as a self-oscillating low-level heterodyne converter. This operation has been demonstrated up to frequencies above 10^{10} Hz by Zimmerman (1970), but the ultimate sensitivity has not been determined.

As a signal generator the device output power is too low for most applications, but it has been suggested that it might be used as a broadly tunable low-level standard source for sensitivity measurements on high-frequency receivers. It has the property, shared by *no* other radiation source so far as we know, of being continuously tunable over eight or 10 decades of frequency just by varying the dc current in the resistive section (Zimmerman *et al.*, 1966). The available power output is determined, except perhaps at very high frequencies, by the loop inductance.

The use of current-biased weak links as a voltage standard is discussed above. However by using voltage-biased weak links in the configuration discussed in Section III.B.8 one can make ultrasensitive voltmeters for low-resistance sources, such as thermoelectric elements.

1. *Tunable resonance spectroscopy*

An application related to radiation detection by heterodyne mixing has been demonstrated by Silver and Zimmerman (1967a). The detection of

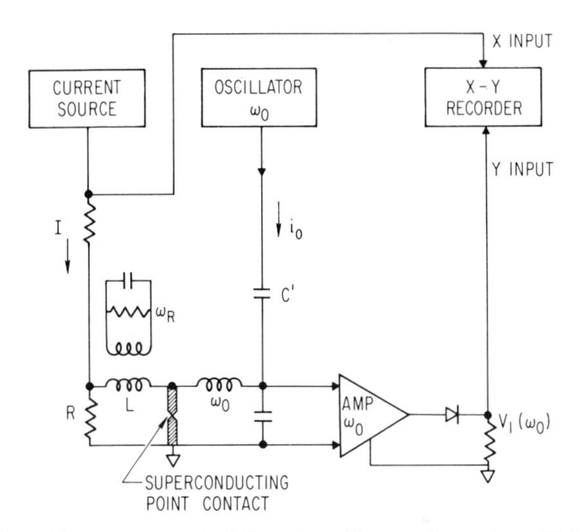

FIG. 48. Schematic arrangement of the absorption spectrometer utilizing a voltage-biased weak-link oscillating detector. The receiver is at a fixed rf frequency ω_0 and the Josephson frequency is scanned by I. Continuous operation over many decades in frequency has been demonstrated.

high-frequency resonance absorption spectra was accomplished using a voltage-biased point contact functioning as an oscillating detector. One observes the transfer of power between the superconducting device and the passive absorber. A description of this technique may be useful in projecting the related process, i.e., the transfer of power between an active source and the superconducting device.

In the reported experiment a voltage-biased point-contact oscillator is inductively coupled to the resonant absorber of frequency ω_R. As the Josephson frequency, Eq. (128), is matched with ω_R one expects an absorption of power from the oscillator. Such power loss must be supplied by the power supply I; hence at the voltage corresponding to ω_R the source current is larger than required by Eq. (128). This excess current or bump in the I–V characteristic near ω_R can be detected by measuring the differential conductance. Figure 48 is the experimental arrangement used, where ω is in the rf region. A small sampling rf current $I_1(\omega_0)$ is added to I and the rf voltage $V_1(\omega_0)$ is measured in a manner similar to that described in Section III.B. Figure 49 shows the measured $V_1(\omega_0)$ as a function of I, hence ω_J. At the two circuit resonances and the ^{59}Co nuclear magnetic resonance one observes a change in V_1 at $\omega_J = \omega_R$. These were interpreted as the differential conductance or derivative of the I–V curve.

Additional striking characteristics of these data can be interpreted as a

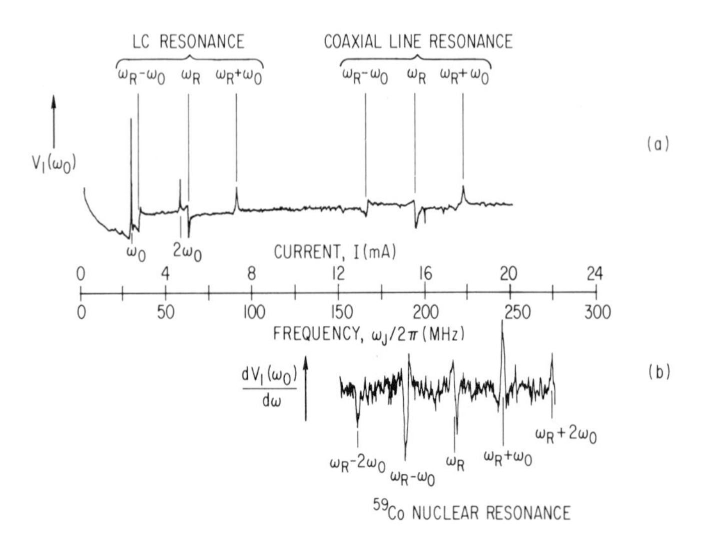

FIG. 49. Experimental spectra for the superconducting point-contact oscillating detector plotted against (a) the current (for a 25 $\mu\Omega$ resistance) and (b) the corresponding Josephson frequency.

parametric mixing effect. Since ω_J and ω_0 are both present in the point contact, one expects to generate frequencies $\omega_J \pm n\omega_0$. Whenever one of these sidebands matches ω_R we also expect power transfer. In this case, however, power is supplied by the generator ω_0 and the level $V_1(\omega_0)$ varies according to elementary expectation. Thus at $\omega_J = \omega_R - \omega_0$, power is extracted at both ω_J and ω_0, and V_1 decreases. For $\omega_J = \omega_R + \omega_0$, power is extracted at ω_J but delivered at ω_0; hence $V_1(\omega_0)$ increases. Similar effects at $\omega_R \pm n\omega_0$ were observed with reduced strength.

2. *Parametric amplifier*

The reversal of the spectroscopy experiment, i.e., the source ω_0 removed and one substituted at ω_R, is the technique being employed for heterodyne detection experiments, as discussed above. Examination of the response at $\omega_J = \omega_R + \omega_0$ in Fig. 49 indicates another possible application. Consider the configuration of Fig. 48 as a parametric amplifier, with ω_0 the signal, ω_J the pump, and ω_R as the idler. If the background level in Fig. 49a is the unamplified signal, one may interpret the voltage for $\omega_J = \omega_R + \omega_0$ as an amplified signal. Although the observed amplification is not large, the circuit Q's (particularly of the idler ω_R) are not high and the coupling of ω_R to the device is not optimum. Amplification is also observed at the degenerate condition, $\omega_0 = \omega_R$, corresponding to the signal and idler at the same frequency. This is observed in Fig. 49, as a strong signal at $\omega_J = 2\omega_0$. This type of parametric operation has not received serious consideration to date, but does not suffer from the disadvantages of the heterodyne down-conversion schemes attempted.

3. *Voltmeter*

We have explicitly discussed utilizing weak-link loop devices to measure flux and current. Voltage can be measured by transforming to flux or current via appropriate circuit elements. However with the rf-biased resistive loop device discussed in Section III.B.8 voltages can be measured directly via the Josephson frequency relation, Eq. (37). Whether one chooses the direct method or that of transforming to a flux or current will depend on source resistance and inductance.

One area where direct voltage measurements are useful is the observation of the superconducting–normal transition of bulk materials. Such experiments were carried out by Borcherts and Silver (1968) on Cu–Nb composites using rf-biased resistive loop devices. The sample material was incorporated into the low-inductance loop and a bias current was introduced. In this manner the critical current and resistance were measured as functions of the temperature.

Another area of application is to cryogenic thermoelectric measurements. By imposing a gradient ΔT across the resistive element one generates the thermoelectric emf of that material with respect to a superconductor. In fact such a thermoelectric device may be a useful incoherent radiation detector. The thermal voltage is proportional to the temperature difference

$$V_\theta = S\,\Delta T \tag{138}$$

and ΔT in the absence of other heat leaks is given by

$$\Delta T = Hl/\mathrm{K}A \tag{139}$$

where H is the power input due to radiant energy, l and A are the length and area of the element, and K is the thermal conductivity. The accuracy with which V can be measured is limited by the thermal fluctuation (Johnson) noise as demonstrated earlier,

$$V_\mathrm{N} = (2k_\mathrm{B}TR/t)^{1/2} \tag{140}$$

Since the thermal and electrical conductivities are approximately related by the Wiedemann–Franz ratio W we have

$$V_\theta/V_\mathrm{N} \approx HS(RT)^{1/2}/W(k_\mathrm{B}t)^{1/2} \tag{141}$$

which for R, T, and t is equal to unity and for $S \approx 10^{-5}$ V K^{-1} is equal to

3×10^{14} H. Thus one could measure 3×10^{-15} W with a S/N = 1 for a 1-sec integration time. Unfortunately at this level one must consider temperature fluctuation noise. This produces temperature fluctuations given by

$$\langle \Delta T^2 \rangle = k_B T^2 / C \tag{142}$$

where C is the thermal capacity of the element. Calculations similar to these are appropriate for other sensors, e.g., superconducting bolometers.

The main advantage offered by the resistive loop technique is in the measurement of *very* low-impedance sources. In these applications standard measuring techniques frequently do not permit measurement to the source-noise limit afforded with the weak-link device.

VI. A Survey of Recent Developments

A. Fundamental Principles

Two new experimental observations relate to the fundamental nature of the Josephson effect. One is the observation of the "pair-quasiparticle interference current" by Pedersen *et al.* (1973). This is a term in the current, predicted by Josephson (1964, 1965), of the form $G_1 V \cos \theta$, which should be added to the three terms given in Eq. (38). Like G, G_1 may also be voltage dependent. In magnitude, G_1 may be comparable to G, but Auracher *et al.* (1973) have shown that in low-capacitance weak links, e.g. point contacts, the effect of the pair-quasiparticle interference current may often be neglected.

The other new observation is the experimental proof of the Riedel peak (Section III.A.1) by Hamilton and Shapiro (1971).

A useful picture of how phase precession takes place in a long, thin (thin compared to the London penetration depth) metal microbridge has been given by Langer and Ambegaokar (1967). They plot the real and imaginary parts of the wave function along the length of the bridge. With no current, the gradient of the phase is zero, and the plot is a straight line denoting constant amplitude and phase. At finite current, a nonzero constant-phase gradient [Eq. (3)] must exist along the bridge (assuming a homogeneous bridge of constant cross section), and the plot becomes a helix of constant pitch around the bridge axis. At higher current the helix twists more tightly, and near critical current it becomes unstable with respect to dropping a loop. That is, the helix collapses in some particular region, as determined by fluctuations or imperfections, in such a way that one particular loop is pulled through the zero axis and ceases to exist. This reduces the total phase shift by 2π. The phase gradient and the current are correspondingly reduced to a lower value. The theoretical dynamics of the process have been investigated by Rieger *et al.* (1972), who emphasize that an instantaneous 2π "phase slip" occurs when the wave function goes to zero, i.e., when the loop of the helix crosses the zero axis.

The above picture is simplified considerably if the bridge is made very short. The limiting case of a short bridge is a bridge in which critical current is reached at a total phase shift of $\frac{1}{2}\pi$ (a quarter-turn of the helix) just as in a tunnel junction. There is considerable theoretical and experimental evidence that such bridges are realizable in practice, and that they, in fact, have sinusoidal current–phase relations (Aslamozov and Larkin, 1969; T. A. Fulton *et al.*, 1972; Gregers-Hansen *et al.*, 1971, 1972).

The value of the above picture is that it further unifies the concepts of the Josephson effect in tunnel junctions and in microbridges. It is a simple exercise in quantum mechanics to show that the wave function near the center of a tunnel barrier also twists around the zero axis, goes through zero at a half-turn ($\theta = \pi$), and undergoes an instantaneous 2π "phase slip" in response to a steadily increasing total phase shift θ across the junction. The whole process is identical to what happens in the limiting case of a short microbridge.

A wave function for an ideal Josephson junction and the electric field derived therefrom were given by Zimmerman (1973), who pointed out further similarities between tunnel junctions and microbridges, e.g., the Bernoulli effect, and also pointed out that the junction capacitance is very different from what would be calculated assuming a uniform electric field across the barrier.

In terms of what we already know about the characteristics of Josephson junctions, some aspects of the qualitative operation of long microbridges are not difficult to understand. The region of the bridge where the wave function collapses as critical current is approached may be thought of as a "short" microbridge (see above) and the rest of the bridge as just so much series inductance. Indeed, the whole repetitive process of slow buildup and rapid collapse of the current (under constant low-voltage bias, for example) is very similar to what happens in any Josephson junction with large series inductance (see, for example, Fig. 16) with $\beta > 1$. For a bridge whose lateral dimensions are small compared to the London penetration depth, the inductance is mainly due to mechanical inertia rather than magnetic flux.

With high current bias there is a complication, namely, when the critical current is exceeded for one part of a long uniform bridge it is exceeded for all parts. Thus, if the bridge current is well above critical, the repetitive collapse of the helical wave function must occur successively in various parts of the bridge; the resulting voltage oscillation might be quite complex. The behavior might resemble the incoherent operation of junctions in series; coherent operation of junctions in series has recently received some attention in the literature (Tilley, 1970; Finnegan and Wahlsten, 1972).

The preponderance of evidence indicates that the microbridges used in practical devices are of the "short" type, as evidenced by sinusoidal or near sinusoidal current–phase relations. Thus, an interest in long microbridges is more academic than practical so far as Josephson weak-link devices are concerned.

An advance in the understanding of fluctuations in rf-biased devices, which goes beyond the simple theory of Section V.A.2, was provided by Kurkijarvi and Webb (1972). For $\beta > 1$, the rf loss that is entailed by traversing hysteresis loops (Fig. 1) results in an effective noise conductance that is greater than the intrinsic junction conductance G [Eq. (102)]. The spectral intensity of the flux fluctuation varies as $\beta^{1/3}$, for β considerably greater than unity. For $\beta < 1$, Eq. (102) should be approximately correct. Experimental measurements of noise were made by Giffard *et al.* (1972).

Mechanical analogs (Sections II and III) have become quite popular both for tutorial purposes and for quantitative analysis of simple circuits (Sullivan and Zimmerman, 1971). Hamilton (1972) and others have made compact electronic analogs of a Josephson junction to which other circuit elements can be connected, just as to a real junction. Mechanical analogs operate in the region of a few hertz and are useful for acquiring an overall intuitive grasp of how a particular circuit functions. Electronic analogs operate at a few kilohertz and are more convenient for extensive quantitative analyses.

B. Weak-Link Developments

1. *Point contacts*

Except for one or two notable innovations, point-contact technology has hardly changed for several years. One useful innovation is to use a thin layer of glass, whose thermal expansion matches that of Nb, as a separator between the two opposing members of a Nb point-contact structure. The composite structure can be cemented together (Buhrman *et al.*, 1971) or the glass can be fused directly to the Nb (deWaele *et al.*, 1972). Such contacts were shown to be quite reproducible under thermal cycling.

In a brief study of "perfectly clean" point contacts, Zimmerman (1972) showed that these approach the ideal of pure Josephson elements more closely than any other type of weak link, in the sense of achieving the theoretical minimum shunt normal conductance G and shunt capacitance C [Eq. (38)].

2. *Thin-film bridges*

Because of the conceptual simplicity and mechanical ruggedness of metal thin-film bridges, considerable effort continues to be devoted to material studies and manufacturing methods. Consadori *et al.* (1971) have made bridges of NbSe by cleaving and notching the material with a razor blade and similar tools. These bridges on a thin mica substrate were used to replace the point contact in a bulk Nb rf-biased loop device.

Notaries *et al.* (1973) have made "overlay" bridges, that is, bridges with a shape similar to that shown in Fig. 30, but with a narrow ($\sim$1 μ) transverse band of a nonsuperconducting metal such as gold overlaid across the narrow part of the bridge. The normal-metal overlay weakens the bridge (reduces the order parameter) by proximity, and so both reduces the critical current of the bridge and also effectively defines its length.

Goodkind and Dundon (1971) have made rf-biased loop devices of sputtered Nb on glass rods. The loops are of the order of 1 mm diam and 1 cm long, and operate down to a few milliKelvin with no bridge at all, that is, with a uniform metallic cross section around the periphery of the loop. The operation of such "non-bridges" seems quite anomalous in view of what we have already learned about the Josephson effect in thin-film bridges, and remains to be understood.

A promising method of making exceedingly narrow bridges in Nb thin films and other metals is by electron-beam photoetching, a technique that has been used by Janocko *et al.* (1971) and others. Bridges narrower than 0.1 μ have been made.

The work of Gregers-Hansen *et al.* has already been mentioned. They have emphasized the importance of making very short bridges, as discussed in Section VI.A. Their bridges were defined by making carefully controlled cuts in soft metal (e.g., Sn) films, using selected extra-sharp razor blades. The criterion was that the end of a cut, which defines one edge of the bridge, should show a microscopically sharp cusp, so that the bridge is geometrically of zero length.

3. *Tunnel junctions*

The large shunt capacitance of most tunnel junctions is disadvantageous both for ultrahigh-frequency detection and for use in rf-biased loop devices (see Section III.B.9). A way of reducing the capacitance is to make small-area junctions with a relatively thick barrier of a low-energy semiconductor. Seto and van Duzer (1971) made Pb–Te–Pb junctions, and Simmonds (private communication) showed that such junctions worked well in rf-biased loops, with no problems from high-order transitions. These particular junctions had a shelf life of only a few months, owing perhaps to interdiffusion of Pb and Te, but they demonstrated the feasibility and utility of low-capacitance tunnel junctions.

C. Device Development

1. *Point-contact devices*

Because of the ease with which good point-contact Josephson junctions can be made, they continue to be used in a variety of research and developmental device applications.

A useful variant of the single- or double-junction quantum interference device (Section IV.C) is the multiple-hole or "fractional-turn" loop device (Zimmerman, 1971). In devices of this type, many loops (as many as 24 in some versions) are connected in parallel in a compact arrangement across a single contact. The idea is adaptable also to double-junction devices and to partly resistive devices. Putting N loops in parallel reduces the device inductance by a factor of N^2 compared to that of a single-hole device with the same hole length and total hole area. This feature can be exploited in two ways. A very low inductance gives a large interference pattern amplitude [Eq. (64) with M proportional to $(LL_1)^{1/2}$], which in turn permits fast flux-quanta counting and improves the lock-on capability of digital–analog magnetometer systems (Section V.D) in noisy environments. Secondly, the fractional-turn concept is a very effective way of increasing magnetometer sensitivity by increasing the effective volume without increasing the inductance (Section V.A). An absolute upper limit on inductance, for a given cryostat temperature, is given by Eq. (98).

D. Applications

The use of Josephson junction devices is spreading rapidly, and it is not possible to discuss everything that has been done. Some examples of recent work are given in the following paragraphs, but this is not an exhaustive survey.

Kanter and Silver (1971) operated a junction as a self-driven degenerate parametric amplifier at 30 MHz. This work was viewed as preliminary to similar operation at millimeter wavelengths. The device configuration used was a fractional-turn (see above) partly resistive loop.

Magnetocardiography using superconducting magnetometers is under active development in several laboratories. A promising approach is to use a gradiometer or field-difference sensor in order to relax the requirement on shielding against environmental interference (Zimmerman and Frederick, 1971; Rosen *et al.*, 1971, and others).

McDonald *et al.* (1972) have carried out a series of remarkable experiments on the use of point contacts for far-infrared spectroscopy, at wavelengths as short as 10 μ. They have achieved efficient four-hundredth-order harmonic mixing of an X-band source and a laser at 3.8×10^{12} Hz, and have obtained good conversion gain when mixing two laser signals in the vicinity of 3×10^{13} Hz to produce an i-f signal at 60 MHz. An interesting feature of their experimental technique is the use of extremely clean and sharp Nb point contacts for reliable operation at the highest frequencies. During and after a final etching, the points are never exposed to the atmosphere, implying that an oxide layer is not only unnecessary but undesirable.

The amplitude of the dc-field pattern (Fig. 13) of an rf-biased loop device is proportional to frequency [Eq. (64)], at least up to the frequency where the analysis is complicated by resonances in the device itself. Thus, it is predictable that the signal-to-noise ratio should be enhanced by operating the device at high bias frequency. This prediction has been confirmed by comparing the results of point-contact devices of nearly the same inductance operated at 30 MHz, 300 MHz (Zimmerman and Frederick, 1971), and at 9 GHz (Kamper and Simmonds, 1972). An ~20-dB improvement in signal to noise was noted for the 9-GHz as compared to the 30-MHz system. Such systems are useful for field measurements in a frequency band from 0 to 1 GHz (a response time of the order of 1 nsec), or for counting flux quanta at rates of 10^7 sec^{-1} or greater. Kamper and Simmonds showed that the quantum-periodic response of the 9-GHz system could be used to calibrate rf attenuators with an accuracy comparable to that of the best piston attenuators, over a frequency range 0–1 GHz.

A new application of the Josephson effect is the generation and detection of phonons (Dayem *et al.*, 1971; Kinder, 1971). A thin-film tunnel junction deposited on one face of an insulating substrate such as sapphire acts as a detector of phonons of energy $2e\Delta$ (Δ = gap voltage; see Fig. 3) as reported earlier by Eisenmenger and Dayem (1967). A similar junction on the other face of the substrate acts as a phonon generator. Sharp increases in the phonon intensity of energy $\hbar\omega \geq 2e\Delta$ are measured by the detector junction whenever the generator bias voltage V_G equals a submultiple of $2e\Delta$, i.e., $V_G = 2e\Delta/n$, $n = 2, 3, 4, \ldots$. This is taken as evidence of multiple quantum processes, since phonons of energy $2e\Delta$ are produced by electron pairs of lower energy, $2e\Delta/n$.

The use of rf-biased loops as noise thermometers (Section V.D.3) has been extended down to the milliKelvin range by Kamper *et al.* (1971), by Soulen and Marshak (1972), and by Giffard *et al.* (1972). Down to the neighborhood of 0.01 K, it appears that absolute thermometry by this method is competitive with CMN (cerium magnesium nitrate) thermometry and other absolute methods.

The use of Josephson junctions in computers has not been discussed in this chapter. Because of the bistable nature of its I–V characteristic (Fig. 3) with current bias, a thin-film tunnel junction could be used as a memory device. The switching speed is given by the time required to charge up the junction capacitance by the bias current, and so is of order $2\Delta/Ci_c$, which may be less than 1 nsec (Matisoo 1966, 1967a, b, 1968, 1969a, b). More recently, the use of an array of parallel junctions as a shift register has been considered (Fulton *et al.*, 1973).

References

Ambegaokar, V., and Halperin, B. I. (1969). *Phys. Rev. Lett.* **22**, 1364.

Anderson, P. W. (1964). "Lectures on the Many-Body Problem" (E. R. Caianiello, ed.), Vol. 2. Academic Press, New York.

Anderson, P. W., and Dayem, A. H. (1964). *Phys. Rev. Lett.* **13**, 195.

Anderson, J. T., and Goldman, A. M. (1969). *Phys. Rev. Lett.* **23**, 128.

Aslamazov, L. G., and Larkin, A. I. (1969). *Zh. Eksp. Teor. Fiz. Pis'ma Red.* **9**, 150.

Auracher, F., Richards, P. L., and Rochlin, G. I. (1973). *Phys. Rev.* **B8**, 4182.

Borcherts, R., and Silver, A. H. (1968). *Bull. Amer. Phys. Soc.* **13**, 379.

Burgess, R. E. (1967). Proc. Symp. Phys. Superconducting Devices, Charlottesville, paper H1 (published by Office of Naval Research).

Buhrman, R. A., Strait, S. F., and Webb, W. W. (1971). *J. Appl. Phys.* **42**, 4527.

Clarke, J. (1966). *Phil. Mag.* **13**, 115.

Cohen, D., Edelsack, E. A., and Zimmerman, J. E. (1970). *Appl. Phys. Lett.* **16**, 278.

Consadori, F., Fife, A. A., Frindt, R. F., and Gygax, S. (1971). *Appl. Phys. Lett.* **18**, 233.

Coon, D. D., and Fiske, M. D. (1965). *Phys. Rev.* **138**, 744.

Dahm, A. J., Denenstein, A., Langenberg, D. N., Parker, W. H., Rogovin, D., and Scalapino, D. J. (1969). *Phys. Rev. Lett.* **22**, 1416.

Dayem, A. H., and Grimes, C. C. (1966). *Appl. Phys. Lett.* **9**, 47.

Dayem, A. H., Miller, B. I., and Wiegand, J. J. (1971). *Phys. Rev. B* **3**, 2949.

Deaver, B. S., and Fairbank, W. M. (1961). *Phys. Rev. Lett.* **7**, 43.

deWaele, A. Th. A. M., Kraan, W. H., deBruyn Ouboter, R., and Taconis, K. W. (1967). *Physica* (*Utrecht*) **37**, 114.

deWaele, A. Th. A. M., and deBruyn Ouboter, R. (1969). *Physica* (*Utrecht*) **42**, 626.

deWaele, A. Th. A. M., Vergouwen, C. P. M., Matsinger, A. J., and deBruyn Ouboter, R. (1972). *Physica* (*Utrecht*), **59**, 155.

Dmitrenko, I. M., Yanson, I. K., and Svistunov, V. M. (1965). *Zh. Eksp. Teor. Fiz. Pis'ma Red.* **2**, 17.

Doll, R., and Näbauer, M. (1961). *Phys. Rev. Lett.* **7**, 51.

Eck, R. E., Scalapino, D. J., and Taylor, B. N. (1964a). "Proc. Int. Conf. Low Temp. Physics, 9th, Columbus," p. 415. Plenum Press, New York.
Eck, R. E., Scalapino, D. J., and Taylor, B. N. (1964b). *Phys. Rev. Lett.* **13**, 15.
Eisenmenger, W., and Dayem, A. H. (1967). *Phys. Rev. Lett.* **18**, 125.
Finnegan, T. F., and Wahlsten, S. (1972). *Appl. Phys. Lett.* **21**, 541.
Fiske, M. D. (1964). *Rev. Mod. Phys.* **36**, 221.
Forgacs, R. L., and Warnick, A. (1966). *IEEE Natl. Convention Rec.*, Pt. 10.
Forgacs, R. L., and Warnick, A. (1967). *Rev. Sci. Instrum.* **38**, 214.
Fulton, T. A. (1970). *Solid State Commun.* **8**, 1353.
Fulton, T. A., Dunkleberger, L. N., and Dynes, R. C. (1972). *Phys. Rev. B* **6**, 832.
Fulton, T. A., Dynes, R. C., and Anderson, P. W. (1973). *Proc. IEEE* **61**, 28.
Giffard, R. P., Webb, R. A., and Wheatley, J. C. (1972). *J. Low Temp. Phys.* **6**, 533.
Goldman, A. M., Kreisman, P. J., and Scalapino, D. J. (1965). *Phys. Rev. Lett.* **15**, 495.
Goodkind, J. M., and Dundon, J. M. (1971). *Rev. Sci. Instrum.* **42**, 1264.
Gregers-Hansen, P. E., Levinsen, M. T., Pedersen, L., and Sjøstrøm, C. J. (1971). *Solid State Commun.* **9**, 661.
Gregers-Hansen, P. E., Levinsen, M. T., and Pedersen, G. Fog (1972). *J. Low Temp. Phys.* **7**, 99.
Grimes, C. C., and Shapiro, S. (1968). *Phys. Rev.* **169**, 397.
Grimes, C. C., Richards, P. L., and Shapiro, S. (1966). *Phys. Rev. Lett.* **17**, 431.
Grimes, C. C., Richards, P. L., and Shapiro, S. (1968). *J. Appl. Phys.* **39**, 3905.
Hamilton, C. A. (1972). *Rev. Sci. Instrum.* **43**, 445.
Hamilton, C. A., and Shapiro, S. (1971). *Phys. Rev. Lett.* **26**, 426.
Hanabusa, M., Silver, A. H., and Kushida, T. (1970). *Phys. Rev.* **B2**, 1293.
Harding, J. T., and Zimmerman, J. E. (1970). *J. Appl. Phys.* **41**, 1581.
Hunt, T. (1966). *Phys. Rev.* **151**, 325.
Ivanchenko, Yu. M., and Zil'berman, L. A. (1968a). *Zh. Eksp. Teor. Fiz. Pis'ma Red.* **8**, 189.
Ivanchenko, Yu. M., and Zil'berman, L. A. (1968b). *Zh. Eksp. Teor. Fiz.* **55**, 2395.
Jaklevic, R. C., Lambe, J., Silver, A. H., and Mercereau, J. E. (1964a). *Phys. Rev. Lett.* **12**, 159.
Jaklevic, R. C., Lambe, J. J., Silver, A. H., and Mercereau, J. E. (1964b). *Phys. Rev. Lett.* **12**, 274.
Jaklevic, R. C., Lambe, J. J., Silver, A. H., and Mercereau, J. E. (1965a). *Low Temp. Phys.* **LT9**, 446.
Jaklevic, R. C., Lambe, J. J., Mercereau, J. E., and Silver, A. H. (1965b). *Phys. Rev.* **140**, 1628.
Janocko, M. A., Gavaler, J. R., Jones, C. K., and Blaugher, R. D. (1971). *J. Appl. Phys.* **42**, 182.
Josephson, B. D. (1962). *Phys. Lett.* **1**, 251.
Josephson, B. D. (1964). *Rev. Mod. Phys.* **36**, 216.
Josephson, B. D. (1965). *Advan. Phys.* **14**, 419.
Kamper, R. A. (1967). Proc. Symp. Phys. Superconducting Devices, Charlottesville, paper M1 (published by Office of Naval Research).
Kamper, R. A., and Simmonds, M. B. (1972). *Appl. Phys. Lett.* **20**, 15.
Kamper, R. A., and Zimmerman, J. E. (1971). *J. Appl. Phys.* **42**, 132.
Kamper, R. A., Mullen, L. O., and Sullivan, D. B. (1969). NBS Technical Note No. 381, Cat. No. C13.46:381. Superintendent of Documents, U.S. GPO, Washington, D.C.
Kanter, H., and Silver, A. H. (1971). *Appl. Phys. Lett.* **19**, 515.
Kanter, H., and Vernon, F. L., Jr. (1970a). *Appl. Phys. Lett.* **16**, 115.

Kanter, H., and Vernon, F. L., Jr. (1970b). *Phys. Rev. Lett.* **25**, 588.
Kanter, H., and Vernon, F. L., Jr. (1970c). *Phys. Lett.* **32**, 155.
Kinder, H. (1971). *Phys. Lett.* **36**, 379.
Kose, V. E., and Sullivan, D. B. (1970). *J. Appl. Phys.* **41**, 169.
Kurkijarvi, J., and Ambegaokar, V. (1970). *Phys. Lett. A* **31**, 314.
Kurkijarvi, J., and Webb, W. W. (1972). 1972 Applied Superconductivity Conference, Annapolis, Md. *IEEE Conf. Rec.* No. 72CH0682-5TABSC, 581.
Lambe, J., Silver, A. H., Mercereau, J. E., and Jaklevic, R. C. (1964). *Phys. Lett.* **11**, 16.
Langer, J. S., and Ambegaokar, V. (1967). *Phys. Rev.* **164**, 498.
London, F. (1950). "Superfluids." Wiley, New York.
Matisoo, J. (1966). *Appl. Phys. Lett.* **9**, 167.
Matisoo, J. (1967a). *Proc. IEEE* **55**, 172.
Matisoo, J. (1967b). Proc. Symp. Phys. Superconducting Devices, Charlottesville, paper N1 (published by Office of Naval Research).
Matisoo, J. (1968). *J. Appl. Phys.* **39**, 2587.
Matisoo, J. (1969a). *J. Appl. Phys.* **40**, 1813.
Matisoo, J. (1969b). *J. Appl. Phys.* **40**, 2091.
McCumber, D. E. (1968). *J. Appl. Phys.* **39**, 3113.
McDonald, D. G., Kose, V. E., Evenson, K. M., Wells, J. S., and Cupp, J. D. (1969). *Appl. Phys. Lett.* **15**, 121.
McDonald, D. B., Risley, A. S., Cupp, J. D., Evenson, K. M., and Ashley, J. R. (1972). *Appl. Phys. Lett.* **20**, 296.
Mercereau, J. E. (1967). Proc. Symp. Phys. Superconducting Devices, Charlottesville, paper No. U1 (published by Office of Naval Research).
Meservey, R. (1965). "Proc. Int'l Conf. on Low Temp. Physics, Columbus," p. 455. Plenum, New York.
Meservey, R. (1968). *J. Appl. Phys.* **39**, 2598.
Nordman, J. E. (1969). *J. Appl. Phys.* **40**, 2111.
Notaries, H. A., Wang, R. H., and Mercereau, J. E. (1973). *Proc. IEEE* **61**, 79.
Parker, W. H., Taylor, B. N., and Langenberg, D. N. (1967). *Phys. Rev. Lett.* **18**, 287.
Parker, W. H., Langenberg, D. N., and Denenstein, A. (1969). *Phys. Rev.* **177**, 639.
Pedersen, N. F., Finnegan, T. F., and Langenberg, D. N. (1972). *Phys. Rev. B* **6**, 4151.
Richards, P. L., and Sterling, S. A. (1969). *Appl. Phys. Lett.* **14**, 394.
Riedel, E. (1964). *Z. Naturforsch* **19a**, 1634.
Rieger, T. J., Scalapino, D. J., and Mercereau, J. E. (1972). *Phys. Rev. B* **6**, 1734.
Rosen, A., Inouye, G. T., Morse, A. L., and Judge, D. L. (1971). *J. Appl. Phys.* **42**, 3682.
Rowell, J. M. (1963). *Phys. Rev. Lett.* **11**, 200.
Scalapino, D. J. (1967). Proc. Symp. Phys. Superconducting Devices, Charlottesville, paper No. G1 (published by Office of Naval Research).
Schroen, W. (1968). *J. Appl. Phys.* **39**, 2671.
Schwidtal, K., and Finnegan, R. D. (1969). *J. Appl. Phys.* **40**, 2123.
Scott, A. C. (1969). *Amer. J. Phys.* **37**, 52.
Seto, J., and van Duzer, T. (1971). *Appl. Phys. Lett.* **19**, 488.
Shapiro, S. (1963). *Phys. Rev. Lett.* **11**, 80.
Shapiro, S., Janis, A. R., and Holly, S. (1964). *Rev. Mod. Phys.* **36**, 223.
Shin, E. E., and Schwartz, B. B. (1966). *Phys. Rev.* **152**, 207.
Silver, A. H., and Zimmerman, J. E. (1967a). *Appl. Phys. Lett.* **10**, 142.
Silver, A. H., and Zimmerman, J. E. (1967b). *Phys. Rev.* **157**, 317.
Silver, A. H., and Zimmerman, J. E. (1967c). *Phys. Rev.* **158**, 423.
Silver, A. H., Zimmerman, J. E., and Kamper, R. A. (1967). *Appl. Phys. Lett.* **11**, 209.

Simmonds, M. B., and Parker, W. H. (1971). *J. Appl. Phys.* **42**, 38.
Soulen, R. J., and Marshak, H. (1972). 1972 Applied Superconductivity Conference, Annapolis, Md., *IEEE Conf. Rec.* No. 72CH0682-5TABSC, 588.
Stephen, M. J. (1969). *Phys. Rev.* **182**, 531.
Stewart, W. C. (1968). *Appl. Phys. Lett.* **12**, 277.
Sullivan, D. B., and Zimmerman, J. E. (1971). *Amer. J. Phys.* **39**, 1504.
Sullivan, D. B., Peterson, R. L., Kose, V. E., and Zimmerman, J. E. (1970). *J. Appl. Phys.* **41**, 4865.
Tilley, D. R. (1970). *Phys. Lett.* **33A**, 205.
Ulrich, B. (1971). *J. Appl. Phys.* **42**, 2.
Vernon, F. L., Jr., and Pederson, R. J. (1968). *J. Appl. Phys.* **39**, 2661.
Zimmerman, J. E. (1970). *J. Appl. Phys.* **41**, 1589.
Zimmerman, J. E. (1971a). *J. Appl. Phys.* **42**, 30.
Zimmerman, J. E. (1971b). *J. Appl. Phys.* **42**, 4483.
Zimmerman, J. E. (1972). *IEEE Conf. Rec. No.* 72CH0682-5TABSC.
Zimmerman, J. E. (1973). *Phys. Lett.* **42A**, 375.
Zimmerman, J. E., and Frederick, N. V. (1971). *Appl. Phys. Lett.* **19**, 16.
Zimmerman, J. E., and Silver, A. H. (1964), *Phys. Lett.* **10**, 47.
Zimmerman, J. E., and Silver, A. H. (1966a). *Phys. Rev.* **141**, 367.
Zimmerman, J. E., and Silver, A. H. (1966b). *Solid State Commun.* **4**, 133.
Zimmerman, J. E., and Silver, A. H. (1967). *Phys. Rev. Lett.* **19**, 14.
Zimmerman, J. E., and Silver, A. H. (1968a). *J. Appl. Phys.* **39**, 2679.
Zimmerman, J. E., and Silver, A. H. (1968b). *Phys. Rev.* **167**, 418.
Zimmerman, J. E., Cowen, J. A., and Silver, A. H. (1966). *Appl. Phys. Lett.* **9**, 353.
Zimmerman, J. E., Thiene, P., and Harding, J. T. (1970). *J. Appl. Phys.* **41**, 1572.

Additional Bibliography

Anderson, P. W. (1967). *Progr. Low Temp. Phys.* **5**, 1.
Anderson, P. W., and Rowell, J. M. (1963). *Phys. Rev. Lett.* **10**, 230.
Aslamazov, L. G., Larkin, A. I., and Ovchinnikov, Yu. N. (1968). *Zh. Eksp. Teor. Fiz.* **55**, 323.
Bura, P. (1966). *Appl. Phys. Lett.* **8**, 155.
Cheishvili, O. D. (1969). *Fiz. Tverd. Tela* **11**, 185.
Claeson, T., Gygax, S., and Maki, K. (1967). *Phys. Kondens. Mater.* **6**, 23.
Clarke, J. (1967). Proc. Symp. Phys. Superconducting Devices, Charlottesville, paper D1 (published by Office of Naval Research).
Contaldo, A. (1967). *Rev. Sci. Instrum.* **38**, 1543.
Dahm, A. J., Denenstein, A., Finnegan, T. F., Langenberg, D. N., and Scalapino, D. J. (1968). *Phys. Rev. Lett.* **20**, 859.
D'Aiello, R. V., and Freedman, S. J. (1966). *Appl. Phys. Lett.* **9**, 323.
Dayem, A. H., and Wiegand, J. J. (1967). *Phys. Rev.* **155**, 419.
deBruyn Ouboter, R., and deWaele, A. Th. A. M. (1970). *Progr. Low Temp. Phys.* **6**, 243.
deBruyn Ouboter, R., Omar, M. H., Arnold, A. P. J. T., Guinan, T., and Taconis, K. W. (1966). *Physica (Utrecht)* **32**, 1448.
deBruyn Ouboter, R., Kraan, W. H., deWaele, A. Th. A. M., and Omar, M. H. (1967). *Physica (Utrecht)* **35**, 335.

deWaele, A. Th. A. M., and deBruyn Ouboter, R. (1969). *Physica (Utrecht)* **41**, 225.
deWaele, A. Th. A. M., Kraan, W. H., deBruyn Ouboter, R. (1968). *Physica (Utrecht)* **40**, 302.
Dmitrenko, I. M., (1969). *Ukr. Fiz. Zh. (Ukr. Ed.)* **14**, 439.
Dmitrenko, I. M., Bondarenko, S. I. (1968). *Zh. Eksp. Teor. Fiz.* **7**, 241.
Dmitrenko, I. M., and Yanson, I. K. (1965). *Zh. Eksp. Teor. Fiz. Pis'ma Red.* **2**, 242.
Dmitrenko, I. M., and Yanson, I. K. (1965). *Zh. Eksp. Teor. Fiz.* **9**, 1741.
Dmitrenko, I. M., Yanson, I. K., and Yurchenko, I. I. (1967). *Fiz. Tverd. Tela* **9**, 3656.
Economou, E. N., and Ngai, K. L. (1968). *Phys. Rev. Lett.* **20**, 547.
Farrell, R. A. (1965). *Phys. Rev. Lett.* **15**, 527.
Farrell, R. A., and Prange, R. E. (1963). *Phys. Rev. Lett.* **10**, 479.
Fetter, A. L., and Stephen, M. J. (1968). *Phys. Rev.* **168**, 475.
Finnegan, T. F., Denenstein, A., Langenberg, D. N., McMenamin, J. E., Novoseller, D. E., and Cheng, L. (1969). *Phys. Rev. Lett.* **23**, 229.
Fulton, T. A., and McCumber, D. E. (1968). *Phys. Rev.* **175**, 585.
Galkin, A. A., and Svistunov, V. M. (1967). *Zh. Eksp. Teor. Fiz. Pis'ma Red.* **5**, 396.
Galkin, A. A., Borodai, B. U., Svistunov, V. M., and Tarasenko, V. N. (1968). *Zh. Eksp. Teor. Fiz. Pis'ma Red.* **8**, 521.
Gandolfo, D. A., Boornard, A., and Morris, L. C. (1968). *J. Appl. Phys.* **39**, 2657.
Gaule, G. K., Breslin, J. T., and Winter, J. J. (1967). Applied Superconductivity Conference, Austin, Texas.
Gaule, G. K., Ross, R. L., and Schwidtal, K. (1967). Proc. Symp. Phys. Superconducting Devices, Charlottesville, paper P1 (published by Office of Naval Research).
Gennes, P. G. de (1963). *Phys. Lett.* **5**, 22.
Giaever, I. (1965). *Phys. Rev. Lett.* **14**, 904.
Goldman, A. M., and Kreisman, P. J. (1967). *Phys. Rev.* **164**, 544.
Golub, A. A. (1968). *Fiz. Tverd. Tela* **10**, 3160.
Goodkind, J. M., and Stolfa, D. L. (1970). *Rev. Sci. Instrum.* **41**, 799.
Gorbonosov, A. E., and Kulik, I. O. (1967). *Fiz. Metal. Metalloved.* **23**, 803.
Harding, J. T., and Zimmerman, J. E. (1968). *Phys. Lett.* **27**, 670.
Ivanchenko, Yu. M. (1966). *Zh. Eksp. Teor. Fiz.* **51**, 337.
Ivanchenko, Yu. M. (1967). *Zh. Eksp. Teor. Fiz.* **52**, 1320.
Ivanchenko, Yu. M. (1967). *Zh. Eksp. Teor. Fiz. Pis'ma Red.* **6**, 876.
Ivanchenko, Yu. M., Svidzinskii, A. V., and Slusarev, V. A. (1966). *Zh. Eksp. Teor. Fiz.* **51**, 194.
Jacobson, D. A. (1965). *Phys. Rev.* **138**, 1066.
Josephson, B. D. (1966). *Wireless World* **72**, 484.
Kamper, R. A., Radebaugh, R., Siegwarth, J. D., and Zimmerman, J. E. (1971). *IEEE Proc. Lett.* **59**, 1368.
Kao, Yi-Han (1968). *Phys. Lett.* **26**, 471.
Kirschman, R. K., Notaries, H. A., and Mercereau, J. E. (1971). *Phys. Lett.* **A34**, 209.
Krasnopolin, I. Ya., and Khaikin, M. S. (1967). *Zh. Eksp. Teor. Fiz. Pis'ma Red.* **6**, 633.
Kulik, I. O. (1965). *Zh. Eksp. Teor. Fiz. Pis'ma Red.* **2**, 134.
Kulik, I. O. (1965). *Zh. Eksp. Teor. Fiz.* **49**, 1211.
Kulik, I. O. (1966). *Zh. Eksp. Teor. Fiz.* **50**, 799.
Kulik, I. O. (1966). *Zh. Eksp. Teor. Fiz.* **51**, 1952.
Kulik, I. O. (1967). *Zh. Tekh. Fiz.* **37**, 157.
Kulik, I. O., Yanson, I. K., and Dmitrenko, I. M. (1967). *Ukr. Fiz. Zh. (Ukr. Ed.)* **12**, 1288.

Kurkijarvi, J. (1972). *Phys. Rev.* B **6**, 832.
Langenberg, D. N., Scalapino, D. J., Taylor, B. N., and Eck, R. E. (1965). *Phys. Rev. Lett* **15**, 294.
Langenberg, D. N., Scalapino, D. J., Taylor, B. N., and Eck, R. E. (1966). *Phys. Lett.* **20**, 563.
Langenberg, D. N., Scalapino, D. J., and Taylor, B. N. (1966). *Proc. IEEE* **54**, 560.
Langenberg, D. N., Parker, W. H., and Taylor, B. N. (1966). *Phys. Lett.* **22**, 259.
Langenberg, D. N., Parker, W. H., and Taylor, B. N. (1966). *Phys. Rev.* **150**, 186.
Larkin, A. I., and Ovchinnikov, Yu. N. (1966). *Zh. Eksp. Teor. Fiz.* **51**, 1535.
Larkin, A. I., and Ovchinnikov, Yu. N. (1967). *Zh. Eksp. Teor. Fiz.* **53**, 2159.
Larkin, A. I., Ovchinnikov, Yu. N., and Fyodorov, M. A. (1966). *Zh. Eksp. Teor. Fiz.* **51**, 683.
Lebwohl, P., and Stephen, M. J. (1967). *Phys. Rev.* **163**, 376.
Lukens, J. E., and Goodkind, J. M. (1968). *Phys. Rev. Lett.* **20**, 1363.
Maki, K. (1963). *Progr. Theor. Phys.* **30**, 573.
Maki, K. (1964). *Phys. Lett.* **10**, 11.
Maki, K., and Griffin, A. (1965). *Phys. Rev. Lett.* **15**, 921.
McCumber, D. E. (1968). *J. Appl. Phys.* **39**, 297.
McCumber, D. E. (1968). *J. Appl. Phys.* **39**, 2503.
Mercereau, J. E. (1969). "Superconductivity" (R. D. Parks, ed.), Vol. 1, p. 393. Decker, New York.
Nakajima, S., and Kuroda, Y. (1968). *Phys. Lett.* **26A**, 106.
Ngai, K. L., Applebaum, J. A., Cohen, M. H., and Phillips, J. C. (1967). *Phys. Rev.* **163**, 352.
Nieto, M. M. (1968). *Phys. Rev.* **167**, 416.
Ohtsuka, T. (1968). *Cryog. Eng.* **3**, 249.
Omar, M. H., and deBruyn Ouboter, R. (1966). *Physica (Utrecht)* **32**, 2044.
Omar, M. H., Kraan, W. H., deWaele, A. Th. A. M., and deBruyn Ouboter, R. (1967). *Physica (Utrecht)* **34**, 525.
Owen, C. S., and Scalapino, D. J. (1967). *Phys. Rev.* **164**, 538.
Papini, G. (1967). *Phys. Lett.* A **24**, 32.
Pritchard, J. P., Jr., and Schroen, W. H. (1968). *IEEE Trans. Magn.* **4**, 320.
Prothero, W. M., Jr., and Goodkind, J. M. (1968). *Rev. Sci. Instrum.* **39**, 1257.
Purna, M., and Deaver, B. S. (1971). *Appl. Phys. Lett.* **19**, 539.
Richards, P. L., and Anderson, P. W. (1965). *Phys. Rev. Lett.* **14**, 540.
Rothwarf, F., Krisch, H. M., and Ford, D. (1968). *J. Appl. Phys.* **39**, 2683.
Rowell, J. M., and Feldmann, W. L. (1968). *Phys. Rev.* **172**, 393.
Scalapino, D. J., and Wu, T. M. (1966). *Phys. Rev. Lett.* **17**, 315.
Schroen, W., and Pritchard, J. P. (1969). *J. Appl. Phys.* **40**, 2118.
Scott, A. C. (1967). *Phys. Lett.* A **25**, 132.
Scully, M. O., and Lee, P. A. (1969). *Phys. Rev. Lett.* **22**, 23.
Shapiro, S. (1967). Proc. Symp. Phys. Superconducting Devices, Charlottesville, paper No. I (published by Office of Naval Research).
Shapiro, S. (1967). *J. Appl. Phys.* **38**, 1879.
Shapiro, S. (1967). *Phys. Lett.* A **25**, 537.
Shigi, T., Saji, Y., Nakaya, S., Uchiho, K., and Aso, T. (1965). *J. Phys. Soc. Japan* **20**, 1276.
Shigi, T., Nakaya, S., and Aso, T. (1966). *J. Phys. Soc. Japan* **21**, 2418.
Silver, A. H. (1967). Proc. Symp. Phys. Superconducting Devices, Charlottesville, paper No. F1 (published by Office of Naval Research.)

Silver, A. H., Jaklevic, R. C., and Lambe, J. (1966). *Phys. Rev.* **141**, 362.
Stephen, M. J. (1968). *Phys. Rev. Lett.* **21**, 1629.
Stewart, W. C. (1969). *Appl. Phys. Lett.* **12**, 392.
Stone, J. L., and Hartwig, W. H. (1968). *J. Appl. Phys.* **39**, 2665.
Svidzinskii, A. V., and Slyusarev, V. A. (1966). *Zh. Eksper. Teor. Fiz.* (*USSR*) **51**, 177.
Svidzinskii, A. V., and Slyusarev, V. A. (1968). *Phys. Lett.* **27A**, 22.
Taylor, B. N. (1968). *J. Appl. Phys.* **39**, 2490.
Thiene, P., and Zimmerman, J. E. (1969). *Phys. Rev.* **177**, 758.
Tilley, D. R. (1966). *Phys. Lett.* **20**, 11.
Urushadze, G. I. (1968). *Phys. Lett.* **27A**, 381.
Vant-Hull, L. L. (1967). *Dissertation Abstracts* **B28**, Paper No. 67-8462.
Vant-Hull, L. L., and Mercereau, J. E. (1966). *Phys. Rev. Lett.* **17**, 629.
Vant-Hull, L. L., Simpkins, R., and Harding, J. T. (1967). *Phys. Lett.* **24A**, 736.
Weinberg, I. (1967). *J. Appl. Phys.* **38**, 3036.
Werthamer, N. R., and Shapiro, S. (1967). *Phys. Rev.* **164**, 523.
Yamashita, T., and Onodera, Y. (1967). *J. Appl. Phys.* **38**, 3523.
Yamashita, T., Kunita, M., and Onodera, Y. (1968). *J. Appl. Phys.* **39**, 5396.
Yanson, I. K. (1967). *Zh. Eksp. Teor. Fiz.* **53**, 1268.
Yanson, I. K. (1967). *Zh. Eksper. Teor. Fiz. Pis'ma Red.* **6**, 729.
Yanson, I. K., and Albegova, I. Kh. (1968). *Zh. Eksp. Teor. Fiz.* **55**, 1578.
Yanson, I. K., Svistvnov, V. M., and Dmitrendo, I. M. (1965). *Zh. Eksp. Teor. Fiz.* **48**, 976.
Yeh, R. H. T., and Mechetti, H. (1968). *Phys. Status Solidi* **25**, K65.
Zawadowski, A. (1967). *Elektrotech. Cas.* **18**, 528.
Zharkov, G. F. (1966). *Usp. Fiz. Nauk.* **88**, 198.
Zimmer, H. (1967). *Appl. Phys. Lett.* **10**, 193.
Zimmerman, J. E., and Mercereau, J. E. (1964). *Phys. Rev. Lett.* **13**, 125.
Zimmerman, J. E., and Mercereau, J. E. (1965). *Phys. Rev. Lett.* **14**, 887.

Chapter 2

Superconductive Switches and Amplifiers

WILLIAM S. GOREE and VICTOR W. HESTERMAN

Superconducting Technology, Inc.
Mountain View, California

I. Introduction

The engineering development and application of superconductive elements to electronic instruments is in its infancy. Yet, many superconductive elements, such as Josephson tunnel junctions and cryotrons, are well understood from a physics viewpoint and their incorporation into superior signal switches and amplifiers should be a straightforward engineering task. Furthermore, construction techniques have been developed to the stage where reproducible and characterizable elements can be built without additional fundamental advances. We have summarized the reported applications of superconducting switches and amplifiers, described their fabrication and operation, and discussed the performance achieved. This chapter should give the reader an understanding of the unique characteristics of superconducting switches and amplifiers, and should be useful as a guide to their practical use.

The chapter is organized according to circuit function, i.e., terminal properties, rather than physical property. Since many of the superconductive elements, such as cryotrons and Josephson devices, have been used both as switches and amplifiers, some redundancy in device description has been necessary to make each major division of the chapter reasonably complete.

II. Signal Switches

Electronic signals can be switched by means of several properties of superconductivity. For example, the change in resistance as a superconducting path is made normal can be used to switch a signal from that path to another path. Resistive signal switching can also be obtained by use of a tunnel junction or weak link by switching the junction or link from the superconducting state to the mixed or flux-flow state. For ac signals (or dc signals for zero-resistance loads), switching can also be accomplished by changing the self-inductance or mutual inductance of coils or strips by switching a nearby superconducting shield into the normal state or mechanically moving a shield relative to the coils.

Mechanical switches, such as sliding contact types, have received little consideration as signal switches because they are slow and it is very difficult to construct a reliable mechanical switch that will be superconducting when closed. A zero-resistance contact switch can be built by depositing a very thin layer of gold (a few hundred angstroms thick) over the superconducting surfaces (Meissner, 1963). The gold prevents oxidation and increases the life of the switch. The superconductor–normal metal–superconductor (S–N–S) structure will be superconducting for transport currents of the order of milliamperes. Interestingly enough this property was reported several years before Josephson's prediction of supercurrent tunneling through dielectric barriers (Josephson, 1962) and well before the work of Clarke on superconductor–normal–superconductor junctions (Clarke, 1966a; Clarke, 1969).

Recent work (Siegwarth and Sullivan, 1971) has shown that a reliable switch can be constructed using sliding contact between a niobium rotor and babbit alloy (4.5% Sn, 10.5% Sb, 85% Pb) fixed contacts. This switch was superconducting for currents up to about 3 A with little change or wear after several hundred switchings at liquid-helium temperatures.

All superconductive materials have well-defined critical values of three parameters: magnetic field H_c (the field required to make the metal go normal), current density J_c, and temperature T_c. For example, the critical value H_c ranges from a few hundred gauss for tin to hundreds of kilogauss for niobium tin (Nb_3Sn); critical current density is the order of 10^3 A/cm^2 for tin and 10^6 A/cm^2 for Nb_3Sn; and transition temperatures are 3.7°K for tin and 18°K for Nb_2Sn. The highest known critical temperature is 21°K for NbAlGe. An excellent summary of critical values of H_c and T_c for many materials has been published (Roberts, 1966, 1969). The critical values are, of course, interrelated and are also dependent on geometry, especially for very thin-film (100 to 1000 Å) superconductors. The selection of the best of these three parameters H_c, J_c, or T_c for use in a switch circuit depends on the specific application. The following examples should clarify many applications:

1. Most signal switches and their associated circuits have made use of the critical magnetic field parameter because they have been generally designed for low-power operation. This application requires switches to be constructed of materials with relatively low critical fields, such as tin, where small currents produce sufficient fields for switching and whose metallurgical properties make for ease of fabrication. This type of switch, called a cryotron, dissipates very little power in the liquid-helium bath since the switch drive element remains superconducting. Sources of power loss are Joule heating in the current leads to the switch element and thermal conductivity down the leads which increases the liquid-helium boiloff rate. The principal disadvantage of the field switch is that the field may induce noise in the signal circuit. This could occur, for example, if the switch were used to switch elements in a sensitive magnetometer circuit.

2. Current switching is used in ryotron (Miller *et al.*, 1964) inductance switches (described in Section II.B.1) to switch a superconducting shield. One disadvantage of the current switch is that the switching circuit is directly coupled to the signal circuit. Tunnel-junction switches also generally use currents to change the junction from the superconducting to the intermediate tunneling state.

3. Little use has been made of thermal properties in signal switches, but they have been used to switch Meissner-effect and quantized-flux modulators in magnetometer circuits (Deaver and Goree, 1967) (Sections III.C.2 and III.C.3). Thermal properties are almost exclusively used in persistent current switches in magnets. In this application, since the switch is not on for long time periods, the power dissipation in the bath is not excessive. The use of a cryotron-type switch for this application would require large currents to switch the high-critical-field magnet material; thus the fixed thermal conduction heat leak down the current leads may

become prohibitive. Thermal switches can, however, be made to obtain almost complete circuit isolation. For example, the heating power can be light coupled to the circuit with a light pipe. This type of switch can be used in circuits where very small fields and/or currents are being measured.

A. Resistance Switches

A superconductor can be switched from zero resistance to the resistive state by means of either a magnetic field, an electrical current, or a temperature change. The field-switched device, i.e., cryotron, has been extensively investigated primarily for computer applications (Newhouse, 1964; Ittner and Kraus, 1961), for amplifiers (Newhouse and Edwards, 1964), and for signal choppers or modulators (Templeton, 1955). Signal choppers are not usually referred to as cryotrons; however, they will be included under cryotrons.

1. *Cryotron switches*

The cryotron, a name coined in 1956 (Buck, 1956), is basically a device in which the magnetic field produced by an entirely superconducting element controls the resistance of a gate element by switching the gate through the superconducting-to-normal transition. Cryotrons have been investigated primarily as computer elements (Newhouse, 1964) and as amplifiers (Section III). Cryotrons will be treated here as they apply to signal switches.

The unique feature of a cryotron switch, compared with conventional transistor or mechanical switches, is the impedance level. The low-resistance state of a cryotron gate is identically zero and the high-resistance state is typically of the order of milliohms. The on–off ratio is infinite; however, it is frequently difficult to obtain a large gate resistance in the normal state.

Early cryotrons were wire wound, consisting of a control winding wound tightly around the central gate wire (Fig. 1). In operation, the magnetic field produced by a current in the control wire drove or switched the gate wire into the normal state. The control was made of a material that would remain superconducting in a field sufficient to drive the gate normal, e.g., niobium control and tantalum gate, or a lead control and a tin gate.

The switching time of a wire-wound cryotron is severely limited by the large inductance of the control winding and the small resistance of the gate wire. Time constants from 50 to 200 μsec are typical (Bremer, 1962), a very slow signal switch by today's standards. The slow switching speed com-

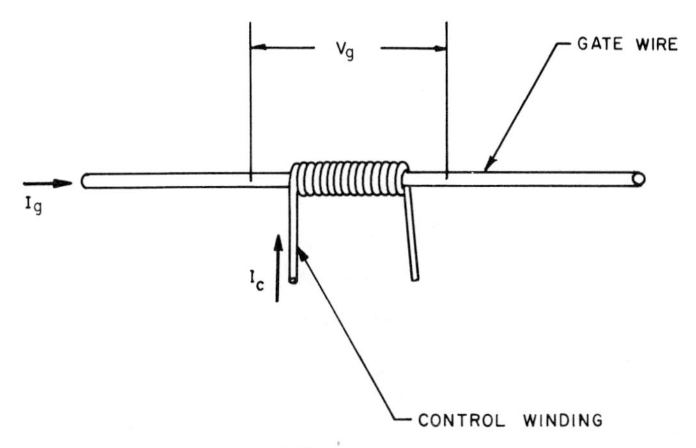

FIG. 1. Wire-wound cryotron.

bined with the difficulty in fabricating a system with many fine interconnecting wires make the wire-wound cryotron of little practical use.

Much better cryotron performance can be obtained by using thin-film fabrication techniques (Slade and McMahon, 1958; Newhouse and Bremer, 1959). Crossed-film cryotrons have been built in the manner shown in Fig. 2. In this configuration, a current in the control film produces a magnetic field on the adjacent portion of the gate film thereby controlling the resistance of that portion of the gate. The performance of thin-film cryotrons can be significantly improved by depositing the entire structure on a superconducting ground plane (Newhouse *et al.*, 1960). The current gain of a cryotron switch is defined as the critical current of the gate, for zero control current, divided by the minimum control current required to drive the gate resistive for very low gate current. If the gate current of one cryotron is to drive the control of another identical cryotron, the current gain must be greater than unity. The current gain of a crossed-film cryotron is proportional to the ratio of the width of the gate to the width of the control film. Switching times as fast as 0.4 μsec can be obtained with a crossed-film cryotron of the type shown in Fig. 2 (Newhouse *et al.*, 1960). Details of the circuit parameters and physical properties of thin-film cryotrons are given in Section III.B.1.

No commercial computer products using cryotrons have yet reached the market (1974). However, such commercial utilization may eventually be made, at least for special applications such as in spacecraft where size and weight are crucial. Application of cryotrons to computer memory is discussed in Chapter 3.

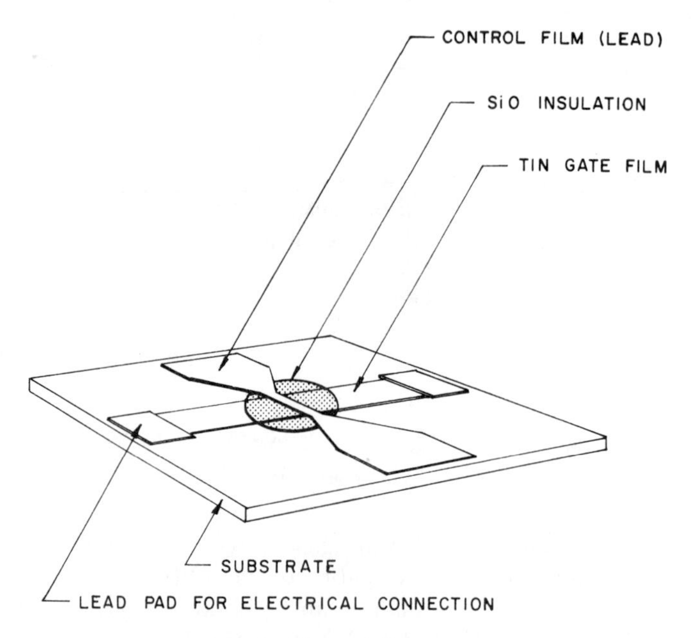

FIG. 2. Crossed-film cryotron.

In most electrical switch applications, it is desirable to have an on–off ratio as large as possible. Since a cryotron has zero off (superconducting) resistance, it appears attractive as far as an on–off ratio is concerned. However, an equally important concern is often that of obtaining a large on, or normal-state resistance. This has the advantage of giving a shorter time constant in switching an inductive load as well as a larger output voltage for driving generally more noisy room-temperature electronic

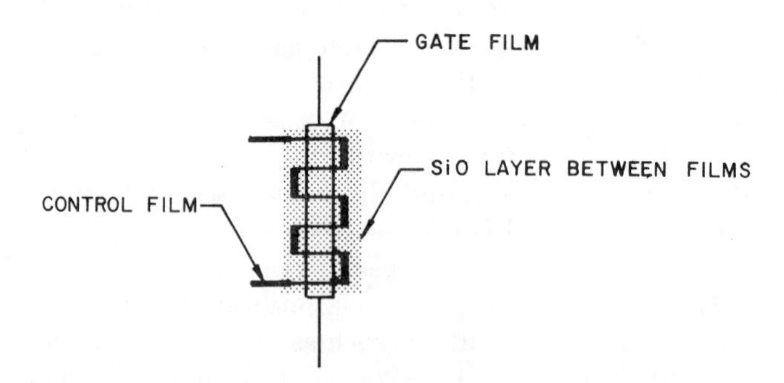

FIG. 3. Cryotron with multiple control crossings.

devices. One method of achieving larger output resistance is to use multiple crossings (Newhouse *et al.*, 1967). This consists of folding the control film so that it passes back and forth across the gate film, thereby making a larger length of gate film go normal and resulting in a larger gate resistance than for a single crossing (Fig. 3). Using multiple control film crossings increases the gate resistance at the expense of higher control inductance. Since the gate resistance and control inductance vary linearly with the number of crossings, the L/R time constant does not change when the gate of one cryotron drives the control of another cryotron. In applications where high current gain is not needed, high gate resistance can be obtained by making the control film about the same width as the gate film, or by using the in-line cryotron (Brenneman, 1963) shown in Fig. 4.

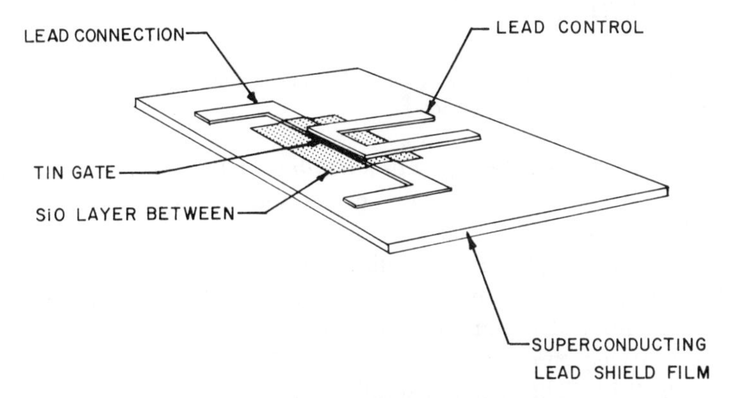

FIG. 4. In-line cryotron.

Little information has been published on signal switching applications of thin-film cryotrons. Therefore, we will discuss the cryotron chopper as an example of a practical circuit making use of a cryotron. The detection of very small dc currents or voltages is often limited by thermal emf's, especially at liquid-helium temperatures, and drift in the dc amplifiers. Also, the resistance of samples at liquid-helium temperatures is frequently so low that the L/R time constant prevents ac resistance measurements at reasonable frequencies (e.g., 1 kHz). These problems can be overcome by use of a resistive chopper to convert the dc signals to ac so that transformers can be used to step up the impedance level and ac amplification can be used to avoid the drift problems. Several superconducting choppers have been discussed in the literature (Templeton, 1955; Kachinskii, 1965; Devroomen and Van Baarle, 1957; and Huebener and Govednik, 1966). Templeton's

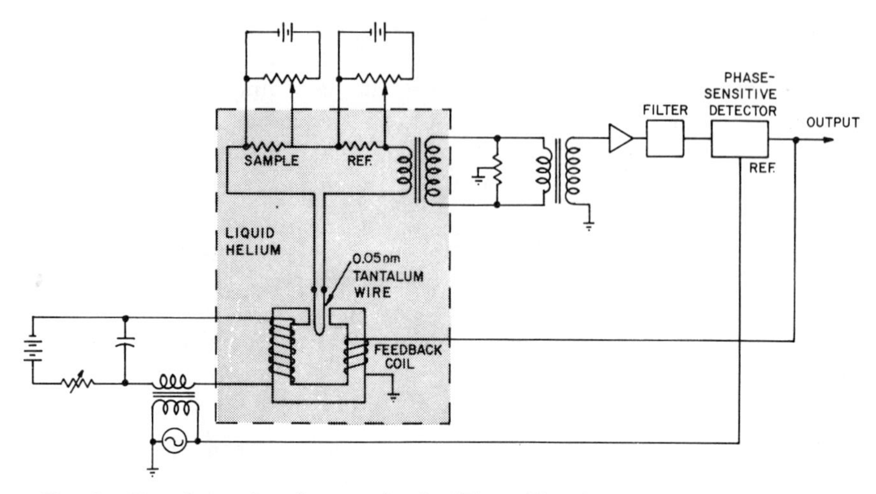

FIG. 5. Tantalum wire chopper circuit. [From Templeton, 1955. Reproduced by permission of the Institute of Physics.]

circuit (Fig. 5) consists basically of a 0.05-mm (0.002-in.-) diam tantalum wire that is switched between the normal and superconducting states by means of a magnetic field. The field is applied by means of a coil and an iron yoke and is supplied by an oscillator via a transformer so that dc bias field can be applied by using the same coil. This bias field is required to hold the tantalum wire near the transition point. Tantalum is used because it has a relatively small critical field (60 G) at 4.2°K. Its normal-state resistance is 0.1 Ω. The signal is amplified first by a 50:1 turns ratio transformer in the liquid-helium bath. The transformer core is mu-metal that, typically, has a permeability at 4°K about one-third that at room temperature (DeVroomen and Van Baarle, 1957).

The signal is amplified at room temperature by a conventional amplifier, filtered with a bandpass filter, and detected by a phase-sensitive detector. The transformer can be tuned with a capacitor to obtain bandpass filtering (Kachinskii, 1965). Templeton's circuit has a resolution of 2×10^{-11} V when used with a sample at 4.2°K having a resistance of 5×10^{-6} Ω. The time constant of the system is 0.1 sec.

Another very simple form of cryotron has been developed that makes use of a ferrite cup core to increase the magnetic field on a gate coil made of tantalum wire (Huebener and Govednik, 1966). The bobbin contains a niobium control coil (Fig. 6). This use of a ferromagnetic material increases the gain at the expense of a larger control inductance. It was designed as a signal chopper and, as such, is similar to the resistive modulator built by

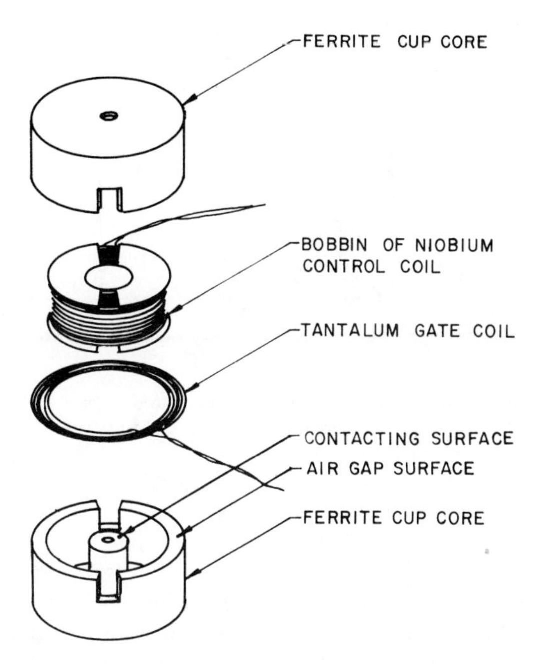

FIG. 6. Signal chopper using ferrite cup core. [From Huebner and Govednik, 1966.]

Templeton (1955). This modulator has been used to detect voltages down to 10^{-9} V.

These resistive modulators or choppers give satisfactory performance for some applications; however, better resolution can be obtained from superconducting tunneling devices and cryotron amplifiers with little or no increase in complexity. For that reason, very little additional development work has been published on these resistive choppers.

2. *Tunneling switches*

Superconducting tunneling devices can exhibit I–V curves that have negative resistance regions. Thus, by the use of a proper load line, two stable states can be obtained and the junction can be switched from a low-resistance state to a high-resistance state. This property can be used for computer memory applications (Chapter 3) or for signal switching. Since tunnel junctions have very low resistance, they cannot directly replace more conventional switches such as diodes or transistors. There are two types of superconducting tunneling junctions; the quasiparticle tunneling junction and the Josephson tunneling junction. Here the terminal properties

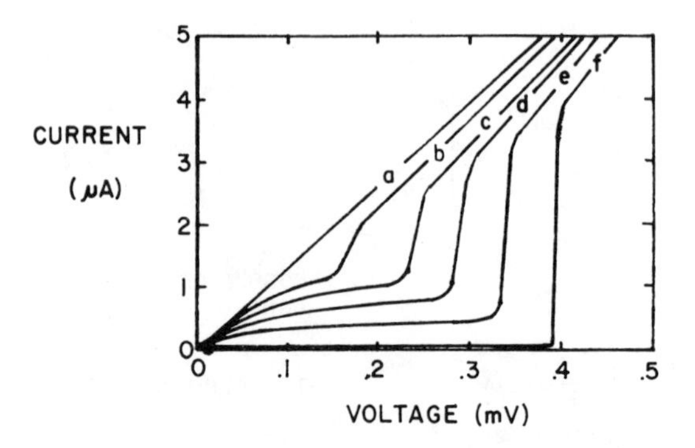

FIG. 7. I–V curve for Al–Al quasiparticle junction—T_c = 1.26°K. Curve a: T = 1.3°K; curve b: 1.2; curve c: 1.1; curve d: 0.9; curve e: 0.8; curve f: 0.3. [After Langenberg *et al.*, 1966].

of tunneling junctions when used as signal switches will be discussed. More details on the physics of tunneling are covered in Chapter 1.

a. Quasiparticle tunneling switches. In a quasiparticle tunneling junction, the insulation thickness separating two superconductors is greater than $\simeq$ 50 Å, i.e., thick enough that superconducting electron pairs cannot tunnel through it. It was demonstrated (Giaever, 1960), prior to Josephson's work on superelectron tunneling (Josephson, 1962), that the tunneling of quasiparticles, i.e., nonpaired electrons, could be influenced by the energy

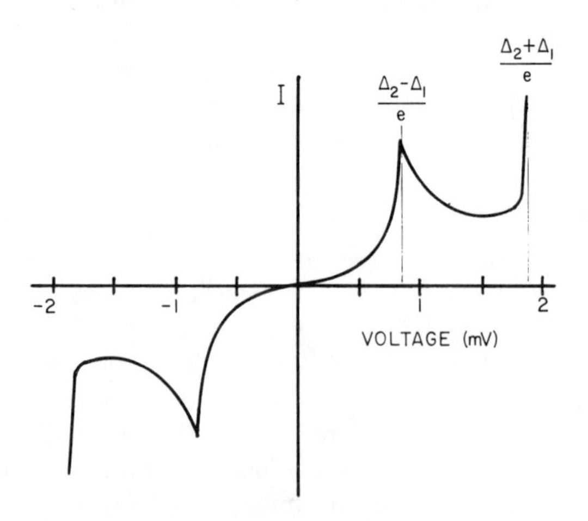

FIG. 8. Characteristic I–V curve for quasiparticle tunneling (T = 2°K, junction type Sn–Sn Oxide–Pb). [After Giaever and Megerle, 1962.]

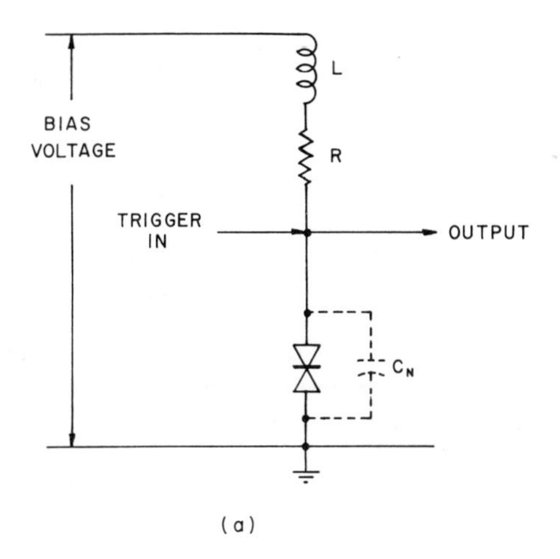

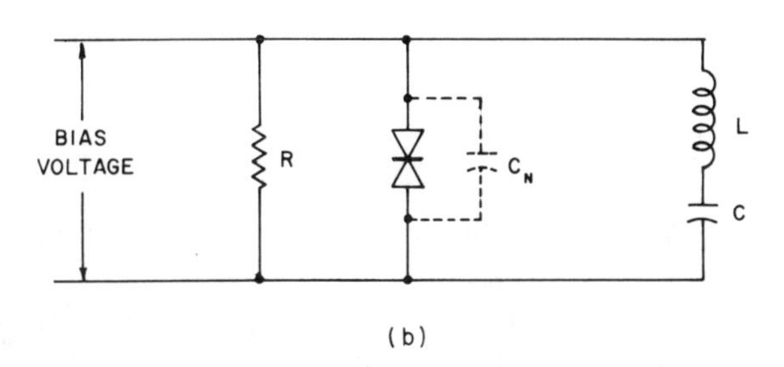

FIG. 9. Quasiparticle tunneling-junction test circuits. (a) Mono and Bi-stable switch and negative resistance oscillator; (b) high-frequency negative resistance oscillator. [After Giaever and Megerle, 1962.]

band gap of the superconductors. The result was an I–V curve with a voltage threshold $V_t = 2\Delta/e$, where 2Δ is the band gap, V_t is the threshold voltage, and e is the electron charge. This is illustrated by the data of Langenberg *et al.* (1966) in Fig. 7. To obtain a negative resistance region, it is necessary that superconductors having different energy gaps, $2\Delta_1$ and $2\Delta_2$, be used to construct the tunneling junction. In this case a threshold is obtained at $V = (\Delta_2 + \Delta_1)/e$. A representative I–V curve for a tin–tin oxide–lead junction (Giaever and Megerle, 1962) is shown in Fig. 8. Fiske and Giaever (1964) give a similar I–V curve for an aluminum–aluminum oxide–lead junction. As $\Delta_1 \to \Delta_2$ the amplitude of the peak voltage decreases and shifts left, finally disappearing in the origin. As $\Delta_1 \to 0$,

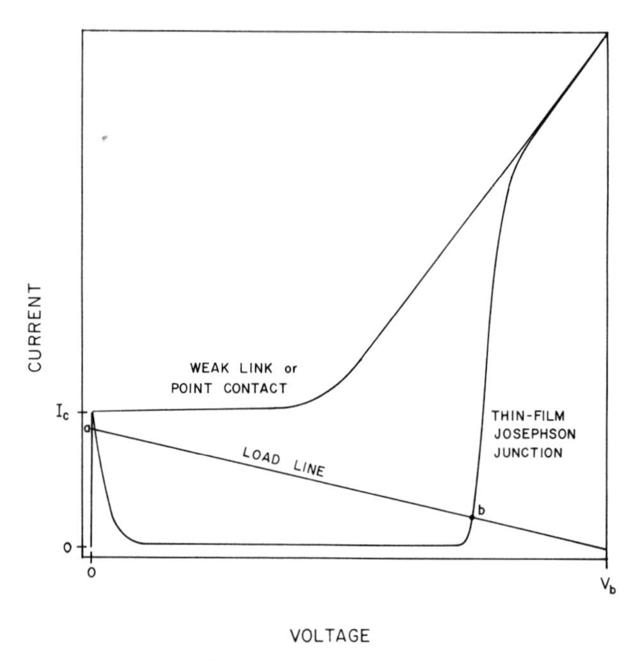

FIG. 10. Supercurrent tunneling I–V characteristic curves.

$(\Delta_2 + \Delta_1) \rightarrow (\Delta_2)$ and the voltage peak moves right and merges with the threshold curve.

This type of tunneling device is constructed by vapor deposition, i.e., by sputtering a strip of superconducting film on a substrate and then allowing the strip to oxidize to form the tunneling barrier. Subsequently, a second superconducting film is deposited crossing the first film. Giaever and Megerle obtained peak-to-valley voltage ratios of 3:1 with their device. They operated switching circuits and negative resistance oscillators using quasiparticle tunneling junctions, as shown in Fig. 9. They obtained oscillation at frequencies up to 4 MHz; however, this does not represent a maximum since the upper frequency was limited by the external electronics used.

b. Superconducting tunneling switches. The discovery of Josephson tunneling junctions and superconducting weak links has stimulated much more device-oriented research than did quasiparticle tunneling. In this case, the barriers are thinner ($\simeq 10$ Å) and a true zero-resistance region is obtained for currents less than the critical current I_c. As discussed in Section III.B the peak-to-valley ratio depends upon the type of junction. Point-contact and weak-link junctions usually exhibit I–V characteristics that have no negative resistance region, i.e., V increases rapidly but continuously for currents slightly in excess of I_c (Fig. 10). Thin-film tunneling junctions, on the other hand, have a very pronounced negative resistance region because of the junction capacitance (McCumber, 1968; Stewart, 1968) as shown in Fig. 10.

When a thin-film Josephson junction is used as a circuit element, an infinite on–off resistance ratio can be obtained, i.e., $R_a = 0$ versus $R_b =$ finite (Fig. 10). Since $R_a = 0$, no power is dissipated in the low-resistance state. The resistance in the high-resistance state R_b is typically of the order 1 Ω. This is much lower than for semiconductor devices. The intrinsic switching time of Josephson tunnel junctions has been shown to be less than 1 nsec (Matisoo, 1967). This measurement was limited by the rise time of the room-temperature electronics. The switching time of very small tunneling junctions (e.g., 0.1×0.1 mm) may be limited by the junction RC time constant, to as high as 50 nsec, as described by Fulton (1971). Additional resistive damping may be required in small junctions to overcome this problem. The experiments of Matisoo (1967) were on larger junctions where the effects described by Fulton were not significant.

In summary, Josephson junctions appear to be ideal switching devices for many applications. The primary problems are reproducibility in the manufacture because of the very thin barrier ($\simeq$10 Å) and the need for liquid-helium temperatures. The reproducibility problem has received some attention (Pritchard and Schroen, 1968), but remains a serious handicap for many potential applications. In applications where the exact value of the critical current is not crucial, as long as it can be measured and is stable, the present technology can produce usable devices.

B. Inductance Switches

The inductance of circuit elements, such as coils, wires, etc., can be changed by switching a nearby superconductor between the normal and superconducting states or by moving a superconducting surface near the circuit element. Various properties of the superconductive state can be used to cause this inductance change. Figure 11 shows several simple superconducting inductance switches. These switches utilize one of the following techniques or properties for operation:

1. A superconductor located near a circuit element is switched between the superconducting and normal states thereby changing the inductance of the circuit element. Circuits of this type are shown in Figs. 11a, 11e, and 11f.
2. A superconducting shield is physically moved relative to the circuit element, as shown in Figs. 11c and 11d.

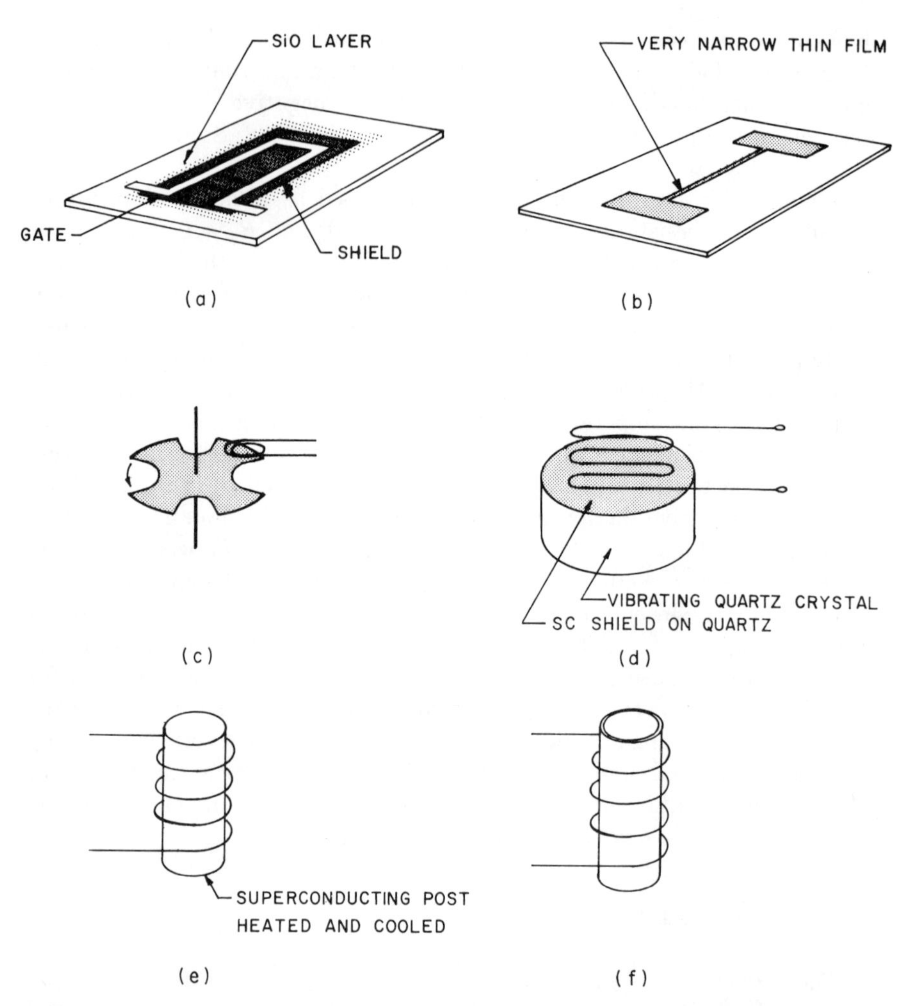

FIG. 11. Self-inductance switches: (a) switched shield, (b) superinductance, (c) moving shield, (d) vibrating shield, (e) Meissner effect, (f) quantized flux.

3. The kinetic inductance of the circuit element is changed as shown in Fig. 11b. See Section III.C.7.

A superconductor can be switched from the superconducting to the normal state by three methods: (1) application of a current in the conductor, (2) application of a magnetic field by means of an adjacent conductor, or (3) thermal switching. The heating required in (3) can be obtained from Joule heating in a thermally coupled resistance element or by application of radiation such as light.

The switching speed of a superconductor is limited by two effects: (1) the thermal time required to heat or cool a superconductor through its transition temperature and (2) the time required for eddy currents in the conductor to decay and allow magnetic flux to penetrate or be expelled. The thermal time constant for heating or cooling depends upon the heat capacity and the thermal conduction to the liquid-helium bath or heat sinks. The time required to heat the superconductor is determined also by how much power is applied to switch the conductor. If the temperature is allowed to go too high, then the conductor will take longer to cool down and go superconducting again. Fortunately, at liquid-helium temperatures, heat capacities are very low and thermal conductivities can be high so that short thermal time constants can be obtained.

The eddy-current time constant is usually a much more serious limitation to switching times than is the thermal time constant, unless a favorable geometry is used, such as a thin-film or a laminated structure. Typically, the eddy currents during the superconducting-to-normal switching are far less serious than during the normal-to-superconducting transition, where very long time constants can result, in some cases up to minutes. The eddy-current time constant depends upon the normal-state resistivity as well as the geometry. A very thin film will give a high normal-state resistance, but this may result in some flux penetration in the superconducting state if the film thickness is comparable to or thinner than the London penetration depth λ. In general, nanosecond switching speeds are possible for a well-designed inductance switch such as a thin-film ryotron (Miller *et al.*, 1964).

In addition to the thermal and eddy-current time constants, some applications involve the L/R time constant of the input circuit. This time constant determines how fast the driving current can be established. The input inductance can be minimized by use of careful design and thin-film deposition techniques.

There are two types of inductance switches: (1) the self-inductance switch and (2) the mutual-inductance switch.

1. *Self-inductance switches*

To indicate the use of self-inductance switches and to provide a background for discussing the performance of various types, consider the simple circuit of Fig. 12 containing two inductance switches (Gange, 1964). In the absence of any resistance, the current I, applied after the circuit is superconducting, will divide or be steered according to the total inductance in each branch, i.e.,

$$I_1 = I[(L_2 + L_4)/(L_1 + L_3 + L_2 + L_4)] \tag{1}$$

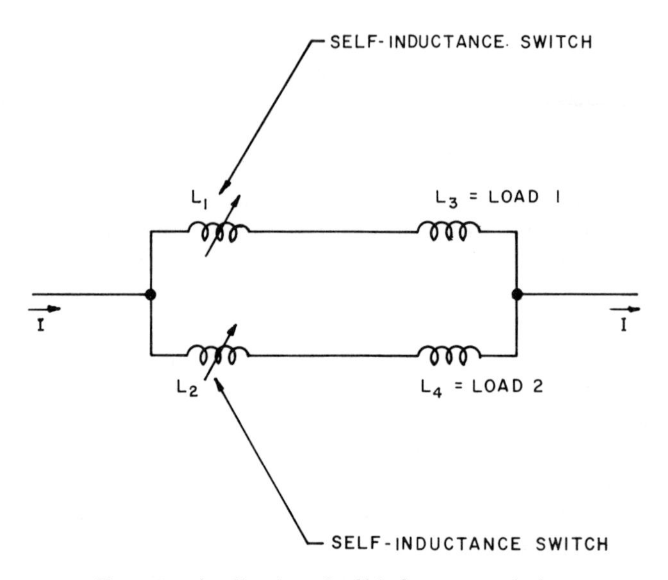

FIG. 12. Application of self-inductance switches.

and

$$I_2 = I[(L_1 + L_3)/(L_1 + L_3 + L_2 + L_4)] \quad (2)$$

A large current steering ratio requires that the load inductance L_3 plus L_4, including lead inductances, be small compared with the high-inductance state of the switch. The above equations then become

$$I_1 \simeq IL_2/(L_1 + L_2) \quad (3)$$

and

$$I_2 \simeq IL_1/(L_1 + L_2) \quad (4)$$

By making L_1 small and L_2 large, the current can be steered almost entirely to branch 1, or by making L_2 small and L_1 large, the current can be steered almost entirely to branch 2. If the inductance is changed after currents have been established in the network, these currents will redistribute. A special case where the variable inductance element traps magnetic flux is discussed in Section II.B.1.a.

Primary applications of self-inductance switches are the switching of purely superconductive loads such as cryotron control lines or superconducting memory drive lines. It is also possible to inductively switch loads having small resistances. In this case, the current will first divide according to the inductance ratio and then redivide, with an L/R time constant, to a new ratio determined by the load-resistance ratio. Thus, current pulses,

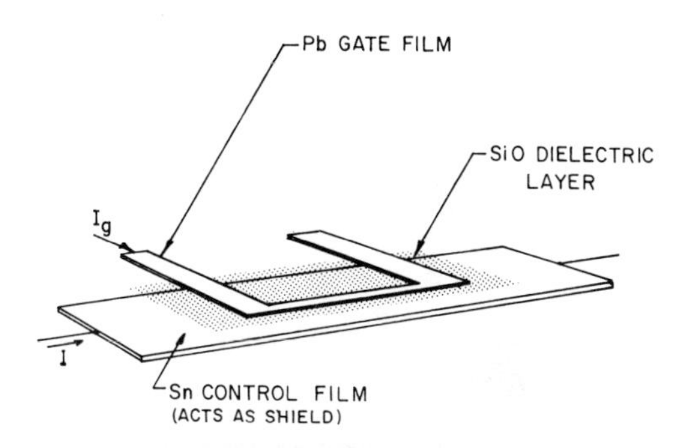

FIG. 13. Simple thin-film ryotron.

short compared with L/R time constant, can be properly steered. If one inductance switch drives the control, i.e., shield film, of another switch (i.e., the ryotron type), we have an example of a load that includes resistance but, in this case, the load resistance develops only after the current exceeds the critical current of the control film.

There are several types of self-inductance switches: (a) ryotron, (b) in-line, (c) crossed-line, (d) thermally switched, (e) superinductance, and (f) mechanical motion switches.

a. Ryotron. The ryotron (Gange, 1964) consists of a narrow-superconducting-gate conductor in close proximity to a wide-shield conductor (Fig. 13). The shield is switched from the superconducting to the normal state by means of a current applied to the shield conductor. The gate conductor always remains superconducting. When the shield is superconducting, magnetic flux due to changes in the gate current cannot pass through the shield, except in the insulation space between the gate and the shield films. This results in a low gate inductance. When the shield is switched normal, flux can penetrate the shield and a large gate inductance is obtained. If the shield is made very thin, for example, a 1000-Å vacuum-deposited thin film, eddy currents in the normal shield will be small. This will allow a large inductance to be obtained at gate-current frequencies up to many megahertz (Gange, 1964).

Experiments have shown that ryotrons, having the simple geometry of Fig. 13, do not have very high ratios between the high- and low-inductance states (Gange, 1964). This results from the return paths of the shielding currents induced in the surfaces of the control film when it is superconducting. A current I_g in the gate film will induce an equal and opposite shielding current (image current) in the surface of the superconducting control film immediately under the gate film. This shielding current must return through the sides of the control film. This will cause magnetic flux to enclose the

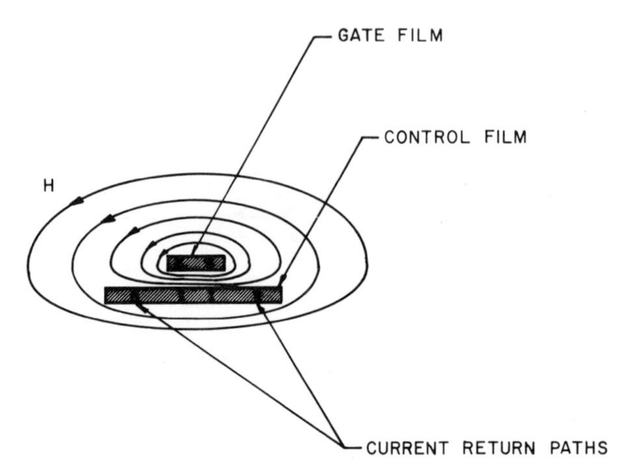

FIG. 14. Magnetic field and current patterns for simple ryotron.

control-film and gate-film combination in addition to just the gate film, as shown in Fig. 14. This will then contribute a significant inductance to the low-inductance state of the gate film. Gange obtained an inductance ratio of only about 1.4 for this type of ryotron.

There are several solutions to the low-inductance ratio problem, and each involves changing the paths of the shielding currents. One method is to put a permanently superconducting shield plane on both sides of and slightly overlapping the control film to prevent magnetic flux from enclosing the control film. Another method is to shape the gate film to have its "go" and "return" paths over the same control film (Fig. 15). This latter construction has been called the horseshoe ryotron (Miller *et al.*, 1964). In this geometry, the induced shield currents can return or close across the bottom end of the horseshoe, as shown in Fig. 15. Miller *et al.* were able to achieve an inductance ratio of 164 with a horseshoe ryotron, and in an array of ryotrons, they achieved switching times of 3.5 nsec.

It is important to note that the inductance change in the ryotron (or other inductance switches using thin-film shields) is a result of the zero-resistance property of the shield film and not of its Meissner-effect or quantized-flux property. A steady current in the gate will not be switched or modulated significantly by an alternating current in the shield. Flux that penetrates the shield when it is normal will be "frozen in" when the shield becomes superconducting, i.e., the Meissner effect is very incomplete for a thin film. If the shield were a thick Type-I superconductor or a small closed ring and the magnetic field due to the gate current were parallel to the shield, then the Meissner-effect and (or) quantized-flux property, would

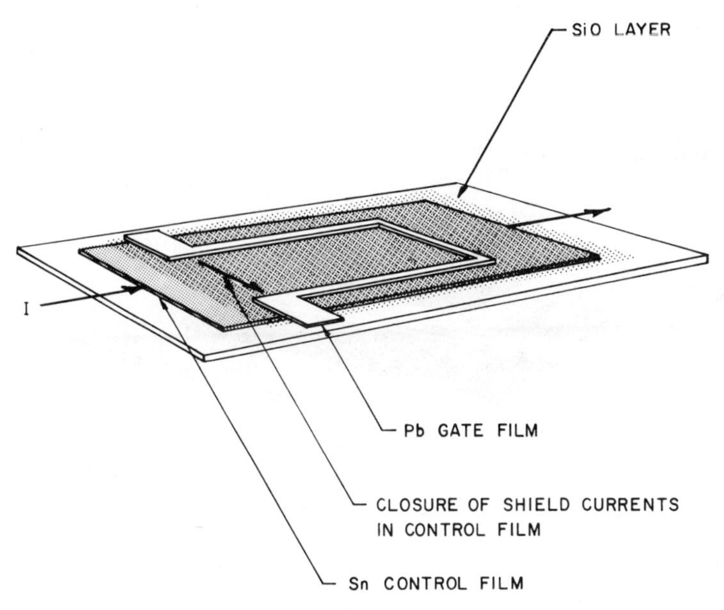

FIG. 15. Horseshoe ryotron.

exclude some of the flux as the shield became superconducting. We then have an inductance modulator, discussed in Sections III.C.2 and III.C.3. Consequently, the ryotron will not operate continuously in circuits in which the current in the shield has a higher frequency than the gate current.

b. In-line inductance switch. The shield in the ryotron, discussed above, was made normal by means of a current in the shield. The shield film can also be switched by means of a magnetic field produced by a current in an adjacent conductor (Meyerhoff *et al.*, 1964). Meyerhoff *et al.*, studied both the in-line and crossed-line versions. The in-line version is shown in Fig. 16. A Pb shield plane contains a hole that is covered with a Sn film. A gate conductor, whose inductance is to be switched, passes over the Sn film. A control film also passes over the Sn film. When there is no current in the control film, the Sn film will be superconducting and the gate conductor will have a small inductance. However, when a current is applied to the control film, a magnetic field is produced on the Sn film making it normal. This allows magnetic flux due to a gate current to penetrate the Sn film so that the gate will have a large inductance. Thus, the gate-film inductance can be switched from a small value to a large value by application of a control current. Meyerhoff *et al.* constructed a circuit similar to that shown in Fig. 12 using in-line switches and measured the branching current I_1

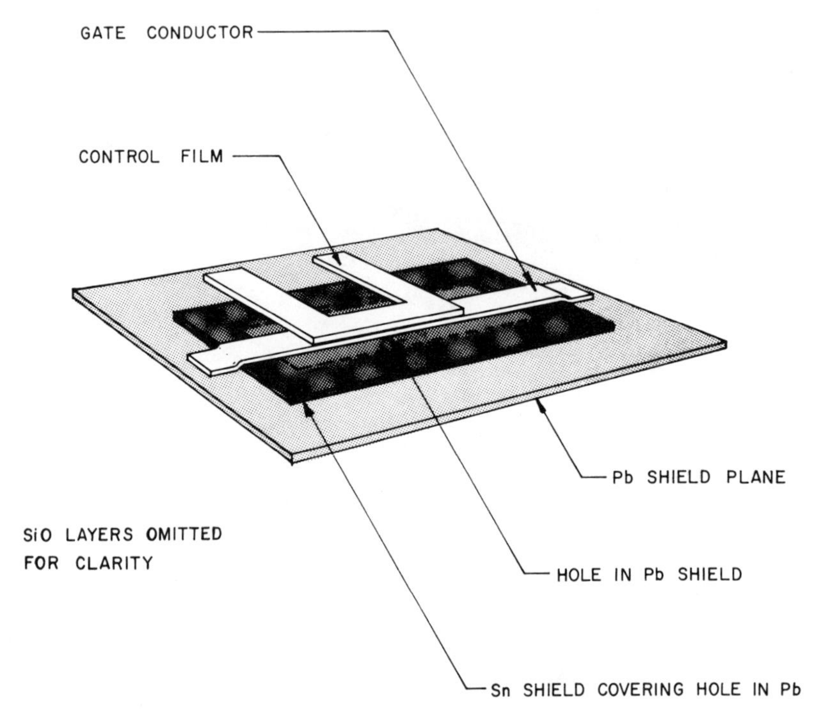

FIG. 16. In-line inductance switch.

and I_2. Their results are shown plotted against input current I_{in} in Fig. 17. Their branch films were 0.005 in. wide in the shield area, the control element was 0.010 in. wide and the hole in the Pb shield was 0.312 in. long. The calculated inductance ratio was 58. The current division, as shown in Fig. 17, was excellent until I_{in} exceeded 0.5 A, at which point the Sn film of the low-inductance switch went normal because of excessive branch current.

c. Crossed-line inductance switch. The crossed-line inductance switch tested by Meyerhoff *et al.* (1964) is shown in Fig. 18. It consists of a Pb shield containing two rectangular holes separated by a narrow Pb bridge. The center of this Pb bridge is cut away and replaced with an Sn shield so that it can be easily switched without switching any of the Pb shield (Pb has a higher T_c and H_c than Sn). A control film passes over and under the Sn bridge; a current in this control film produces a magnetic field on the Sn shield section of the bridge thereby making it go normal. When no control current flows, the Sn shield is superconducting and magnetic flux

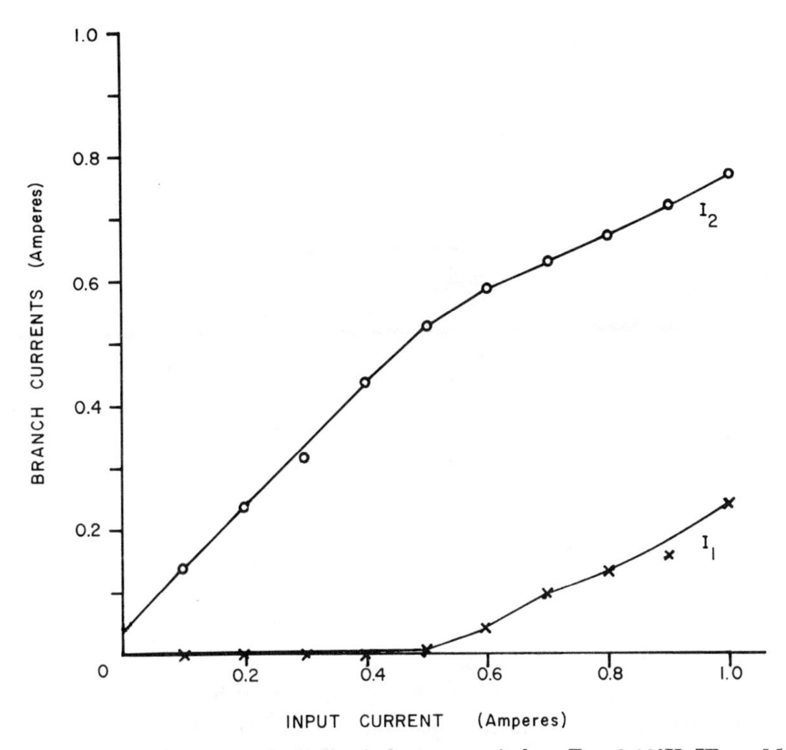

FIG. 17. Branch currents for in-line inductance switches; $T = 3.60°K$. [From Meyerhoff *et al.*, 1964, courtesy of Plenum Publishing Corp.]

from the gate conductor cannot penetrate the holes in the Pb shield. This gives a very low inductance for the gate conductor. Flux cannot penetrate the holes because supercurrents can flow around each hole to prevent penetration. However, when a control current is applied to make the Sn shield in the bridge go normal, then flux due to the gate current can penetrate the holes in the Pb shield since shielding currents passing through the bridge decay to zero because the bridge is resistive. A large inductance is thus obtained. Using this type of crossed-line switch in the circuit of Fig. 12, Meyerhoff *et al.* obtained a current division ratio of 8; however, this ratio was limited by the load inductances L_3 and L_4 in Fig. 12. Thus, the inductance ratio of the switch without loads was greater than 8.

d. Thermally switched inductance switch. No experimental data have been published for the thermally driven inductance switches except data related to the quantized-flux and Meissner-effect modulators, as discussed

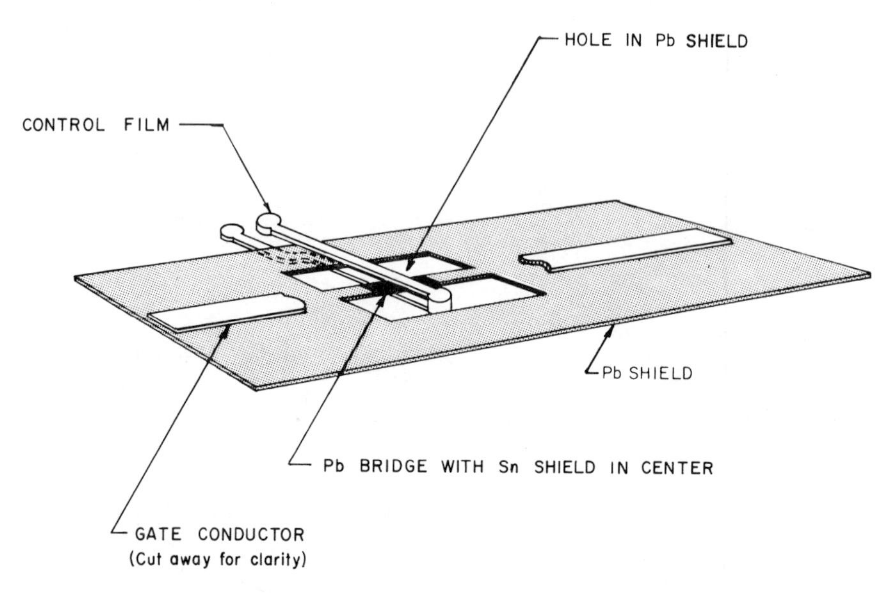

FIG. 18. The crossed-line inductance switch.

in Sections III.C.2 and III.C.3. Thermally switched inductance switches have been used only as magnetometers. Quantized-flux modulators were thermally switched at frequencies as high as 40 kHz (Deaver and Goree, 1967). The Meissner-effect modulators could only be switched up to about 500 Hz before the percentage modulation dropped below about 20%; however, these modulators were not optimized for speed, and much better performance should be possible.

e. Superinductance switch. The superinductance (or kinetic inductance) effect (see Fig. 11b and Section III.C.7) can be used to build an inductance switch (Little, 1967). This inductance effect is a result of the momentum of the superelectron pairs in a conductor having a very small cross-sectional area. It is not associated with external magnetic flux which is usually distinguished by the term geometric inductance. The superinductance of a thin film $\simeq 50$ Å thick, 20 μ wide, and 3 cm long has been measured by Little to be as much as 50 times the geometric inductance. The superinductance is not very sensitive to magnetic field; however, it is very sensitive to temperature, approaching infinity as T approaches T_c. Thus, a superinductance switch need only consist of a very thin, narrow superconducting film deposited on a resistive heater film or heated by means of incident radiation such as light.

f. Mechanical-motion inductance switch. Mechanical-motion inductance switches consist of a superconducting shield that can be moved relative to a conductor. One possible geometry is a solenoid with a close-fitting superconducting shield cylinder that can be inserted into the bore of the solenoid. When the shield cylinder is inside the solenoid, the inductance will be low because the magnetic flux cannot change through a closed superconducting path if the cylinder is thick compared with the London penetration depth (about 1000 Å). Another geometry is that of a planar shield placed close to a thin-film conductor (Fig. 11d). The inductance can be changed by varying the spacing between the shield plane and the conductor plane. An example of this type of switch is the vibrating crystal magnetometer (Opfer, 1970) discussed in Section III.C.2. The superconducting shield plane is vibrated at about 100 kHz by means of a quartz crystal and produces about 10% modulation in the inductance. This modulation is not very good for a signal switch, but it may be excellent for magnetometer applications because of the high Q possible, i.e., Q's up to 10^4 were reported. Several factors limit the usefulness of mechanical motion switches. One is their slow operating speed. Another limiting factor is the necessity of generating mechanical motion without electrical noise or of coupling the switch into the low-temperature environment. Still another problem is that of obtaining a large inductance ratio. Since a low self-inductance requires close spacing between shield and conductor, mechanical motion may result in damage to the switch and poor reproducibility of the low-inductance value.

2. *Mutual-inductance switches*

A mutual-inductance switch can consist of two coils or conductors separated by a superconducting shield that can be made normal or mechanically removed from between the coils. Any of the techniques mentioned under self-inductance switches (i.e., current, field, or temperature) can be used to switch the superconducting shield to the normal state. Since nearly perfect magnetic shielding can be obtained with superconductors, very large inductance ratios should be possible. Close conductor to shield spacing is not required. Meyerhoff *et al.* (1964) discuss briefly a mutual-inductance switch constructed with a technique similar to that shown in Fig. 18 for their self-inductance switch. A primary was added on the side of the shield opposite the gate conductor. A crossed-line control was used to switch the Sn shield in the bridge. No performance data were given. Another possible geometry is that of a cylindrical shield with the primary winding wound on the outside and the secondary winding on the inside. If a thermally switched shield is used, the heater can be magnetically shielded with a superconductor to eliminate noise from the heater current.

A mutual-inductance switch can be made using mechanical motion of the shield (Figs. 11c and 11d). For Fig. 11c, a second coil can be added below the rotating disk. This circuit has been considered (Buchhold, 1963) for use as a parametric amplifier (up-converter) (see Section III.C.1). The switch of Figure 11d can be generalized to a mutual-inductance switch by replacing the vibrating superconducting plane with a vibrating coil. This type of modulator has been built and tested by Ries and Satterthwaite (1967) as a parametric amplifier (up-converter type) (Section III.C.1).

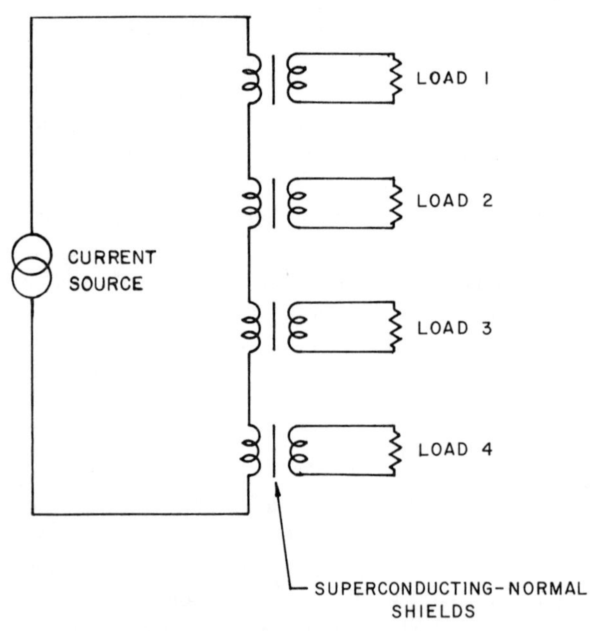

FIG. 19. Signal distribution with mutual-inductance switches.

Little application has been reported for mutual-inductance switches to date. However, as superconducting circuits come into more common use, applications will surely be found. One possible application is in the distribution of signal power from one source to a number of loads, as shown in Fig. 19. The shields shown in Fig. 19 could be made of one common superconducting sheet or cylinder with switchable windows under each secondary. The great advantage of mutual-inductance switches over self-inductance switches is the very high isolation possible in the off state.

III. Amplifiers

A number of different superconducting properties can be used to amplify electrical signals. These properties have been divided into two general classes: (1) resistive modulation and (2) inductance modulation. Each of these two classes will be discussed after first considering the general input circuit requirements.

Superconducting amplifiers have two unique features: an unusually low input-noise level and low input resistance (zero in most cases). These two features make superconducting amplifiers attractive for special applications. How these features can be used in special applications is discussed in Section III.A.

A. Input Circuits

Since superconductive amplifiers have very low noise properties, it is important to examine the noise properties of the general input circuits to determine the conditions necessary to make maximum use of these low-noise properties. We will, in general, follow the noise analysis developed by Rorden (1965) and discussed with respect to superconducting circuits by Goodman *et al.* (1973). Special cases will be considered to more simply illustrate the basic ideas. Comparison will be made with the treatment by Radhakrishnan and Newhouse (1971) in which the uncorrelated voltage noise term V_u is not included.

In addition to very low noise, another unique feature of all the amplifiers, except the negative resistance amplifiers of Section III.B.2, is the true zero-input resistance and low-input inductance.

1. *Input circuit without a transformer*

The IRE Subcommittee 7.9 on Noise (1960) has shown that the noise of any amplifier can be characterized by four parameters: (1) the input current noise, (2) the input voltage noise, (3) the real part of the correlation factor, and (4) the imaginary part of the factor correlating the current and voltage noise. This model is also discussed by Motchenbacher and Fitchen (1973). Each of these four parameters is, in general, a function of frequency. Figure 20 shows a current generator for the equivalent amplifier current noise per root hertz, I_a, and a voltage generator for the voltage noise per root hertz, V_a, in the input circuit of a superconducting amplifier. The amplifier symbol in Fig. 20 includes the superconducting sensor and whatever electronics that are necessary to obtain an amplified

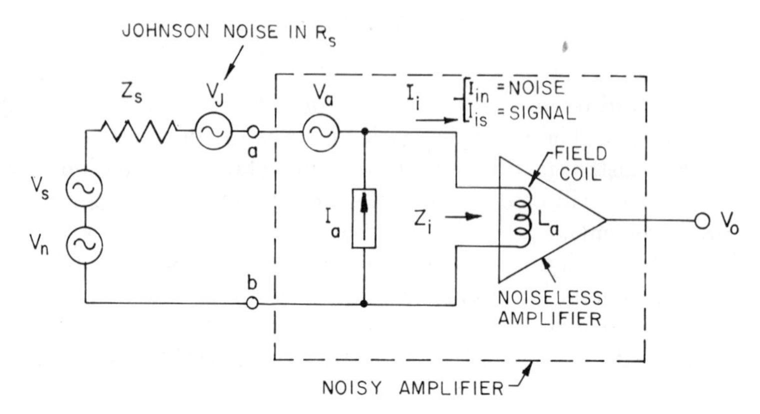

FIG. 20. Input circuit of a superconducting amplifier (V_n = equivalent input voltage noise due to V_a and I_a).

reproduction of the input current. When the input (a,b) is short circuited or the source impedance is very low, the voltage noise will predominate, provided that V_a and I_a are uncorrelated. When the input is open circuited or the source impedance is very large, then the current noise will predominate, provided that V_a and I_a are uncorrelated. When the predominant noise originates from somewhere beyond the input stage, the output noise is relatively independent of source impedance. In that case V_a and I_a have to be strongly correlated. This model is then somewhat artificial, but does correctly describe the amplifier's noise performance. Alternatively, one could add an extra noise term in the output to take care of noise sources beyond the input stage.

The signal voltage observed on the output of the amplifier of Fig. 20 is given by

$$V_o = Z_t I_i \tag{5}$$

where Z_t is the transimpedance, $\partial V_o/\partial I_i$, and I_i is the input current (I_{is} will be used for input signal current, and I_{in} for input noise current). The source signal voltage is V_s. If V_s is zero, the rms input current noise per root hertz as a function of frequency, is given by

$$I_{in} = (V_J + V_a + I_a Z_s)/(Z_s + Z_i) \tag{6}$$

where V_J is the Johnson noise voltage per root hertz in R_s, $V_J^2 = 4k_B T R_s$, k_B is Boltzmann's constant (1.38×10^{-23} J °K^{-1}), T is the absolute temperature of R_s in degrees Kelvin, Z_s is the source impedance ($R_s + JX_s$), and Z_i is the amplifier input impedance. Note that Eq. (6) is a *vector equation*, where V and I are Fourier transforms of $v(t)$ or $i(t)$ when the term is aperiodic and Fourier amplitudes when the term is periodic (IRE

Subcommittee 7.9 on Noise, 1960). The bandwidth does not appear in the Johnson-noise term because all current and voltage terms are given per root hertz.

The noise current I_{in} of Eq. (6) can be considered as originating from an equivalent input noise generator V_n in series with V_s given by

$$V_n = I_{\text{in}}(Z_s + Z_i) \tag{7}$$

Using Eq. (6) we obtain

$$V_n = V_J + V_a + I_a Z_s \tag{8}$$

The degree of correlation between V_a and I_a can be expressed as

$$V_a = V_u + I_a Z_\gamma \tag{9}$$

where V_u is the part of V_a uncorrelated with I_a, and Z_γ is the complex correlation factor.

The equivalent input voltage noise V_n of Eq. (8) can now be written as

$$V_n = V_J + V_u + I_a(Z_s + Z_\gamma) \tag{10}$$

Since Eq. (10) is a vector equation, the random uncorrelated noise terms must be added as the square root of the sum of the squares. Therefore the time average of V_n^2 is given by

$$\bar{V}_n^2 = 4k_B T R_s + \bar{V}_u^2 + \bar{I}_a^2(Z_s + Z_\gamma)(Z_s + Z_\gamma)^* \tag{11}$$

where * means complex conjugate.

The value of I_a for superconducting amplifiers typically varies from $\sim 10^{-10}$ A Hz$^{-1/2}$ for rf-driven tunneling amplifiers to 10^{-8} A Hz$^{-1/2}$ for superconducting low-inductance undulating galvanometer (SLUG) amplifiers. Still smaller values of I_a can be obtained by increasing the number of turns in the input field coil of the amplifiers. However, this will increase L_a and, therefore, increase the L/R response time of the amplifier input circuit. Typical values of V_a are not available in the literature; however, several very small voltage measurements have been published which give an upper bound for V_a. The actual values of V_a were obscured by Johnson noise in R_s or by $I_a Z_s$. Clarke (1966b) obtained a voltage resolution of 10^{-14} V using a SLUG (see Section III.B.1.b). Zimmerman and Silver (1968) obtained a voltage resolution of 4×10^{-17} V Hz$^{-1/2}$ using a point-contact rf-driven tunneling amplifier (see Section III.C.4).

In experiments by the authors to measure V_a of a thin-film rf-driven tunneling amplifier, a source inductance L_s of about 5×10^{-11} H was connected to the input field coil. The field coil had an inductance of 540 nH. A signal voltage was introduced into L_s through a known mutual inductance. It was verified that the transimpedance Z_t of the amplifier (ratio of output voltage to input current) was not changed from that observed with a high-impedance source. The output noise voltage up to 1 Hz was also observed to be unchanged, indicating that the current noise I_{in} in Fig. 20 is also unchanged.

For the case of the high-source impedance $I_{in} = I_a$. For the very small source inductance, Eqs. (6) and (9) give

$$I_{in} = [V_u + I_a(Z_\gamma + Z_s)]/(Z_s + Z_i) \tag{12}$$

By equating these two values of I_{in} we obtain

$$I_a Z_i = V_u + I_a Z_\gamma \tag{13}$$

where Z_i is $j\omega L_a$. It is not possible to determine both Z_γ and V_u explicitly from Eq. (13), unless additional information is available. However, for example, if we know from the physics of the amplifier operation, or by independent measurements, that V_u is constant with frequency, then V_u has to be small compared to $I_a Z_\gamma$. In that case we can determine from Eq. (13) that $Z_\gamma \simeq Z_i$, and this makes $V_a \simeq I_a Z_i$. In the above experiment I_a was 10^{-10} A Hz$^{-1/2}$, $\omega/2\pi$ was 1 Hz, and Z_i was $j\omega L_a$. Therefore, if V_u is frequency independent, $V_a \simeq 3.4 \times 10^{-16}$ V Hz$^{-1/2}$, and $V_u \ll 3.4 \times 10^{-16}$ V Hz$^{-1/2}$.

For the sake of illustration of Eq. (8) assume that $I_a = 10^{-10}$ A Hz$^{-1/2}$, $V_u = 10^{-16}$ V Hz$^{-1/2}$, $Z_s = R_s$, $T = 4.2$°K, and the correlation is zero, i.e., $Z_\gamma = 0$. The plot of V_n versus R_s is shown in Fig. 21. Several important factors are made clear by Fig. 21. First, for large R_s, the current noise $I_a R_s$ predominates, and, for small R_s, the voltage noise predominates. In the center region of R_s, the Johnson noise in R_s predominates unless the temperature is very low. Second, extremely low-voltage resolution is possible provided that R_s is low enough, e.g., at $R_s = 10^{-12}$ Ω, $V_n = 10^{-16}$ V Hz$^{-1/2}$. The center region of R_s, where Johnson noise predominates, is bounded on the lower side by

$$R_s = \bar{V}_a^2/4k_B T \tag{14}$$

and on the upper side by

$$R_s = 4k_B T/\bar{I}_a^2 \tag{15}$$

This Johnson-noise region disappears if

$$4k_B T < |I_a|\,|V_a| \tag{16}$$

This condition of Eq. (16) will not normally be met for superconducting amplifiers, e.g., if $T = 4.2$°K then $4k_B T = 2.3 \times 10^{-22}$ J. If I_a is as high as 10^{-8} A Hz$^{-1/2}$, then V_a would have to be as large as 2.3×10^{-14} V Hz$^{-1/2}$. As mentioned above, V_a is probably much less than 10^{-14} V Hz$^{-1/2}$ for most

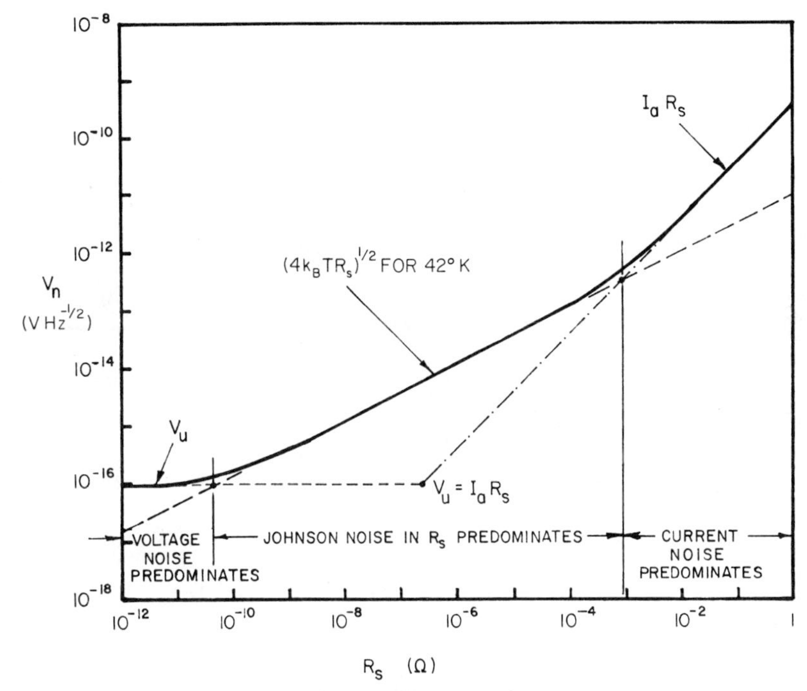

FIG. 21. Equivalent input voltage noise versus source resistance [assumed values: $I_a = 10^{-10}$ A $Hz^{-1/2}$; $V_u = 10^{-16}$ V $Hz^{-1/2}$; $T = 4.2$°K; $V_n = (4k_BTR_s + I_a^2R_s^2 + V_u^2)^{1/2}$].

superconducting amplifiers. Of course if T is very low (e.g., 10^{-3} °K), then it is possible to obtain a condition in which Johnson noise is not important for any value of R_s. In this situation the voltage resolution will always be limited by either I_a or V_a, or both.

The signal-to-noise ratio V_s/V_n obtained using Eq. (8) is a maximum for $R_s = 0$. This means that the lowest noise voltage possible is limited by V_a, which requires that R_s be less than the value given by Eq. (14).

2. *Noise match using an input transformer*

If we assume that we have a given source impedance and a given amplifier, and want to use an ideal transformer to obtain an optimum noise match (largest signal-to-noise ratio), then we can derive the required turns ratio N. By an ideal transformer we mean one having a unity coupling coefficient (no leakage inductance), zero resistance in both the primary and secondary, and primary reactance very large compared with the primary circuit impedance. The circuit and its reduced equivalent are shown in Fig. 22. It should be emphasized that the optimum noise match is more important

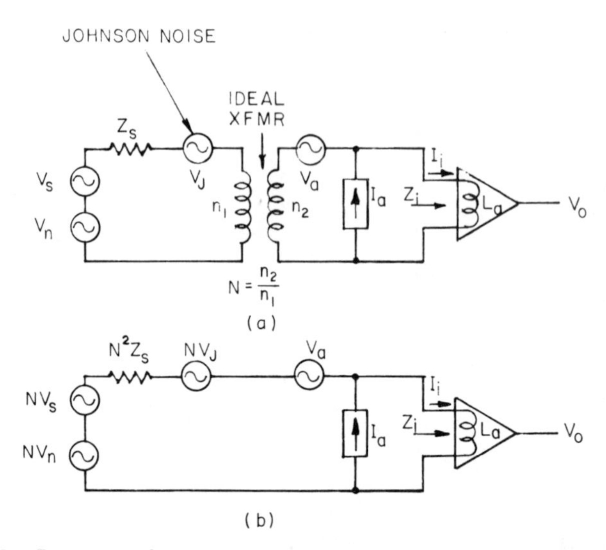

FIG. 22. Input transformer for noise matching: (a) circuit; (b) equivalent.

than maximum power transfer if we wish to obtain a maximum signal-to-noise ratio. Note that the equivalent circuit in Fig. 22b is the same as that given in Fig. 20 except that Z_s is replaced by N^2Z_s, V_s by NV_s, and V_J by NV_J. Thus the input current noise will be

$$I_{in} = (NV_J + V_a + I_aN^2Z_s)/(N^2Z_s + Z_i) \tag{17}$$

NV_n can be obtained from Eq. (17) by using Eq. (7), modified to include the transformer, i.e.,

$$NV_n = I_{in}(N^2Z_s + Z_i) \tag{18}$$

Therefore

$$V_n = V_J + (V_a/N) + I_aNZ_s \tag{18a}$$

The noise in the output will be given by

$$V_{on} = Z_tI_{in} \tag{19}$$

The output signal is given by

$$V_{os} = NV_sZ_t/(N^2Z_s + Z_i) \tag{20}$$

The signal-to-noise power ratio is then

$$\begin{aligned} P_{os}/P_{on} &= V_{os}^2/V_{on}^2 \\ &= N^2V_s^2/[N^2V_J^2 + \bar{V}_u^2 + \bar{I}_a^2(N^2Z_s + Z_\gamma)(N^2Z_s + Z_\gamma)^*] \end{aligned} \tag{21}$$

Note that the signal-to-noise ratio is independent of Z_i. By taking the partial derivative of (P_{os}/P_{on}) with respect to N and equating it to zero, we find that the signal-to-noise power ratio of Eq. (21) has a maximum when

$$N^2|Z_s| = \{[\bar{V}_u^2 + \bar{I}_a^2(R_\gamma^2 + X_\gamma^2)]/\bar{I}_a^2\}^{1/2} \tag{22}$$

or

$$N^2|Z_s| = |V_a|/|I_a| \tag{22a}$$

For this value of $N^2|Z_s|$ Eq. (18a) becomes

$$V_n = V_J + 2V_a/N \tag{18b}$$

Equation (22a) is the optimum noise match. When the Johnson-noise term predominates (center region in Fig. 21), there is little advantage in using a transformer as far as signal-to-noise ratio is concerned. This can be seen by inspection of Fig. 22 in which it is seen that the transformer affects V_s and V_J equally. There may be an advantage or disadvantage as far as the input time constant is concerned, depending upon whether N is greater or less than unity, as will be discussed later.

To illustrate the advantage of using a transformer when V_a predominates, assume the values: $V_s = 10^{-14}$ V, $V_u = 10^{-14}$ V Hz$^{-1/2}$ (this is probably much larger than V_u for most superconducting amplifiers), $I_a = 10^{-10}$ A Hz$^{-1/2}$, $Z_s = R_s = 10^{-10}$ Ω, $Z_\gamma = 0$, and $T = 4.2$°K. The optimum turns ratio N for a noise match computed from Eq. (22) is 10^3. The noise voltage V_n computed using Eqs. (17) and (18) is then 1.5×10^{-16} V Hz$^{-1/2}$. This gives a signal-to-noise ratio of 67. As a comparison, the noise voltage for no transformer as computed from Eq. (8) is 10^{-14} V Hz$^{-1/2}$, and the signal-to-noise ratio is unity.

3. *Transformer design*

The pertinent question here is "can a sufficiently ideal transformer be built?" The answer rests on obtaining zero resistance, high coupling coefficient, adequately large reactances, and minimum stray capacitance. Zero resistance can be readily achieved using superconducting wire. Using care in construction, one can obtain a coupling coefficient k of about 0.85 to 0.95. When the turns ratio is small a coupling coefficient of this magnitude gives nearly ideal transformer operation. For a turns ratio of 10^2 or more, however, a coupling coefficient greater than about 0.95 is desirable. A coupling coefficient greater than 0.95 can probably be obtained by careful design, making use of the shielding properties of superconductors. For example, the primary (or secondary) can be a very wide 1-turn strap with generous overlap. The secondary (or primary) can then be many turns of fine wire wrapped tightly on the primary. The requirement that the transformer reactances be large compared with other circuit impedances can be a problem if small transformer size is required. However, small input impedances are usually used so that very small transformer reactances may be adequate. For example, in amplifiers using thin-film sensors, the inductance of the input field coil is typically of the order of 100 nH. Therefore, the transformer secondary must be larger than 100 nH to obtain ideal transformer operation. This is easy to achieve if the transformer is a voltage step-up type, i.e., $N \gg 1$. On the other hand, if a very small N is required (i.e., $N \simeq 10^{-3}$) then the secondary should be one or at most a few turns to avoid having a prohibitive number of turns on the primary. With only a few turns on the secondary the transformer size may have to be larger than the superconducting sensor to get a large enough secondary inductance. Note that amplifier response down to $\omega = 0$ (i.e., dc) can be achieved with a superconducting transformer because the entire secondary circuit resistance is zero.

The circuit where R_s is zero is typically used in superconducting magnetometers where only emf's due to magnetic flux changes can be introduced. In this case inductance matching is required for optimum magnetometer performance, as discussed later in this section.

The last transformer consideration is stray capacitance of the winding. Since R is zero, high-Q windings are obtained which can result in large peaks in the frequency response curve. The extent of this problem depends upon the details of the transformer design and upon the resonant frequencies relative to the high frequency cutoff obtained with the L_a, N, and R_s of the input circuit.

In general, adequately ideal transformers can be built for many applications; however, some applications will most likely require sacrifices in performance compared to that of an ideal transformer. Superconducting shields can be used very effectively to shield these transformers to prevent stray pickup and coupling with other circuits.

Either air cores or ferromagnetic cores can be used; however, ferromagnetic cores have the potential problems of low-frequency drift and Barkhausin noise. Mu-metal cores, which have a permeability at helium temperatures of about one-third their room-temperature value (DeVroomen and Van Baarle, 1957), have been used in sensitive superconducting circuits by Clarke *et al.* (1971) and Templeton (1955) with no direct evidence of Barkhausin noise.

To improve the current resolution (large source resistance) of the SLUG, Clarke *et al.* (1971) have built and tested a superconducting transformer in which a coupling coefficient k of 0.94 was achieved with a 1000-turn primary and a 1-turn secondary. The primary inductance was 8×10^{-10} H, and the secondary inductance was 8×10^{-4} H. A mu-metal core with an effective permeability of 15 was used to improve the coupling. This transformer improved the current resolution of a SLUG from 5×10^{-8} A to 9×10^{-11} A, a factor of about 560.

4. *Input time constant*

So far little has been said about the important problem of the input circuit time constant τ. In the circuit in Fig. 20 with $Z_s = R_s$, and no transformer, we see that the input time constant is L_a/R_s. Both L_a and R_s are typically very small for superconducting amplifiers, but they may vary by orders of magnitude. In many applications a 1-sec time constant is adequate. When this is the case an L_a of 100 nH, for example, would require that R_s be greater than 10^{-7} Ω. Such a large R_s will seriously limit the voltage resolution achievable, as seen by Fig. 21. If a transformer is used, as in Fig. 22, the input time constant will be

$$\tau = L_a/N^2R_s \tag{23}$$

If the optimum N^2R_s from Eq. (22) is used, we obtain

$$\tau = L_a|I_a|/|V_a| \tag{24}$$

Using the same values as in most of our previous examples, $L_a = 10^{-7}$ H, $I_a = 10^{-10}$ A Hz$^{-1/2}$, and $V_a = 10^{-16}$ V Hz$^{-1/2}$, results in a τ of 0.1 sec. This corresponds to an upper-frequency 3-dB-point, $1/(2\pi\tau)$, of only 1.6 Hz. At higher frequencies, still for $Z_s = R_s$, Eqs. (17), (19), and (20) show that the signal and noise output will both decrease at 3 dB per octave. Equation (21) shows that the signal-to-noise power ratio will not change with frequency. Flat frequency response to higher frequencies can be obtained by sacrificing signal-to-noise ratio. This can be done by using a larger value of N than given by Eq. (22). The signal-to-noise ratio is not very sensitive to N near the optimum given by Eq. (22), whereas τ varies as N^2. Thus, the frequency response can be improved substantially with only a small loss in signal-to-noise ratio. This is especially true when the Johnson-noise term predominates (center region in Fig. 21). For example, assume $T = 4.2$°K, $Z_\gamma = 0$, $\Delta\nu = 0.01$ Hz, $L_a = 10^{-7}$ H, $I_a = 10^{-10}$ A Hz$^{-1/2}$, $V_u = 10^{-16}$, $Z_s = R_s = 10^{-6}$ Ω, and $V_s = 10^{-12}$ V. The optimum N from Eq. (22) is 10^{-1}, i.e., a current step-up transformer is required. The value of τ, as computed above, is 10 sec. From Eqs. (17) and (18) we obtain a Johnson-noise term of 1.5×10^{-14} V Hz$^{-1/2}$ and an amplifier noise term $(V_u^2/N^2 + I_a^2N^2R_s^2)^{1/2}$, of 1.4×10^{-16} V Hz$^{-1/2}$. Thus, Johnson noise in R_s predominates and V_n is 1.5×10^{-14} V Hz$^{-1/2}$. If we increase N from 10^{-1} to 10^2, then τ from Eq. (23) is 10^{-5} sec and the frequency response is flat to 16 kHz, provided that the amplifier itself has flat response to at least 16 kHz. A 100:1 voltage step-up transformer will increase the voltage noise V_n as obtained from Eq. (18) to only 1.8×10^{-14} V Hz$^{-1/2}$.

5. *Servo feedback*

The input time response τ and the amplifier input impedance Z_i can both be greatly improved (i.e., τ is decreased and Z_i is increased) by coupling the servo-feedback signal of the amplifier to the input circuit rather than coupling it internally in the amplifier. This can be done by applying the feedback voltage across a very small resistance in series with the input of the amplifier (Fig. 23). An additional Johnson-noise term is introduced by

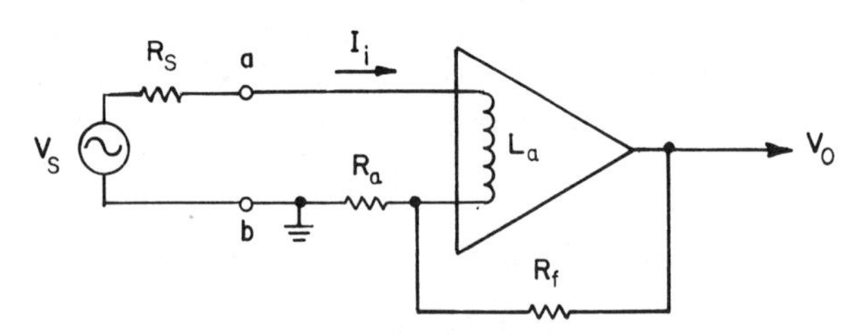

FIG. 23. Servo feedback to the input circuit of an amplifier.

this series resistance (R_a). The response time τ will be decreased by a factor approximately equal to the system excess gain and the input impedance will be increased by the same factor. This improvement can be substantial and will generally far more than offset any disadvantages. When necessary a transformer may also be used in the circuit of Fig. 23.

6. *Noise temperature*

The performance of low-noise amplifiers is frequently specified in terms of noise temperature T_n. Noise temperature can be defined as follows. Consider an amplifier properly noise matched to a source resistance that is at a temperature of absolute zero, i.e., there is no Johnson noise in the source. Define the resulting output noise power of the amplifier as P_n. Next, raise the temperature of the source resistance until the output noise power is doubled. The source temperature required is the noise temperature T_n.

Consider the equivalent circuit of Fig. 22b using a transformer input. The amplifier output noise power can be written as

$$P_n = V_o^2/R_L \tag{25}$$

where R_L is the load resistance in the output circuit. The output voltage V_o is given in terms of the transimpedance Z_t by Eq. (5). Therefore,

$$P_n = Z_t^2 \bar{I}_{in}^2 / R_L \tag{26}$$

When T is zero, $\bar{I}_{in}^2$ will be given by Eq. (17) as

$$\bar{I}_{in}^2 = [\bar{V}_u^2 + \bar{I}_a^2(N^2Z_s + Z_\gamma)(N^2Z_s + Z_\gamma)^*]/(N^2Z_s + Z_i)^2 \tag{27}$$

The Johnson-noise current from Eq. (17) at temperature T is

$$\bar{I}_J^2 = 4k_B T N^2 R_s/(N^2Z_s + Z_i)^2 \tag{28}$$

The output noise power will be doubled when T is large enough to make $\bar{I}_J^2$ equal to the $\bar{I}_{in}^2$ given by Eq. (27). Solving Eqs. (27) and (28) for temperature results in

$$T = [\bar{V}_u^2 + \bar{I}_a^2(N^2Z_s + Z_\gamma)(N^2Z_s + Z_\gamma)^*]/4k_B N^2 R_s \tag{29}$$

Minimizing Eq. (29) with respect to N gives the minimum noise temperature

$$T_n = (|V_a| + \bar{I}_a^2 Z_\gamma^2/|V_a|)|I_a|Z_s/2k_B R_s \tag{30}$$

For the special case where $Z_\gamma = 0$, the noise temperature becomes

$$T_n = Z_s|V_a|\,|I_a|/2k_B R_s \tag{31}$$

Using the noise-match condition of Eq. (22) in Eq. (29) also gives the minimum noise temperature.

It should be noted that Eq. (30) is derived assuming that the source reactance and resistance are not separately adjustable, i.e., only N is varied to obtain an optimum. If, in addition to an optimum choice of N, the source reactance is tuned out with a reactance of the opposite sign, than a somewhat lower T_n is possible. That case is considered by Rorden (1965) and Goodman *et al.* (1973).

If a noise source beyond the input stage completely dominates the noise sources in the input stage, we must have I_{in} in Eq. (6) be independent of Z_s. This result can be obtained if we set $V_u = 0$ and $Z_\gamma = Z_i$, where $|Z_i| = \omega L_a$. In this special case Eq. (31) reduces, with the aid of Eq. (9), to

$$T_n = \omega L_a \bar{I}_a^2/2k_B \tag{32}$$

Equation (32) gives the same result obtained by Radhakrishnan and Newhouse (1971). In their noise analysis the uncorrelated voltage noise term V_u was not included.

Note that lower noise temperature can be obtained by reducing ω. The minimum value of T_n is limited eventually by V_u.

The noise temperature of the amplifier discussed in connection with

Eq. (13), i.e., $V_a = 3.4 \times 10^{-16}$ V Hz$^{-1/2}$, and $I_a = 10^{-10}$ A Hz$^{-1/2}$, is 1.2×10^{-3} °K. The time constant τ from Eq. (24) for this case, if we assume L_a is 5.4×10^{-7} H, is 0.18 sec.

7. *Room-temperature sources*

The possibility of using superconducting amplifiers with room-temperature sources is not to be overlooked. In most amplifiers the signal-to-noise ratio is limited by noise in the room-temperature source rather than by noise in the amplifier. However, when the signal source is inductive in character and has low resistance, e.g., the signal is generated from a magnetic flux change, then amplifier noise may be a limit. In this instance use of a superconducting amplifier may improve the signal-to-noise ratio, especially if very low-frequency response is desired where flicker noise is high in transistors and vacuum tubes. In this example the source resistance is likely to be of the order of 1 Ω or more. The leads into the cryostat are typically 0.1 Ω to 10 Ω and must be included. Hence the input time constant τ will be quite short, e.g., if L_a is 100 nH and R_g is 1 Ω then τ will be 10^{-7} sec.

Consider the case where X_s is zero, R_s is 10 Ω, V_u is 10^{-18} V Hz$^{-1/2}$, I_a is 10^{-10} A Hz$^{-1/2}$, Z_γ is zero, and T is 300°K. The value of R_s given by Eq. (15) that gives the boundary between the Johnson-noise region and the current-noise region (Fig. 21) is 1.65 Ω. This example, therefore, falls in the right-hand region of Fig. 21, i.e., the current noise predominates over Johnson noise in R_s. Therefore, a superconducting transformer will improve the voltage resolution of the amplifier. The turns ratio N given by Eq. (22) is 3×10^{-5}, i.e., a noise match requires a current step-up transformer (see Fig. 22). Actually, a turns ratio of 10^{-1} would be adequate to reduce the amplifier current-noise term in Eq. (17) below the Johnson-noise term by a factor of 40.

To better illustrate the case of large room-temperature source resistances, it is helpful to make a comparison with a room-temperature transistor amplifier. The signal-to-noise ratio of each can be plotted versus the turns ratio N of the input transformer. Equation (18) can be used to make these plots. The comparison will be made for a frequency of about 1 Hz. For the superconducting amplifier assume that $V_a = 10^{-18}$ V Hz$^{-1/2}$, $I_a = 10^{-10}$ A Hz$^{-1/2}$. Typical values for a good transistor amplifier at a frequency of 1 Hz are $V_a \simeq 3 \times 10^{-9}$ V Hz$^{-1/2}$, $I_a = 3 \times 10^{-12}$ A Hz$^{-1/2}$. These transistor amplifier values correspond to an optimum source resistance of V_a/I_a, from Eq. (22) of 1 kΩ and a noise temperature of 325°K (at higher frequencies much lower noise temperatures can be obtained). Let $T = 300$°K for the temperature of the source resistance of each amplifier, and let R_s be 1 Ω. As a reference, to get signal-to-noise ratios, assume the signal V_s is 1 μV.

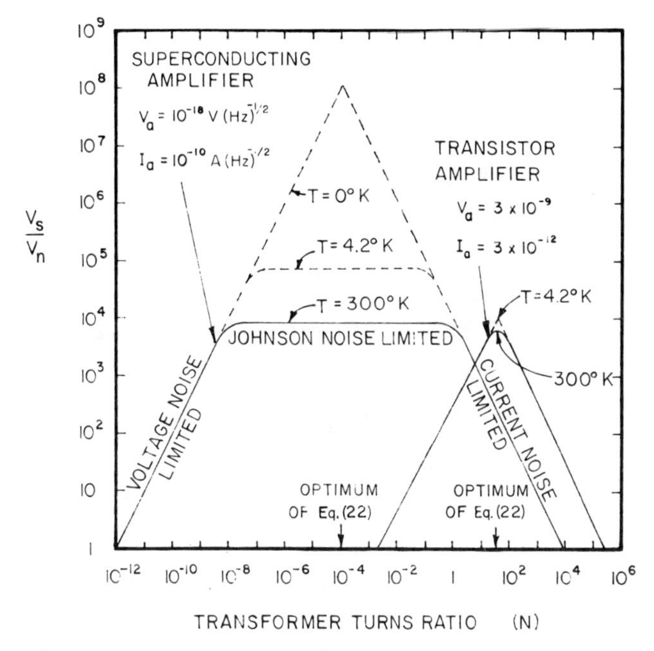

FIG. 24. Signal–to–noise ratio comparison of a superconducting amplifier and a transistor amplifier (assumed values: $R_s = 1\ \Omega$, $V_s = 1\ \mathrm{V\ Hz^{-1/2}}$, $f = 1$ Hz).

The resulting curves are shown in Fig. 24. Note that for $N = 1$ (no transformer required) the superconducting amplifier gives an 18.5 times better signal-to-noise ratio. The curves of Fig. 24 cross at $N \simeq 2$. When N is larger than 2 the transistor amplifier gives a better signal-to-noise ratio. The flat tops of the curves of Fig. 24 result from the Johnson noise predominating over the current noise and voltage noise. Dashed curves were included for a source temperature of 4.2°K to demonstrate the effect of Johnson noise. Note that the flat top disappears at $T = 4.2$°K for the transistor amplifier. The optimum N computed from Eq. (22) is also shown in Fig. 24. This demonstrates how the optimum value of N gives a maximum signal-to-noise ratio for very low source temperature. When the Johnson noise predominates (flat top region), N can be made closer to unity than the value given by Eq. (22) without loss in signal-to-noise ratio.

In the above case where R_s is 1 Ω and T is 300°K, one has two choices: (1) use of a transistor amplifier with $N \simeq 32$ to get a signal-to-noise ratio of 5400, or (2) use a superconducting amplifier with $N \leq 10^{-1}$ to get a signal-to-noise ratio of 7800. The advantage of the superconducting amplifier is not great but in some situations the inconvenience of a cryogenic amplifier may be justified. If, however, R_s is less than 1 Ω, or the temperature of the source resistance is less than 300°K, then the superconducting amplifier can have a substantial advantage over a transistor amplifier.

8. *Applications*

It should be clear from the discussion that the most outstanding performance of superconducting amplifiers is obtained with low-temperature input circuits and low resistance sources. At first this may seem to limit severely the application of superconducting amplifiers; however, there are many areas in which these amplifiers can be very beneficial. The following will give an idea of the extent of these applications: (1) thermal emf's; (2) Hall voltages; (3) measurement of very small resistances and inductances; (4) electrical noise in devices and materials; (5) readout for superconducting bolometers for infrared detection; (6) measurement of V–I curves of tunneling junctions; (7) the study of flux-flow properties of superconductors; and (8) the measurement of very small magnetic fields. It is interesting to note that in item (3) an inductance change as low as 10^{-13} H has been measured at a frequency of 10 Hz (Lukens *et al.*, 1971).

To illustrate one application, assume that the resistivity of very pure metals at low temperatures is to be measured. The circuit of Fig. 23 can be used by applying an accurately measured dc current I_b, to terminal *a*. This will produce a voltage drop V_s across R_s that can be measured by the amplifier.

Assume R_a is 10^{-9} Ω, T is 4.2°K, I_a of Fig. 20 is 10^{-10} A Hz$^{-1/2}$, $\omega/2\pi$ is 1 Hz, L_a is 2×10^{-7} H, $R_s \ll R_a$, V_u is 10^{-18} V Hz$^{-1/2}$, and $|Z_\gamma| = \omega L_a$. Then, using Eq. (11), V_n is about 5×10^{-16} V Hz$^{-1/2}$. If we allow a maximum I_b of 0.1 A, then the minimum R_s that can be detected is given by

$$R_{s_{\min}} \cong V_n/I_b = 5 \times 10^{-15}\ \Omega \tag{33}$$

Thus we could detect a change in resistance of about 5 fΩ. The time constant τ is given by

$$\tau = L_a/(R_s + R_a)G \tag{34}$$

where G is the excess gain of the amplifier. Using the above values and a gain of 10^4, we obtain a τ of 0.02 sec.

Another important application of superconducting amplifiers is in the measurement of magnetic fields, e.g., the amplifier becomes a magnetometer. In most superconducting magnetometers, the source resistance R_s will be zero so that the frequency response to magnetic fields will continue down to dc. In the circuit of Fig. 25 we obtain an induced emf due to magnetic field change ΔB in the sensing coil. Since R_s is zero, the $(L_a + L_s)/R_s$ time constant is infinite, and currents induced by a step change in the field, ΔB,

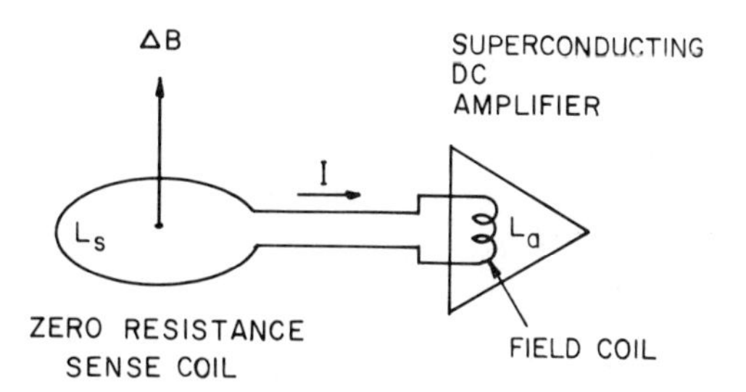

FIG. 25. Use of a superconducting amplifier as a magnetometer.

do not decay from their initial value. The current i is proportional to ΔB not to $\partial B/\partial t$.

The applied field ΔB may be due to an ambient magnetic field or it may be produced by the magnetic moment of a sample placed in the sense coil. The latter procedure can be used to build superb magnetometers to measure the magnetic moments of rocks or other samples or to build a very sensitive instrument for measuring susceptibility by having a very stable magnetic field present in the sense coil before the sample is inserted. The field will magnetize the sample as the sample is inserted into the sense-coil region.

We want to determine the value of L_s to obtain the best signal-to-noise ratio for measuring the magnetic field changes in L_s. For a sense coil having a given area, we can adjust L_s by adjusting the number of turns N_s of the sense coil. Alternatively, we can use a superconducting transformer having a turns ratio N to obtain a noise match with a given L_s, or we can adjust the number of turns on the field coil to obtain the required L_a, V_a, and I_a.

The noise-match condition can be obtained from Eq. (24). Letting $|Z_s| = \omega L_s$, and $|Z_\gamma|^2 = R_\gamma^2 + X_\gamma^2$, we obtain

$$N^4 L_s^2 = (\bar{V}_u^2/\omega^2 \bar{I}_a^2) + (|Z_\gamma|^2/\omega^2) \qquad (35)$$

If no transformer is used then N can be set equal to unity. If Z_γ is written as $R_\gamma + j\omega L_\gamma$, then we obtain one frequency independent term L_γ^2, i.e.,

$$N^4 L_s^2 = (1/\omega^2)[(\bar{V}_u^2/\bar{I}_a^2) + R_\gamma] + L_\gamma^2 \qquad (36)$$

Note that this noise-match condition depends upon ω. At low frequencies a larger source inductance is required than at higher frequencies.

In the special case where the predominant noise source is beyond the input stage, or if the sensor field coil is very loosely coupled to the sensor, the output voltage noise V_{on} will be independent of source impedance. Since $V_{on} = Z_t I_{in}$, then I_{in} must also be independent of source impedance and equal to I_a. From Eq. (6) we see that these conditions can be satisfied if $|Z_\gamma| = |Z_i| = \omega L_a$, and $V_u \ll I_a|Z_s + Z_i|$. For this special case then,

$R_\gamma = 0$, and $L_\gamma = L_a$, and Eq. (36) reduces to

$$N^2 L_s = L_a \tag{37}$$

This is the familiar result often derived on the basis of maximum energy transfer (Zimmerman, 1971), and is used in the following treatment of field amplifiers. It is correct only when $Z_\gamma = Z_i$ and V_u is negligible compared to $I_a\omega(L_s + L_a)$. If V_u is 10^{-20} V Hz$^{-1/2}$, I_a is 10^{-10} A Hz$^{-1/2}$, and $(L_s + L_a)$ is 10^{-6} H, this condition on V_u requires that ω be much greater than 10^{-4} rad sec^{-1}.

9. *Magnetic field amplifier*

The dc magnetometer input circuit of Fig. 25 can be considered a field amplifier because a pickup or sense coil is used to sense a very small magnetic field change in a large area and essentially compress this flux and apply it to the sensor as a more intense field in a smaller area.

The complete field amplifier circuit is made of superconductive material. Once this circuit is cooled below its critical temperature, the total magnetic flux linking the circuit must remain constant. This is because of the zero-resistance property of superconductivity. Thus, if the ambient field is changed at the sense coil, a current will flow so that the internal field at the sense coil will change and so will the field at the field transfer coil, thereby satisfying the condition that the total flux linkages remain constant.

The resulting field amplification will now be derived. It must be emphasized that the following derivation is based on *maximum energy transfer* from the sense coil to the field coil. It does not take into account variations in input noise as a function of the sense-coil inductance L_s or of any frequency dependence of the input noise as given by Eq. (6). These noise effects, however, are only significant when the field coil is very tightly coupled to the magnetometer sensor.

Coil s is a large field sense coil and coil a is a field transfer coil coupled to the magnetometer. If the magnetic flux linking coil s is changed by an amount $N_s\,\delta\phi_s$, where N_s is the number of turns of coil s and $\delta\phi_s$ is the applied flux change, then a circulating current will be induced in the coils, as given by

$$I = N_s\,\delta\phi_s/(L_s + L_a) \tag{38}$$

where L_s and L_a are the geometric inductances of coils s and a, respectively. The resulting field at the sensor (field transfer coil) is a function of the induced current I and the geometry of the field coil, i.e., dimensions and turns. The field amplification will then be the magnetic induction field at the sensor B_a, divided by the applied induction field B_s.

Maximum transfer of field energy from the applied field at the pickup coil to the sensor occurs when the inductance of the two coils is equal, $L_s = L_a$. This can be shown by equating the final energy

$$E = \tfrac{1}{2}LI^2 \tag{39}$$

of the two coils and realizing that the maximum energy that can be transferred is one-half the applied energy. The inductance relation $L_s = L_a$ can also be derived by equating the current in coil a to the current induced by the flux change $\delta\phi_s$ at coil s:

$$N_a\phi_a/L_a = N_s\,\delta\phi_s/(L_s + L_a) \tag{40}$$

Solving for the flux in coil a:

$$\phi_a = N_s\,\delta\phi_s L_a/N_a(L_s + L_a) \tag{41}$$

In general, the inductance of a coil is a function of the square of the number of turns; thus we can write

$$L_a = N_a^2 L \tag{42}$$

where L is a function of the remaining geometrical parameters. Thus,

$$\phi_a = N_s\,\delta\phi_s N_a L/(L_s + N_a^2 L) \tag{43}$$

For maximum flux in coil a, set $\partial\phi_a/\partial N_a = 0$ and solve for L_s. This gives the relation $L_s = L_a$. For special cases in which the coils are long solenoids,

$$B_a = \mu_0 N_a I/l_a \tag{44}$$

and

$$I = N_s\,\delta\phi_s/(L_s + L_a) \tag{45}$$

Therefore,

$$B_a = \mu_0 N_s N_a\,\delta\phi_s/l_a(L_s + L_a) \tag{46}$$

Using $L_s = L_a$ for maximum energy transfer, the applied field as $B_a = \delta\phi_s/A_s$, and

$$L_s = \mu_0 N^2 A_s/l_s \tag{47}$$

we obtain the field amplification

$$B_a/B_s = N_a l_s/2N_s l_a \tag{48}$$

By using Eqs. (43) and (47) this can be rewritten as

$$B_a/B_s = N_s A_s/2N_a A_a \tag{49}$$

or as

$$B_a/B_s = \tfrac{1}{2}(V_s/V_a)^{1/2} \tag{50}$$

where V_s and V_a are the volumes of the solenoids $A_s l_s$ and $A_a l_a$, respectively.

If the coils are not long solenoids, we must replace Eq. (44) with the appropriate field and inductance expressions for the particular coil to derive the field amplification. In general, the field transfer coil should be small while the pickup coil should be as large as possible.

A circuit has been tested (Deaver and Goree, 1967) in which the pickup coil was one turn of 0.005-cm niobium wire wound on a 1-cm-diam form. The single-layer field transfer coil was about 180 turns of 0.005-cm niobium wire on a 0.005-cm-diam form. The field amplification was a factor of 50.

Another example of a field amplifier is one in a rock magnetometer system designed to measure the magnetic moment of lunar rock specimens (Goree, 1972). This system has a 16-cm-diam 1-turn Helmholtz-pair pickup coil placed around a 10-cm-diam room-temperature sample access region. The field resolution of the pickup coils is about 2×10^{-11} G rms $\mathrm{Hz}^{-1/2}$ and is limited by environmental fluctuations inside the system's superconducting shield (e.g., Johnson noise in nearby conducting surfaces) rather than by intrinsic sensor noise. The measured magnetic field amplification factor between the pickup coil pair and the sensor is $\simeq 500$ which is near the theoretical estimate obtained by using Eq. (50).

B. Resistance Modulation Amplifiers

Resistance modulation amplifiers use an input signal to change the resistance of a superconductive element that is biased near the midpoint of its resistive transition. Superconductive circuits can utilize either positive resistance or negative resistance changes to obtain signal amplification.

Positive resistance modulation amplifiers use the magnetic field produced by a signal current in a superconducting control element to vary the resistance of a biased gate element. The gate is either a thin-film superconductor, as in a cryotron, or a Josephson tunnel junction. The amplifier is usually a four-terminal device with the input (control) inductively coupled to the output (gate).

Negative resistance amplifiers are two-terminal devices in which the signal is applied across the same terminals as the bias and load. The element utilizes the tunneling of normal electrons (quasiparticles) or of superelectron pairs through a dielectric barrier separating two superconductors. Power gain is possible because of the negative resistance property of the device.

1. *Positive resistance amplifiers*

a. Cryotron amplifiers. The most widely studied superconductive amplifiers use the basic cryotron circuit. The gate of the cryotron is a

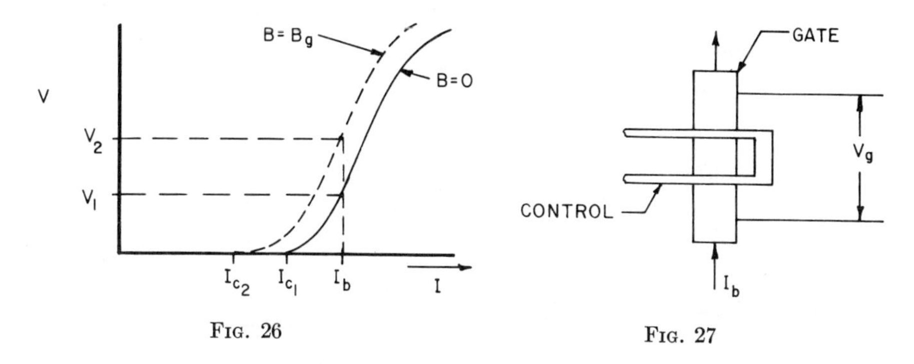

FIG. 26 FIG. 27

FIG. 26. V–I curve for a cryotron gate.
FIG. 27. Cryotron with double control crossing.

thin film or wire that is biased with a current source near the midpoint of its resistive-superconductive transition. Figure 26 shows a typical V–I curve for a cryotron gate. If the external magnetic field at the gate is increased then the critical current of the gate will be decreased, also as shown in Fig. 26. Thus, the magnetic field produced by a signal current in the control of a cryotron will change the effective resistance of the cryotron gate.

Cryotrons have been constructed in many different varieties as noted in Section II on switches and more completely by Bremer (1962), Newhouse (1964), and Newhouse and Edwards (1965). The wire-wound and in-line cryotrons have not been used in amplifier circuits to our knowledge; most of the studies have been devoted to crossed-film cryotrons.

A typical crossed-film cryotron is shown in Fig. 27. The gate is biased into its transition region as shown in Fig. 26 and is usually made of tin or indium. The control is made of a high-temperature superconductor (relative to the gate) such as lead or niobium and always remains superconducting. Experience with cryotron switches (Section II) has shown that thin-film circuits have advantages such as rapid switching times and small size over wire-wound types. All cryotron amplifiers discussed in this section have been constructed with thin films.

The transimpedance of a cryotron element is given by

$$R_t = \left(\frac{\partial V_g}{\partial I_c}\right)_{I_g} = I_g\left(\frac{\partial R_g}{\partial I_c}\right)_{I_g} \simeq I_g\left(\frac{\partial R_g}{\partial B}\right)_{I_g}\left(\frac{\partial B}{\partial I_c}\right) \tag{51}$$

where R_t is the transimpedance, V_g is the gate voltage, R_g is the differential gate resistance, I_c is the control current, I_g is the gate bias current, and

B is the magnetic induction field produced at the gate by the control current. Equation (51) is approximate because the magnetic induction field B is not uniform over the entire region of the gate where the control crosses. However, the relation is useful to illustrate the effects occurring in a cryotron.

It is desirable that R_t be as large as possible so that large signal output voltages can be obtained for very small signal input currents. If R_t is too low then the noise of the room-temperature amplifier used to read the output might predominate over the input current noise of the cryotron. In that case several cryotrons could be cascaded at the expense of added complexity to obtain the full performance potential of the first stage.

The power gain of a cryotron cannot be given because the input resistance is zero and, therefore the input power is also zero (Section III.A). The dc current gain per stage is given by

$$G = \left(\frac{\partial I_g}{\partial I_c}\right)_{V_g} \tag{52}$$

Since the differential gate resistance R_g is given by

$$R_g = \frac{\partial V_g}{\partial I_g} \tag{53}$$

then the current gain can be expressed in terms of the transimpedance and gate resistance as

$$G = \frac{R_t}{R_g} = \left(\frac{\partial V_g}{\partial I_c}\right)\left(\frac{\partial V_g}{\partial I_g}\right) = \frac{\partial I_g}{\partial I_c} \tag{54}$$

Several other crucial parameters of a cryotron affect their performance. These are the following:

L_c is the inductance of the control and determines the time constant of the input circuit.

$\partial B/\partial I_c$ is the change in magnetic field at the gate (∂B) with respect to a change in the control current (∂I_c). This parameter is a function of the control geometry, i.e., a coil, wire, thin film, etc., and of the coupling between the control and the gate.

$\partial R_g/\partial B$ is the change in gate resistance for a given change in magnetic field at the gate, for constant gate current. This factor is particularly important in determining the transimpedance R_t of the cryotron.

R_g is the differential resistance of the gate. This is important in determining the output signal level.

R_0 is the gate resistance (V_g/I_g) that determines the time constant of a series connected cryotron and the power dissipation in the gate film.

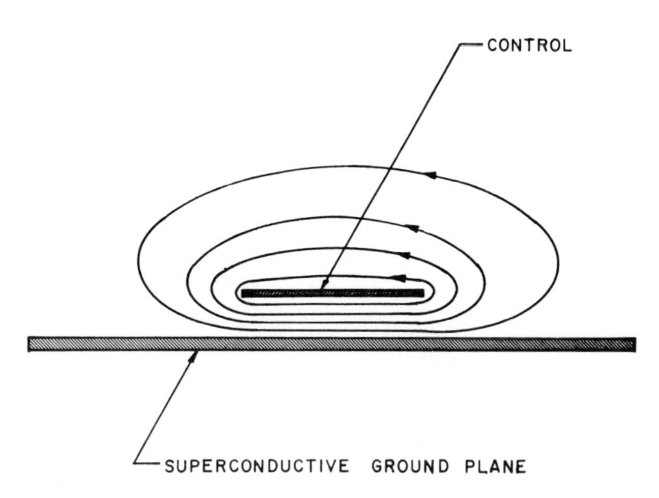

FIG. 28. Magnetic field about a strip conductor near a superconducting ground plane.

L_g is the inductance of the gate. This parameter determines the minimum time constant of the output circuit, i.e., $\tau = L_g/R_0$.

$\partial/\partial_t(\partial R_g/\partial I_c)$ is the time dependence of the change in gate resistance under the influence of a field produced by the control. When a magnetic field is applied to the gate it takes a finite time for the gate to react. This is due to effects such as flux motion and thermal time constants and often is a more severe limit on the frequency response than is the gate or control inductance, or the gate resistance.

Control circuit. As noted above the crucial cryotron control circuit parameters are the inductance of the control L_c and the magnetic field produced on the gate by a current in the control, $\partial B/\partial I_c$. How each of these parameters is affected by the construction of the control circuit is described below.

The inductance of the control circuit is important because it determines the time constant of the input circuit of the cryotron. Low-inductance thin-film circuits can be obtained by placing the circuit in close proximity to a superconducting ground plane. The inductance in henrys per meter of length of a film of width W separated by a distance s from a superconducting ground plane is given by (Young, 1959)

$$L/l = \mu_0(s/W) \tag{55}$$

The general effect of the ground plane is shown in Fig. 28. The magnetic field beneath the control is confined by the ground plane and is, therefore increased by the presence of the ground plane. The current distribution in the control is also more uniform because of the presence of the ground plane. This enhanced current homogeneity results in an increase in the control field by as much as a factor of 2.

Thus, the control film should be wide enough (W) and close enough to a superconducting ground plane (s) so that its total inductance L_c is low enough so that it will not limit the desired time response of the cryotron. Reductions of the induction by two orders of magnitude using a ground plane are readily achieved. Of course, there are compromises to be made. Since the control is normally driven by the gate of another cryotron, the overall time constant is given by

$$\tau = (L_c + L_g)/R_0 \tag{56}$$

Thus the gate inductance must also be considered. Furthermore, the field produced by the control is strongly related to the control geometry as will be discussed in the following section.

Coupling between control and gate. The major function of the control is to produce a field change at the gate as a function of the applied signal. The geometry sought is one that will give a maximum $\partial B/\partial I_c$. The original cryotron of Buck (1956) used a solenoid wrapped around a wire which made the field parallel to the gate surface. Thin-film cryotrons can be constructed so that the field is predominantly normal or parallel to the gate film and a choice should be made based on performance.

The critical magnetic field of thin films with the field applied normal and parallel to the film has been the subject of many investigations. The general result of these studies indicates that a control field normal to the gate is *undesirable.* The major reason for this is the large demagnetization factor for fields normal to a thin film as indicated in Fig. 29. This causes the film transition to occur first near the edges, and to then move inward as the field or transport current is increased. This, in effect, broadens the magnetic transition and decreases the slope of the V_g–I_g curve thereby decreasing the sensitivity. A parallel field has a demagnetization factor of nearly zero and the entire film surface is exposed to the same field. Thus, the field-induced transition is much sharper and easier to control if the field is parallel to the gate film.

A rather complete study of the magnetic field between the control and ground plane was reported by Newhouse *et al.* (1960). When the gate film is placed between the control and ground plane, the field configuration changes when the gate allows flux penetration. With a gate bias current I_g less than the critical value the gate behaves as a ground plane. As the control current is increased the gate will eventually become normal in the

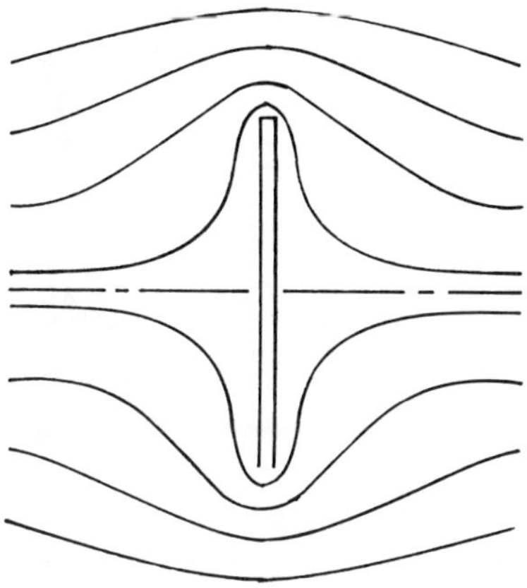

FIG. 29. Thin superconducting film in a perpendicular magnetic field.

vicinity of the control. The design parameters that are important here are the width of the control film and the separations to be used between films to give a maximum change in the gate voltage for a given change in control current, i.e., a maximum transimpedance. Newhouse and Edwards (1964) obtained the V_g–I_c data shown in Fig. 30. We want to maximize the transimpedance, the slope of the V_g–I_c curve, as a function of control width. If the field due to the control current is parallel to the gate at all points, the transition should occur abruptly when this field H_c equals the parallel critical field of the gate film, $H_{c||}$. The width of this transition in control current $I_{c2} - I_{c1}$ as shown on Fig. 30, should be such that the slope of the transition $\partial V_g/\partial I_c$ is independent of the width of the control film W_c. This

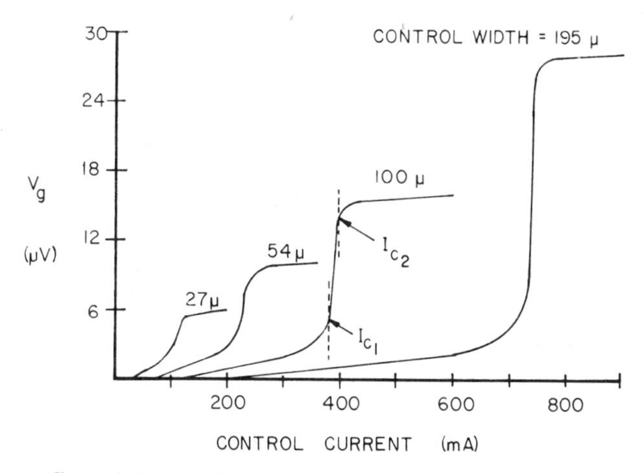

FIG. 30. Control characteristics as a function of control width for a shielded crossed-film cryotron. [After Newhouse, 1964, courtesy John Wiley & Sons, Inc.]

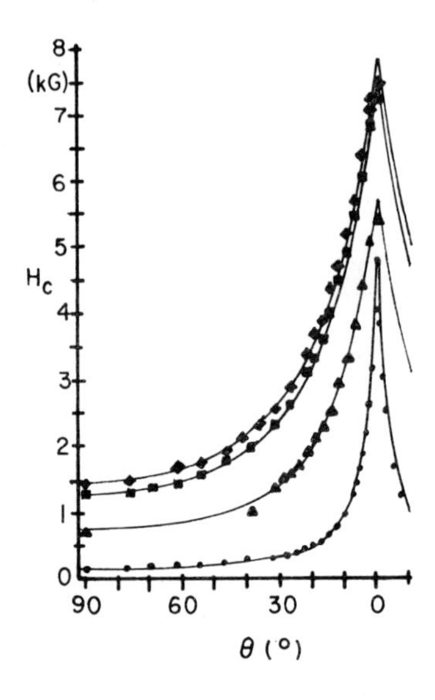

FIG. 31. Critical magnetic field for thin superconducting films as a function of field orientation (◆: 5°K; ■: 34°K; ▲: 163°K; ●: 300°K). [After Schiller and Bulow, 1969a.]

is clear since the gate voltage V_g and the control current I_c necessary to produce a given field are both proportional to the control width W_c therefore the slope $\partial V_g/\partial I_c$ is independent of control width.

Experimentally, this is not the case. If we plot the slope at the midpoint of the V_g–I_c curves, we find the slope becomes constant only for rather wide control films (200 μ in the experiments of Newhouse and Edwards). This can be explained if we consider the following fact: The critical magnetic field for thin films is a strong function of the angle between the plane of the film and the field. Experiments on tin films by Schiller and Bulow (1969a, b) present striking evidence of this, and are in excellent agreement with the theoretical work of Tinkham (1963) in which the effect is explained in terms of a fluxoid structure in the film. Some of the results obtained by Schiller and Bulow (1969a, b) are summarized in Fig. 31. The critical field for $\theta = 90°$ (normal field) is seen to be a factor of 8 or more greater than for parallel fields. Similar measurements for lead films 1950 Å thick showed $H_{c\perp}/H_{c\parallel} \simeq 2$.

A model for the control-gate film geometry would look as shown in Fig. 32. At some control current I_c the normal component of the control field penetrates the gate film, starting at the edges of the gate at the corners of the gate-control intersection, where the normal fields due to the gate current and control current add. This field penetration introduces a flux-flow voltage in the two small parallel regions of the gate film under the edges of the control. As long as the control width is several times the separation distance between the control and the ground plane, these regions will be

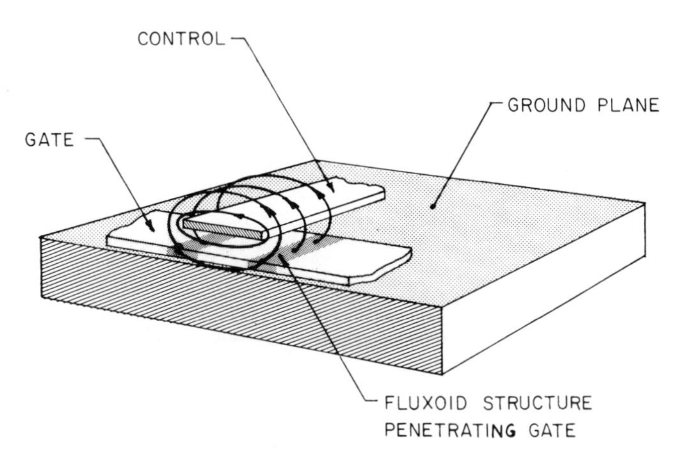

FIG. 32. Field distribution model for control and gate films.

separated and the flux-flow voltage should be independent of control width. For narrow control widths these parallel flux-flow regions merge and the flux flow voltage will depend upon the control width W_c.

As the control current is increased beyond this value for initial flux penetration, a point is finally reached at which the parallel field beneath the control switches the gate film to the flux-flow state. This should be an abrupt transition, but for narrow control films the presence of the flux-flow region decreases the control current required to initiate the transition of the gate and broadens the transition. Thus for narrow films the trans-impedance is not independent of control width. As the control film width increases up to $\simeq 200\ \mu$ the maximum slope of the V_g–I_c curves also increases, an effect exhibited in the samples studied by Newhouse and Edwards (1964). For films wider than 200 μ the maximum slope remains essentially constant.

For narrow control films the amount of normal field at the edges is increased. As the ratio W/d becomes larger, more of the field is parallel to the gate and the slope then becomes more nearly constant. For the sample studied by Newhouse and Edwards (1964), control-film–ground-plane separation is about 1 μ; thus, it is only when the control width is several hundred micron that the field under the control is parallel. This is in agreement with the film width of 200 μ required to give constant slope.

Thus, the design of optimum control films should satisfy the following guidelines.

1. The insulation between films should be as thin as possible and still avoid electrical shorts. Generally films of silicon dioxide from 1000 to 3000 Å have been used.

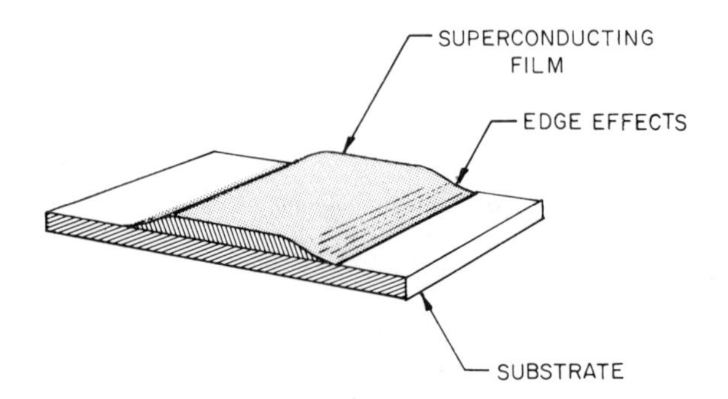

FIG. 33. Penumbra at the edge of evaporated superconductive films.

2. The control film width should be from 100 to 200 times the separation distance between the control film and the ground plane.

3. The control film should be thick enough so that flux flow does not occur in the film; e.g., at temperatures well below T_c this film need only be a few hundred angstroms thick.

Gate circuit. The cryotron gate circuit largely determines the performance of a cryotron amplifier. Amplifier noise is due primarily to fluctuations in the gate resistance; the output impedance and signal levels are also determined by the gate. In this section the details of the gate circuit will be related to the design parameters noted earlier, i.e., $L_g(\partial V_g/\partial I_c)_{I_g}$, material, thickness, width, etc.

The inductance of the gate film is made very low by placing the gate near a superconducting ground plane as discussed earlier for the control. The spacing s between gate film and ground plane, should be as small as possible and still eliminate electrical shorts between the superconductive films. Most researchers have used 1500 to 5000 Å silicon monoxide (SiO) dielectric separations. Recent techniques such as reactive sputtering of metal oxide insulating films make it now possible to obtain good insulating films down to 100 Å thickness. Since the inductance in henrys per meter of a film of width W separated by the distance s from a superconducting ground plane is given by $L/l = \mu_0 s/W$, thinner dielectrics could appreciably reduce the gate-film inductance.

A further complication is introduced by edge effects. The cross section of a typical evaporated film is shown in Fig. 33. The tapered edges result from the penumbra effect of evaporating through a mask and cause an additional broadening of the transition. The magnetic flux penetration into the edges causes their critical field to be higher than that of the main film.

The gate thickness should be as small as practical and at least thinner than $\sqrt{5}\lambda$, where λ is the London penetration depth, to minimize hysteresis

(Newhouse and Edwards, 1964). This hysteresis effect has been determined experimentally and is presumably caused by flux pinning in the gate film. Most cryotrons reported in the literature have used gate films of 2000-to 5000 Å thickness.

The width of the gate film is chosen to optimize the gate inductance and resistance. Since the gate inductance is inversely proportional to the gate's width W, the film should be made as wide as possible for low inductance. The gate resistance is inversely proportional to the gate width requiring a narrow film for high resistance. For some amplifier designs Newhouse and Edwards (1964) have used a single dc bias supply for series-connected control and gate films and as a consequence the width and thickness of these two films are restricted so that the same bias current in each will give satisfactory operating characteristics.

The actual requirements for the inductance depend on the desired frequency response of the cryotron element. Newhouse (1969) notes that typical cryotron amplifiers are usually limited to frequencies in the 1- to 2-MHz range because of thermal effects, i.e., Joule heating in the gate film and poor heat transfer to the liquid-helium bath. Thus, there is no need to make the gate much wider than that required to give, say, 10-MHz response, unless better thermal response can be achieved.

The electrical response time was given by Eq. (56). For the typical cryotron amplifier, in which the gate of one stage is in series with the control of the following stage, L_c is usually much higher than L_g (the control is usually longer and narrower). Thus L_g does not play a dominant role in determining the response time.

The transimpedance R_t is not a function the width of the gate film. This was discussed in the control circuit section of this chapter. The dc current gain, as given by R_t/R_g from Eq. (54), is thus directly proportional to the gate width.

The two parameters that determine the optimum gate width are therefore the required gate bias current I_g and the gate resistance R_g. If a single bias supply is to be used, the gate must be chosen wide enough to give the proper bias with the same current used to bias the control. The gate must also be narrow enough to give the desired level of output (gate) resistance since $R_g \propto 1/W_g$.

Note also that R_g and I_g are both functions of gate film thickness, R_g increasing with decreasing thickness, and I_g decreasing with decreasing thickness. Thus the choice of thickness and width offers a wide range of R_g and I_g. Since the inductance of the gate is independent of thickness, the resistance can be increased by decreasing the thickness without changing the width or the inductance. However, the minimum thickness is set by the steepness of the transition as mentioned above.

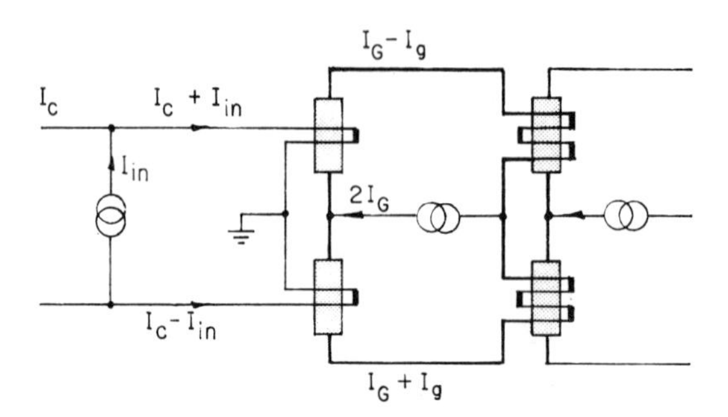

FIG. 34. Balanced-bridge cryotron amplifier. [After Newhouse and Edwards, 1964.]

The final gate parameter to be considered is the temperature dependence of the gate-film transition with both gate and control bias currents. In reported cryotron amplifiers a bias current is used on the control (bias field) as well as on the gate film to achieve maximum gain. The choice of operating temperature is then dictated by the slope and stability of the V_g–I_c curve for constant gate bias current. The general feature of all reported circuits is that the operating temperature has to be very near the critical temperature of the gate for low noise, stable performance.

For gate films of tin or indium, operation is usually best within 50 m°K below T_c. As the bath temperature is additionally decreased the steep slope region of the V_g–I_c curve becomes very unstable due to Joule heating in the gate film. This heating problem can be reduced by using high-thermal-conductivity substances such as single-crystal quartz or sapphire rather than glass, which is unsatisfactory for amplifier operation (Newhouse and Edwards, 1964). Even with a high-conductivity substrate a major source of thermal impedance is in the dielectric insulating films. This thermal time response has limited the frequency response of cryotron amplifiers to a few megahertz.

Performance of linear cryotron amplifiers. The first successful amplifiers were built with series stages of the single gate, single control cryotron that has been the subject of discussion so far. Thermal instabilities in the bath generally made it impractical to keep more than two stages properly biased.

A technique used by Gygax (1961) and by Newhouse and Edwards (1964) dramatically reduces these temperature effects. The cryotrons are connected in a balanced bridge circuit as shown in Fig. 34. A current signal I_{in} (Fig. 34) increases the control field on the upper cryotron and reduces the control field on the lower cryotron. This changes the resistance of the gates, increasing the resistance of the upper gate and decreasing the resistance of the lower gate. The gate bias current will then redistribute itself to compensate for the resistance change with more current flowing through the lower gate since its resistance is lowered by the signal current I_{in}.

Temperature fluctuations will affect both cryotrons of a balanced pair equally, thereby not affecting the difference current I_G but changing the distribution of the bias current I_g. This assumes, of course, that the cryotron is biased in a linear region and that the fluctuations are not large enough to drive the cryotron out of this region. The dc current gain per stage, $\partial I_v/\partial I_c$, will remain constant as will I_g itself. Any number of these balanced pairs of cryotrons may be connected in series to increase the overall gain by the number of pairs.

The output characteristics of multistage cryotron amplifiers can be tailored, within limits, to meet varied requirements. The output resistance is determined entirely by the resistance of the last stage while previous stages contribute only to the current gain. The circuit shown in Fig. 34 illustrates the technique of multiple crossings of the control film on the last stage for increasing the gate resistance and output voltage of a cryotron amplifier. The maximum output voltage is proportional to the sum of the control widths crossing the gate since each crossing causes more of the gate length to change resistance for a given current input to the control. If the width of the control is also increased in the same proportion as the number of crossings, then the total inductance of the final control stage remains unchanged since the control inductance is inversely proportional to control width. It should also be noted from Fig. 34 that the control and gate interconnections are made wider than the control to minimize stray inductance.

Newhouse and Edwards (1964) have found that multistage crossed-film cryotrons have a large low-frequency noise probably caused by local temperature fluctuations within the cryotron itself. This noise is *not* removed by the balanced-pair configuration. They have eliminated the noise using negative feedback stabilization in which the signal from the output cryotron is coupled back to the input cryotron control as shown in Fig. 35. The feedback inductors are chosen to provide negative feedback up to frequencies around 100 Hz without attenuating signals above this frequency. Thus the feedback-stabilized cryotron amplifier is not a dc amplifier. Each stage of the amplifier has a small phase lag associated with its inductance and gate resistance, and after a sufficient number of stages this phase lag can reach 180°. At this point, the feedback circuit produces positive feedback, and the circuit will oscillate unless the gain at the feedback frequency is less than unity. The oscillation can be minimized by reducing the magnitude and (or) phase shift of the feedback signal. The magnitude can be reduced by (1) increasing the feedback inductance, (2) decreasing the resistance driving the feedback loop, and (3) decreasing the closed-loop gain by

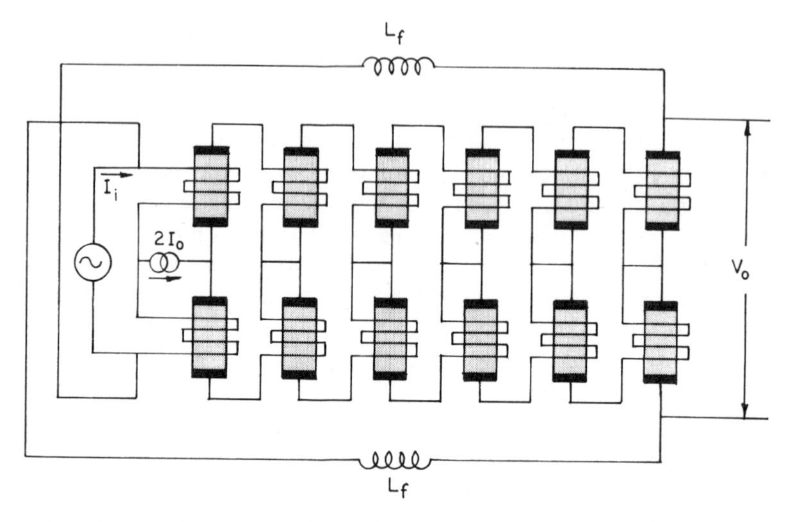

FIG. 35. A crossed-film cryotron amplifier with five cascaded stages and a negative feedback stage to eliminate low-frequency noise. (One current source supplies the biasing current for all the controls and gates.) [After Newhouse and Edwards, 1964.]

decreasing the number of stages over which the feedback occurs. Number (3) is doubly effective since it reduces the magnitude and phase shift of the feedback signal. This oscillation problem restricts the number of stages permissible in one feedback loop, but does not limit the number of feedback-stabilized loops that can be connected in series.

The circuit shown in Fig. 34 used a separate current generator for each gate and each control pair. A much simpler circuit uses only one bias supply as shown in Fig. 35. As noted earlier, this circuit requires that the gate-film width and thickness as well as the control-film width be chosen so that the same current through the gate and control will produce the desired operating bias condition. Feedback and gain are the same as for the previous amplifier with individual current supplies.

Multicrossover cryotron. An even simpler and possibly more useful cryotron amplifier is the multicrossover, single-stage circuit of Newhouse *et al.* (1967). The circuit consists essentially of a very long gate film with 512 crossovers by the control film. A separate current bias supply is used for the gate and control film. To achieve a large number of crossovers the gate as well as the control film is folded many times as shown schematically in Fig. 36. The gate and control films were 200 μ wide (Newhouse *et al.*, 1967). The gate was a 3000-Å-thick film of either tin or indium and the control was a 7500- to 10,000-Å-thick lead film. The ground plane was lead evapo-

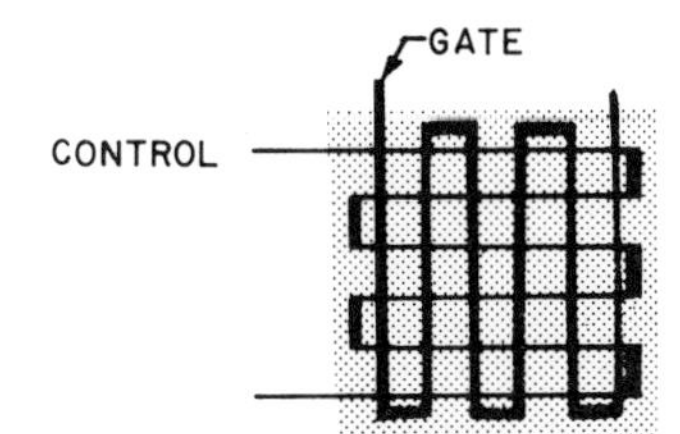

FIG. 36. Cryotron with multiple control and gate crossings.

rated on glass, single-crystal quartz, or anodized niobium. Silicon monoxide insulating layers about 4000 Å thick were used between all metal films.

The multicrossover cryotron amplifier does not require negative feedback since the gate and controls are directly driven by current sources and thus the bias currents are fixed. Local thermal fluctuations do produce low-frequency variations in the output voltage because the film properties change with temperature, but this noise is not amplified by successive stages as with multistage cross-film cryotrons; thus it may be filtered at the room. The noise is generally below 1 kHz.

Summary. Table I summarizes the performance of the different types of cryotron amplifiers described, i.e., balanced-pair multistage cryotrons and multicrossover cryotrons.

Of these types of cryotron amplifiers it appears that the multicrossover type is by far the simplest and the one that most easily lends itself to optimization using state-of-the-art deposition techniques.

We conclude that cryotron amplifiers have already achieved impressive performance, and there appears to be room for further development, especially in view of modern techniques of vacuum deposition. Very little developmental work has been reported recently; however, as superconducting devices become more common, interest in cryotron amplifiers may again be generated. The multicrossover cryotron, in particular, deserves further development. Once a high-performance cryotron amplifier has been developed, the potential for low-cost mass production appears favorable.

b. Supercurrent tunneling amplifiers. The operation of a supercurrent tunneling junction, point contact, or weak-link device, as an amplifier is similar to that of the cryotron discussed in Section III.B.1.a. The junction resistance of a supercurrent tunneling amplifier varies with the magnetic field produced by the input current. Similarly, the gate resistance of a cryotron varies with the field imposed by the control current. The physics involved is, of course, quite different.

A typical tunnel junction is shown in Fig. 37. The V–I curves may have many different shapes as shown in Fig. 38, depending on the nature of the junction, the source used to drive the junction, and the environment (especially high-frequency electromagnetic noise). The voltage–current

TABLE I

PROPERTIES OF THREE TYPICAL AMPLIFIERS

Property	Sample C: Eight-stage high-gain amplifier	Sample D: Four-stage high-bandwidth amplifier	Multicross-over cryotron
Maximum current gain			
of amplifier	22,000 (=86 dB)	16.8 (=24.5 dB)	100
per stage	3.5 (=10.8 dB)	2.02 (=6.1 dB)	
Upper cutoff frequency			
of amplifier (measured)	45 kHz	700 kHz	900 kHz
of one stage (calculated)	150 kHz	1680 kHz	
Gain-bandwidth product			
of amplifier	9.9×10^8	1.18×10^7	9×10^7
per stage	5.25×10^5	3.4×10^6	
Minimum detectable input current (rms)	0.068 μA	3 μA (limited by vacuum tube preamp)	—
Calculated Johnson noise referred to input	0.017 μA	0.088 μA	—
Temperature of operation	3.658°K	3.615°K	3.38°K
Critical temperature of gates	3.702°K	3.698°K	3.42°K
Resistance per square of gates at 4.2°K	6.1 mΩ	13.3 mΩ	—
Dynamic transimpedance of output gate	3.2 mΩ	60 mΩ	—

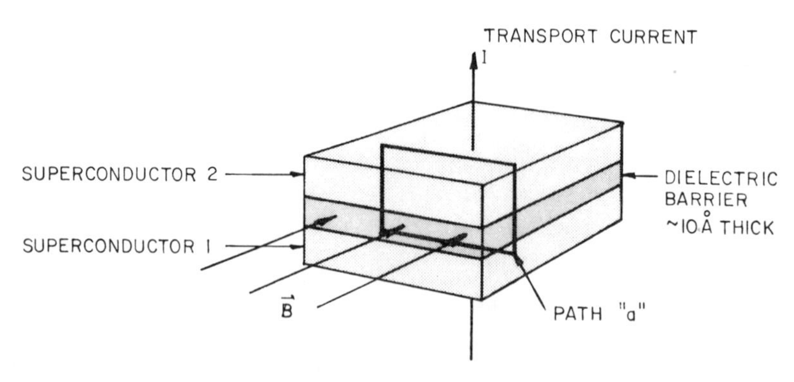

FIG. 37. Supercurrent tunnel junction.

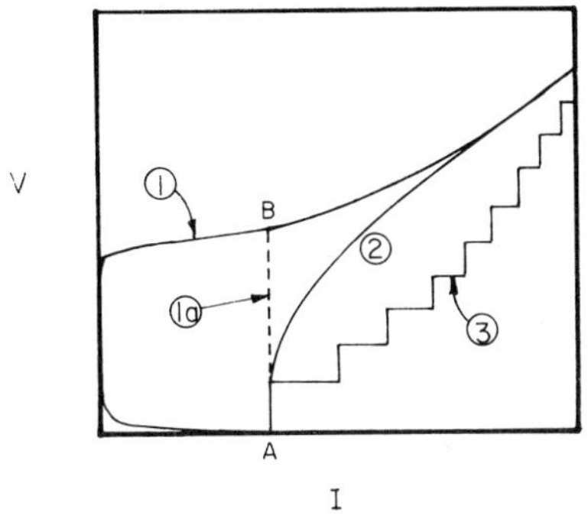

FIG. 38. Sketch of four characteristic V–I curves for tunneling junctions.

(V–I) characteristic is discussed in Section II.A.2.

Curve 1 in Fig. 38 is for a Josephson tunneling junction (dielectric barrier) driven with a voltage source. The negative resistance region can actually be measured if a true voltage source is used, and this characteristic has amplifier applications as discussed in Section III.B.2. Curve 1a is for the same junction driven with a constant current source. The switching process from point A to B is in reversible and occurs in less than 1 nsec. This Josephson tunneling junction is ideal for computer and other switching applications, but it is not generally useful for amplifiers. Curve 2 is a typical curve for a narrow bridge or point-contact device. The curve may not be as smooth as sketched in the figure but it is usually monotonic. Curve 3 is for a narrow bridge or point-contact device subjected to high-frequency radiation. The induced constant voltage steps are caused by coupling between the applied radiation and the induced radiation at the Josephson frequency ν_J, i.e.,

$$V = nh\nu_J/2e \tag{57}$$

where V is the voltage across the junction at the nth step, h is Planck's constant, and e is the electron charge. The quantity $2e/h$ is equal to 483.6 MHz μV^{-1}.

McCumber (1968) and Stewart (1968) have described the different V–I curves for tunneling junctions in terms of equivalent circuits for the junctions that include shunt capacitance and series inductance. They show that junctions with a large capacitance and low inductance, such as thin-film Josephson junctions, will exhibit the negative resistance region and rapid switching. On the other hand, junctions with very low capacitance can exhibit continuous, monotonic V–I curves.

Voltage-field relations for tunnel junctions. The basic property of a resistive-modulation, tunnel-junction amplifier is the variation of junction voltage (gate voltage) with respect to applied flux (control current). Josephson (1962) predicted that the current density j in a dielectric barrier separating two superconductors would be

$$j = j_{\max} \sin \psi$$

where $j_{\max}$ is the maximum value of the current density for zero flux in the junction and ψ is the gauge invariant phase difference across the junction. When a magnetic field is present, ψ is given by

$$\psi = \gamma_1 - \gamma_2 - (2e/h) \int \mathbf{A} \cdot d\mathbf{l} \tag{58}$$

where γ_1 and γ_2 are the phases of the electronic wave functions in the superconductors on either side of the barrier and **A** is the magnetic vector potential. The closed-line integral of the vector potential around path "a" as shown in Fig. 37, gives the total magnetic flux linking the junction. Thus

$$j = j_{\max} \sin \left(\gamma_1 - \gamma_2 - 2e/h \int_{\text{junction}} \mathbf{A} \cdot d\mathbf{l} \right) \tag{59}$$

If Eq. (49) for the current density is integrated over the cross-sectional area of the junction, the transport current as a function of magnetic flux linking the junction is obtained. Furthermore, by setting the phase difference across the junction (γ_1–γ_2) equal to $\pi/2$, the dependence of the junction critical current on magnetic flux is determined, i.e.,

$$I_c = I_{\max} \left| \frac{\sin \pi\phi_j/\phi_0}{\pi\phi_j/\phi_0} \right| \tag{60}$$

where $I_{\max}$ is the maximum current that can flow through the junction without developing a voltage across the junction, ϕ_j is the magnetic flux linking the junction, and ϕ_0 is the flux quantum $= 2.07 \times 10^{-7}$ G cm^2. This is the well-known diffraction effect for Josephson junctions, which is shown in Fig. 39.

The area of a typical tunnel junction is very small, e.g., 10^{-6} cm^2. A magnetic field of the order of 0.1 G is therefore required to produce one quantum of flux linking the junction; thus the junction is not very sensitive to magnetic fields.

The variation of I_c with respect to magnetic flux may be greatly increased by connecting two junctions in parallel, as shown in Fig. 40. The current is now given by $j = j_{\max_1} \sin \psi_1 + j_{\max_2} \sin \psi_2$. The gauge invariant phases, ψ_1 and ψ_2, across junctions 1 and 2, respectively, are coupled by the continuous superconducting paths connecting the junctions. This phase coupling results in the quantum interference effects (see Chapter 1, for two

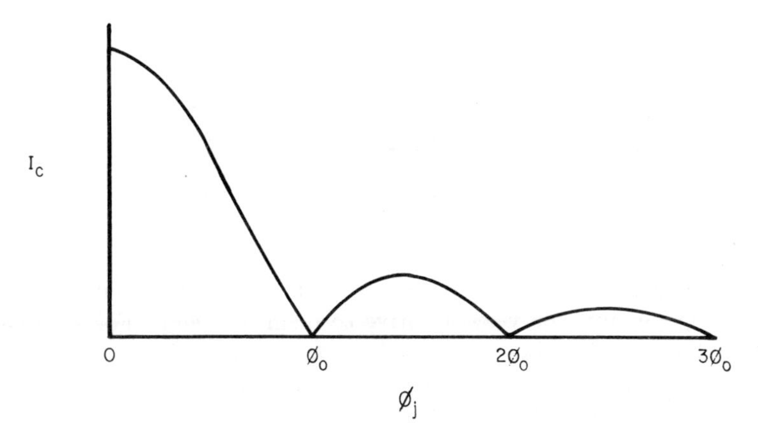

FIG. 39. Diffraction effect for Josephson junctions.

identical junctions connected in parallel,

$$I_c = 2I_{\max}\left|\frac{\sin \pi\phi_j/\phi_0}{\pi\phi_j/\phi_0}\right| \cos \pi \frac{\phi_A}{\phi_0} \tag{61}$$

where I_c is the critical transport current of the parallel combination of junctions, $I_{\max}$ is the maximum critical current of each junction, and ϕ_j is the magnetic flux linking the junctions. The quantity ϕ_A is the magnetic flux linking the area enclosed by the two junctions and the superconducting connections. This equation is similar to the equation for a two-slit optical interferometer. With point-contact junctions the area of each junction is orders of magnitude less than the area of the central area, which is usually from about 0.01 to 0.1 cm^2. Thus the variation in critical current with flux

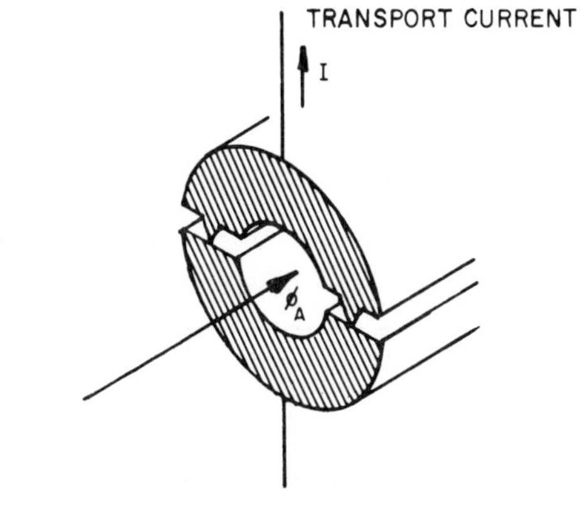

FIG. 40. Interferometer.

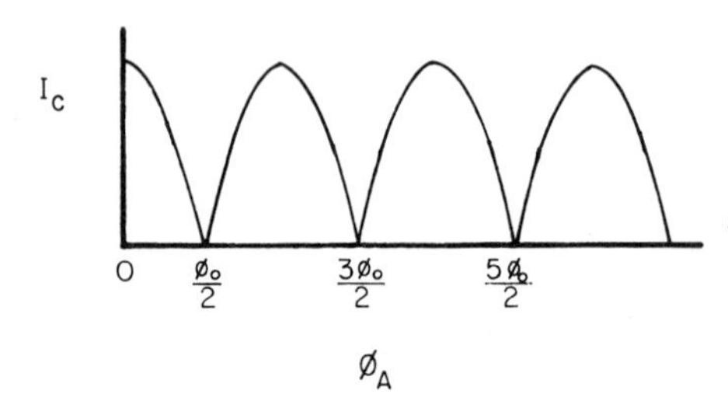

FIG. 41. Critical current versus flux for an interferometer.

is essentially given by $|\cos \pi\phi_A/\phi_0|$, as shown in Fig. 41. The double-junction device can, therefore, be very sensitive to magnetic fields. For a 0.01-cm² area, the flux quantum $\phi_0 = 2 \times 10^{-7}$ G cm² corresponds to a field change of 2×10^{-5} G. It is usually possible to resolve a flux quantum to at least 1 part in 1000, thus resolution of field changes of 10^{-8} G is possible. Resolution to greater than 10^{-9} G has often been attained with still no clear definition of intrinsic noise limits.

To construct an amplifier one needs only to connect the signal to a circuit arranged to produce a change in magnetic flux linking the junction device. This flux change will then change the critical current of the device. The operation is best described from the idealized V–I curve of Fig. 42. The following operation is valid for single-junction (diffraction dependence of I_c on ϕ) as well as the double-junction (interferometer) devices to be considered here.

Assume that the device is initially in a zero magnetic field. The critical current I_c is then $2I_{max}$. As a magnetic field is applied, the critical current

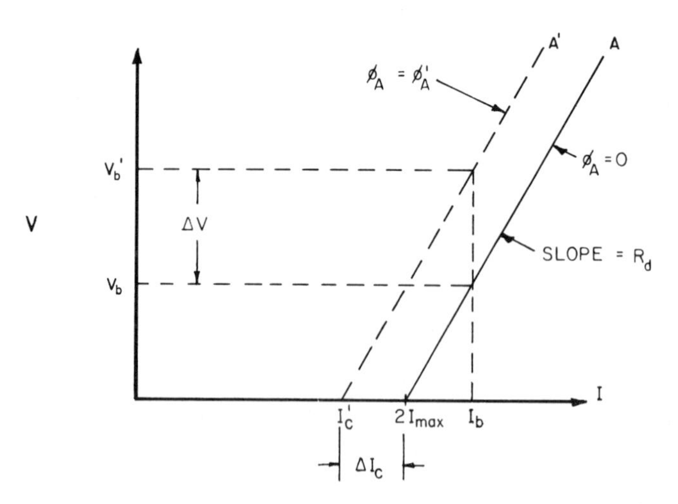

FIG. 42. Idealized V–I curve of an interferometer.

will decrease, finally reaching zero if full modulation depth is achieved. To transform the device to an amplifier, a bias current I_b larger than $2I_{max}$ is applied to obtain a finite voltage drop across the device. As a field is applied I_c decreases from $2I_{max}$ to I_c' and the V–I curve translates to the left. If the V–I curve is really linear, then for fixed bias current I_b the voltage drop across the device will increase as the critical current decreases according to

$$\Delta V = -R_d(2I_{max} - I_c') \tag{62}$$

where R_d, the differential resistance, is the slope of the V–I curve. The voltage V will reach its *maximum* value when the critical current is at a *minimum* and vice versa. We have seen that the critical current is proportional to $|\cos \pi\phi_A/\phi_0|$ for an interferometer. Thus, we have a simple functional form for the change in gate voltage ΔV with respect to the flux ϕ_A linking the junction device:

$$\Delta V = -R_d I_0(1 - |\cos \pi\phi_A/\phi_0|) \tag{63}$$

Now we need only to relate the flux linking the device, ϕ_A, to the external flux ϕ_x produced by the signal current.

For a two-junction interferometer, the magnetic flux that links the device, ϕ_A, differs from the external flux ϕ_x because of persistent circulating currents that are induced in the device. In general, $\phi_A = \phi_x + Li$, where L is the inductance of the current. A thorough discussion of the relationship between ϕ_A, ϕ_x, L, and I_c is given in Chapter 1 (see also Silver and Zimmerman, 1967).

The relationship is such that the critical current of the interferometer has the same periodicity with respect to the *external* flux as to the internal flux. The period is equal to one flux quantum ϕ_0; however, the modulation depth depends on the circuit inductance and zero-field critical current I_{max} of each junction. The modulation depth, i.e., the maximum change in critical current produced by the changing flux divided by the zero-field critical current, is a crucial device parameter since it largely determines the maximum slope of the I_c versus ϕ_x relationship and therefore the current gain of the device. Maximum modulation will be obtained (Silver and Zimmerman, 1967) when $I_{max} = \phi_0/2\pi L$.

The minimum value of the critical current I_{min} is set by thermal fluctuations (Dahm *et al.*, 1969) and is typically about 10^{-8} A for 4.2°K. Thus, the circuit inductance must be less than about 3×10^{-8} H for maximum modulation of the critical current. This requirement has restricted the cross-sectional area of the device to the order 0.1 cm^2 in the typical geometries that have been built.

The "terminal properties" of a tunnel junction and a double-junction

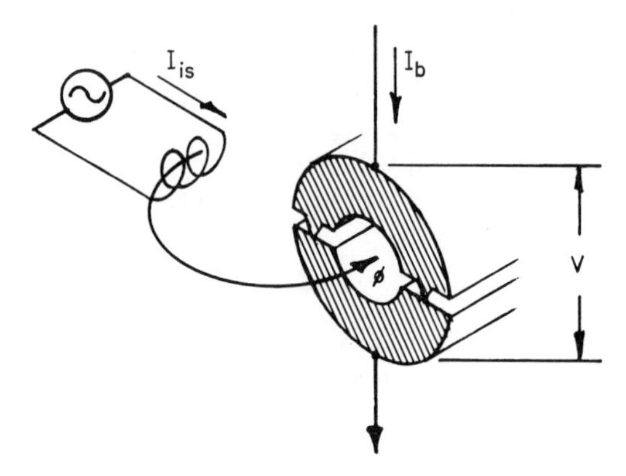

FIG. 43. Interferometer amplifier.

interferometer exhibit a variable resistance as a function of magnetic field. For that reason tunnel-junction devices are included under Section III.B. However, the field sensitivity is actually due to parametric up-conversion and down-conversion, or detection, all in one deceptively simple device, i.e., the physical process within the device is parametric amplification. The effective pump frequency is the Josephson frequency (see Section III.C) given by Eq. (57).

An interferometer amplifier can be constructed as shown in Fig. 43. A signal current I_{is}, produces a change in the magnetic flux linking the device, thereby causing a change in the voltage across the device. The required bias currents and corresponding voltage drop depends on the junction material, contact area, and operating temperature. The junction resistance of point contact devices is found to vary from about 10 to 10^{-3} Ω. Critical currents may vary from $\sim 10^{-1}$ to $\sim 10^{-6}$ A with the lower currents corresponding to the higher junction resistance. A representative bias level would be $\sim 10^{-4}$ A with a junction resistance of 0.1 Ω, giving a bias voltage of $\sim 10^{-5}$ V.

The double-junction interferometer has been used as a variable resistance amplifier with several different geometries, i.e., the Clarke SLUG, the double-point-contact interferometer, and the vacuum-deposited S–N–S junction interferometer. In each of these amplifiers the input or signal current produces a magnetic flux that links the interferometer device thereby altering the resistance of the device to transport currents.

SLUG amplifier. A simple amplifier circuit, developed by Clarke (1966a), is the superconducting low-inductance undulating galvanometer (SLUG). The SLUG is constructed by casting a bead of solder around a

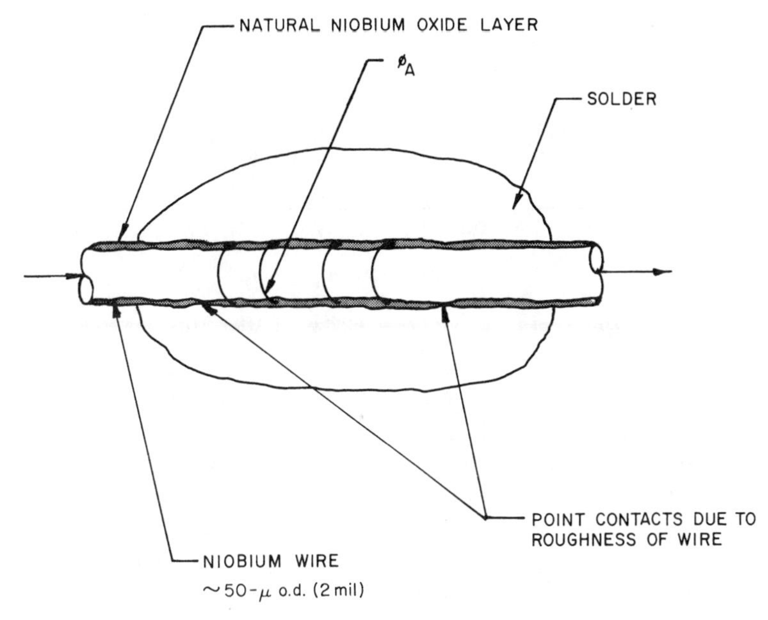

FIG. 44. SLUG junction.

niobium wire that is coated with a thin natural niobium oxide layer. It is believed that the rough surface of the wire presents many point-contact-like regions in which tunneling can occur between the wire and the solder. Figure 44 is a greatly magnified sketch of how the junction may look.

The V–I characteristic is obtained by passing a current I into the niobium center wire and out of the solder bead, as shown in Fig. 45. The voltage drop V occurs across the thin niobium oxide dielectric and also across any

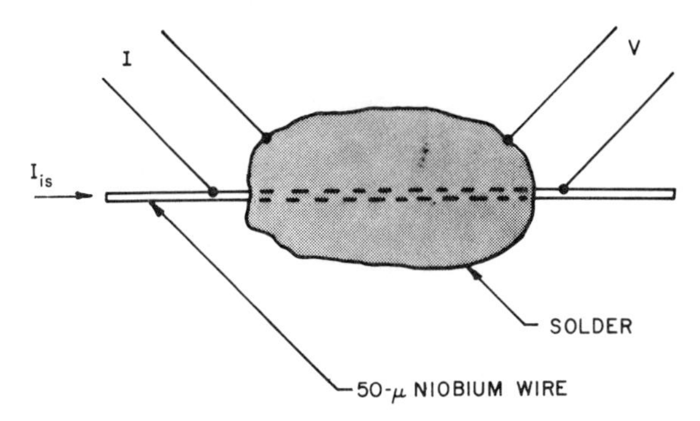

FIG. 45. SLUG circuit.

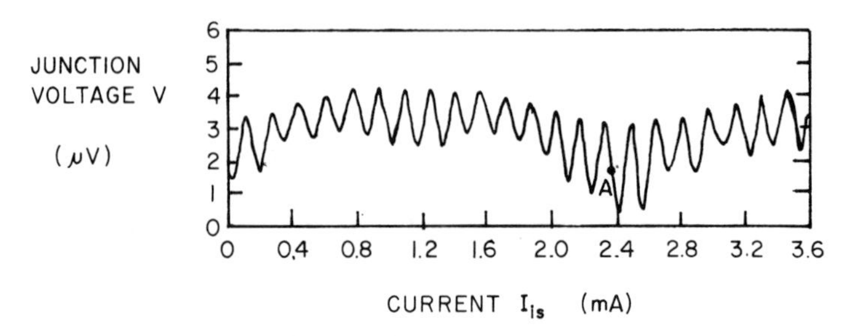

FIG. 46. Typical response of a SLUG; (T = 4.22°K). [After McWane *et al.*, 1966.]

point contacts that exist. A magnetic field is coupled to the junctions by passing a current I_{is} through the 50-μ niobium center wire. The field about the wire is, of course, circumferential, linking the area between any point contacts or other tunneling regions. In practice, the critical junction current I_c is usually found to vary smoothly with flux (i.e., with the current I_{is} through the center wire). Also, the junction voltage V for fixed bias current is usually a continuously varying function of the flux. Typical response of the Clarke SLUG is shown in Fig. 46 (McWane *et al.*, 1966).

If the SLUG is to be used as an amplifier, the control current would be biased to some fixed point, such as A, in Fig. 46, giving a steep slope for V–I_{is}. Then, a small signal I_{is}, in the niobium wire will change V. The response can be made more linear and the dynamic range can be increased by using a servo loop to maintain the operation at a fixed point (such as point A in Fig. 46).

In his first experiments Clarke (1966a) used the circuit shown in Fig. 47. The input inductance L was approximately 10^{-8} H and R_s was $5 \times 10^{-8}\ \Omega$. This gave an L/R_s time constant of 0.2 sec. The signal voltage V_s corresponding to a unity signal-to-noise ratio was 5×10^{-14} V. (The Johnson noise in R_s would be about 3×10^{-15} V rms.) The corresponding current I was 10^{-6} A. More recent experiments by Clarke *et al.* (1971) using improved r-f shielding have permitted measurements of current changes as small as 10^{-8} A.

The transimpedance R_t of a SLUG is defined by $R_t = \partial V/\partial I_{is}$. Typically R_t is about 0.1 to 1 Ω for a SLUG. The differential junction resistance

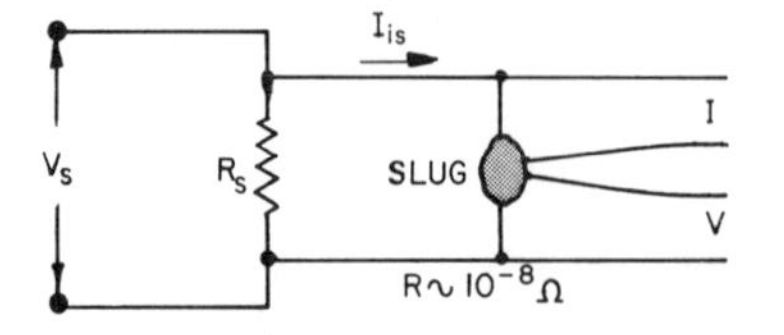

FIG. 47. SLUG amplifier.

$R_d = \partial V/\partial I$ is normally from 0.1 to 5 Ω. The current gain G is defined as $\partial I/\partial I_{is}$, and is given by

$$G = R_t/R_d \tag{64}$$

From the above numbers for R_t and R_d it can be seen that the current gain is roughly unity. These values make possible very large voltage gains. The voltage gain for the circuit of Fig. 47 is given by $(\partial V/\partial I_{is})(\partial I_{is}/\partial V_s)$, or

$$\text{voltage gain} = R_t/R_s \tag{65}$$

Note that this voltage gain is not unique to the SLUG, i.e., it depends upon the resistance of the source. If R_s is 10^{-8} Ω, then a voltage gain of 10^8 is obtained, and the time constant is 1 sec. An input-current resolution of 10^{-8} A would then correspond to an input voltage of 10^{-16} V and an output voltage of 10^{-8} V. However, the input-voltage resolution would be limited by the Johnson noise to 1.5×10^{-15} V. The output voltage can be measured with a low-noise af amplifier using an input transformer to match the amplifier to the $\sim$1-Ω junction resistance of the SLUG.

The intrinsic response time of a SLUG has not been determined, but it would probably respond at least up to microwave frequencies. Usually the response of a SLUG amplifier is limited by the L/R response time of the input circuit (see Section III.A) or of the readout electronics used to measure the SLUG voltage. If the servo feedback is applied to the input circuit, the response time of the input circuit can be grossly improved, as discussed in Section III.A.5.

Note that the input coupling factor of a SLUG cannot be altered very much since it is built into the device. The intrinsic input inductance of the center niobium wire is less than 10^{-12} H from one end of the ball of solder to the other. Usually much higher inductance is obtained (e.g., $\simeq 10^{-8}$ H) because of the leads going from the SLUG to the signal terminals.

A superconducting transformer can be used to obtain a better match of the source to the very low inductance of the SLUG; however, great care must be used in that case to maintain a very low inductance in the transformer secondary and the SLUG input circuit. The use of an input transformer is important for large source resistances to prevent amplifier current noise (I_a in Fig. 22) from limiting the current resolution. This is discussed in Section III.A and by Clarke *et al.* (1971).

Double-point-contact interferometer amplifier. Many researchers have used double-point-contact interferometers successfully as amplifiers (Silver and Zimmerman, 1967; Beasley *et al.*, 1969; Forgacs and Warnick, 1967). Microbridges, using the same circuitry, should also lend themselves to this application although this has not been reported to our knowledge.

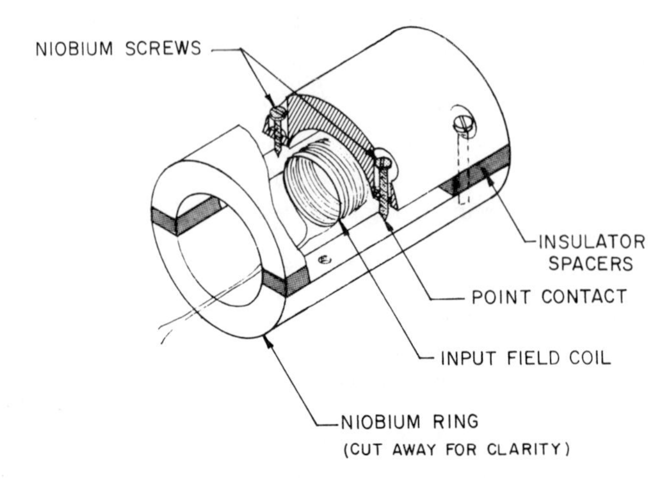

FIG. 48. Double-point-contact interferometer.

The basic circuit of the interferometer is shown in Fig. 48. The input signal is usually inductively coupled to the device with a close-fitting solenoid inserted in the bore of the device. Maximum coupling between the input-field-coil inductance and the interferometer-ring inductance requires that the ring gaps be as small as possible and that the points be short and fat.

Few data are available on the current gain, transimpedance, and input inductance of these interferometers. In the main, they have been used as magnetometers and most of the published data relate to the sensitivity of the device to applied magnetic field. Magnetic field resolution of $\simeq 5 \times 10^{-8}$ G $\mathrm{Hz}^{-1/2}$ has been reported (Forgacs and Warnick, 1967).

Typical dimensions of the device are a length of from 1 to 5 cm with a 1-mm- or 2-mm-hole diam. The point-contact screws are generally number 000–120. The loop inductance is then of the order of 10^{-10} H. The critical current of each junction should then be about $\phi_0/2\pi L$ or about 5 μA. The junction resistance, which determines the output impedance of the device, can vary widely but generally is around 1 Ω. Such a low impedance makes it difficult to match the device to a low-noise, room-temperature amplifier whose input impedance is typically at least 50 Ω. A mu-metal core transformer is frequently used to improve the matching. The output voltages for input signals equal to the noise level are typically between a few nanovolts and 1 μV. This output is high enough so that signal-to-noise ratio is not limited by the room-temperature amplifier or by the input transformer in a well-designed system.

A typical circuit diagram for a double-point-contact interferometer is

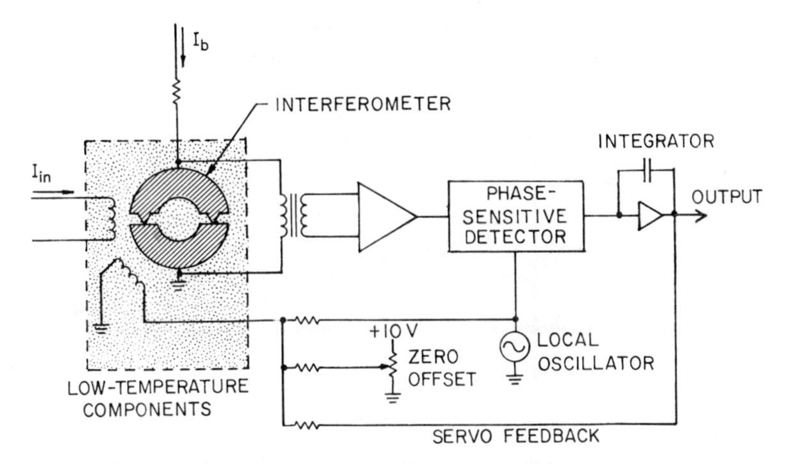

FIG. 49. Circuit diagram for a double-point-contact interferometer amplifier.

shown in Fig. 49. The two point contacts can be adjusted to obtain a wide range of junction critical currents. It is sometimes difficult to maintain constant performance as the point contacts are handled and temperature cycled. Some workers have used linkages to adjust the point contacts in the liquid-helium bath. It is unwise to consider placing these devices in complex cryogenic systems where they are not accessible for adjustment. In spite of their need for occasional adjustment, these devices can perform satisfactorily.

Superconductor–normal metal–superconductor amplifier. Thin-film Josephson junctions with dielectric barriers have not often been used as practical devices because of the difficulty in fabricating the very thin ($\simeq$10 Å) barriers. Much more reproducible junctions can be built using normal metal barriers because the barrier thickness may be as much as 10,000 Å and still exhibit supercurrent tunneling. This type of junction (S–N–S for superconductor–normal metal–superconductor) has very low resistance $\sim 10^{-6}$ Ω and critical currents of the order of 10^{-3} A, therefore the junction voltage is in the 10^{-9}-V range. The low voltage limits the practical use of S–N–S devices, although they are ideal for basic studies of tunneling effects. Most measurements of the S–N–S junctions have been made using the Clarke SLUG as a voltmeter to read the output, in fact, the SLUG was initially conceived for this purpose (Clarke, 1966a). The *I–V* curve of an S–N–S junction is given in Fig. 50. A negative resistance region is not obtained as in a thin-film Josephson junction because the junction capacitance is low due to the thick normal metal barrier, and because of the low shunt resistance of the normal metal (McCumber, 1968).

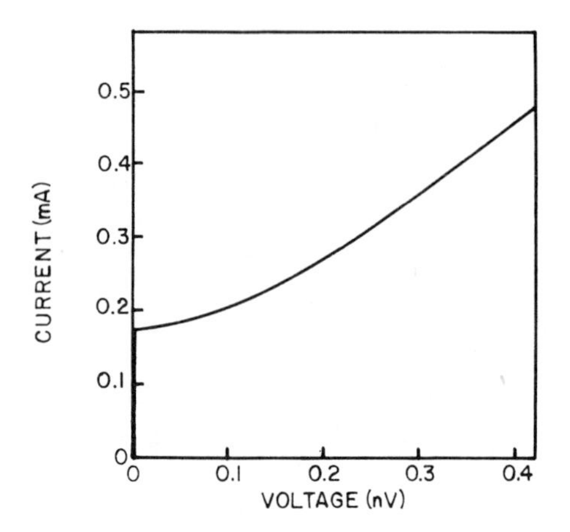

FIG. 50. I–V curve for S–N–S junction; (T = 2.94°K). [After Clarke, 1971.]

Clarke and Paterson (1971) constructed Pb, Cu–Al, Pb interferometers as shown in Fig. 51 to study asymmetrical current inputs. A magnetic field was applied to the interferometer by means of the current I_{in} in the top Pb film. Since the top film was about one-tenth the width of the bottom film, the current I_{in} was tightly coupled to the interferometer inductance. The junction areas were 2.5×10^{-4} cm² and the junction critical currents I_{max} at 4.2°K were about 1 mA. The junction resistance was larger than typical, about 1.5×10^{-5} Ω, and the interferometer inductance was about 2×10^{-11} H. The output voltage V_o was monitored with a Clarke SLUG.

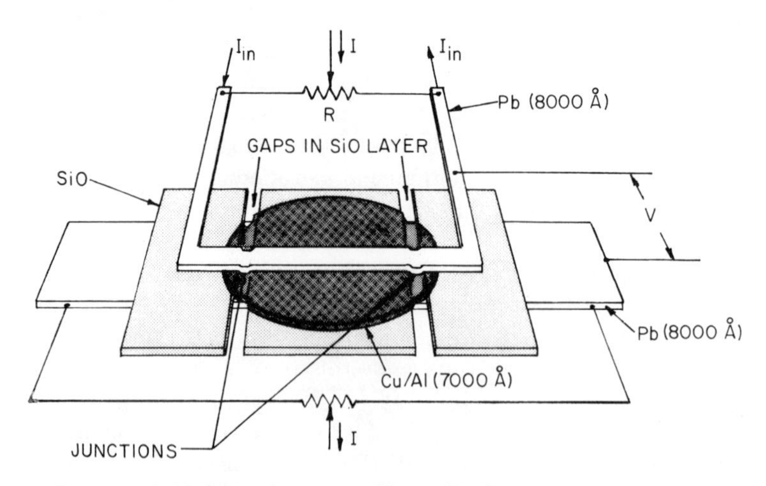

FIG. 51. S–N–S interferometer. [After Clarke and Paterson, 1971].

The current period for I_{in}, corresponding to a one quantum flux change, was roughly 100 μA.

The input-current noise at $T = 1.5°K$ for $2LI_{max}/\phi_0 = 60$, $I = 4$ mA, and $V = 10^{-9}$ V, was determined to be 10^{-8} A in a 1-Hz bandwidth. This is in good agreement with the current noise of $(2k_BT/R)^{1/2}$ derived by Dahm *et al.* (1969) for $eV \ll k_BT$. The value of R, the static resistance of the junction, was 5×10^{-7} Ω per junction. This current noise is no better than that possible using only a SLUG. The problems of high current noise and low output voltage of an S–N–S interferometer can be eliminated by using very small junction areas, but then the reproducibility is reduced because of the small junction dimensions necessary to obtain high resistance.

Clarke and Paterson (1971) state that a junction resistance of 1 Ω would reduce the input current noise to 10^{-11} A Hz$^{-1/2}$ and result in an output voltage of about 10^{-9} V for an input of 10^{-11} A. This voltage is detectable with a room-temperature amplifier using a cooled transformer, thereby eliminating the SLUG readout. This would require junction dimensions of about 2.5 μ. Such a device might compete favorably with a thin-film, rf-driven, tunneling amplifier (Section III.C.4), which typically requires a junction width of less than 1 μ.

The asymmetric feature of Clarke and Paterson's S–N–S interferometer is significant. By inserting the transport current I off center and removing it off center using the resistance dividers shown in Fig. 51, they obtain a positive feedback effect and consequent asymmetry in the critical transport current I_c versus I_{in} curves. One side of the I_c–I_{in} curve becomes flatter and the other side steeper. By biasing the input current I_{in} to the steep side a large increase in current gain, $\Delta I_c/\Delta I_{in}$, can be achieved. This will also result in a larger transimpedance $\partial V/\partial I_{in}$. The maximum current gain is obtained when I is inserted and withdrawn entirely on one side (100% asymmetry) and when $2LI_{max}/\phi_0$ is very large. This same asymmetric technique can be applied to other double-junction interferometers, although the geometry in some cases may not lend itself to 100% asymmetry. The transport current leads must be attached to a point on the interferometer loop very close to the junction, i.e., all the inductance of the loop should be on one side of the attachment points.

The output voltage of an interferometer is an amplified reproduction of the input current. Thus the output of an interferometer can be directly connected to the input of another similar device, i.e., one that can be staged without first bringing leads out of the cryostat. This is similar to a cryotron amplifier. In contrast, the rf-driven tunneling amplifier (Section III.C.4) is basically an upconverter. Its output is at a higher frequency than the input and must be amplified and detected if the amplified input signal is to be recovered for multiple-stage amplifiers.

2. *Negative resistance amplifiers*

Any negative resistance device can, in principle, be used to obtain power amplification. Historically, relatively little use has been made of negative resistance amplifiers because the vacuum tube and transistor amplifiers have generally proven superior. The primary disadvantage of negative resistance amplifiers is gain instability and a tendency to oscillate when out of adjustment. When Esaki tunnel diodes became available in the late 1950s, much interest developed in negative resistance amplifiers, especially for microwave frequencies, because of the fast time response of the diodes.

There are two superconducting devices that exhibit negative resistance in their current–voltage (I–V) characteristic: the quasiparticle tunneling junction (Giaever tunneling) and the supercurrent tunneling junction (Josephson tunneling). No experimental information on superconducting negative resistance tunneling amplifiers has been published, to our knowledge.

We will briefly consider the principles of the low-frequency operation of a negative resistance amplifier and then examine the conditions required to use a Giaever or Josephson junction as an amplifier. The physics of these junctions are discussed briefly in Section II.A.2.

A schematic I–V curve for a negative resistance device is shown in Fig. 52. For the device to operate as an amplifier it is necessary that the bias supply load line be steeper ($R_b < R$) than the slope, $-1/R$, of the

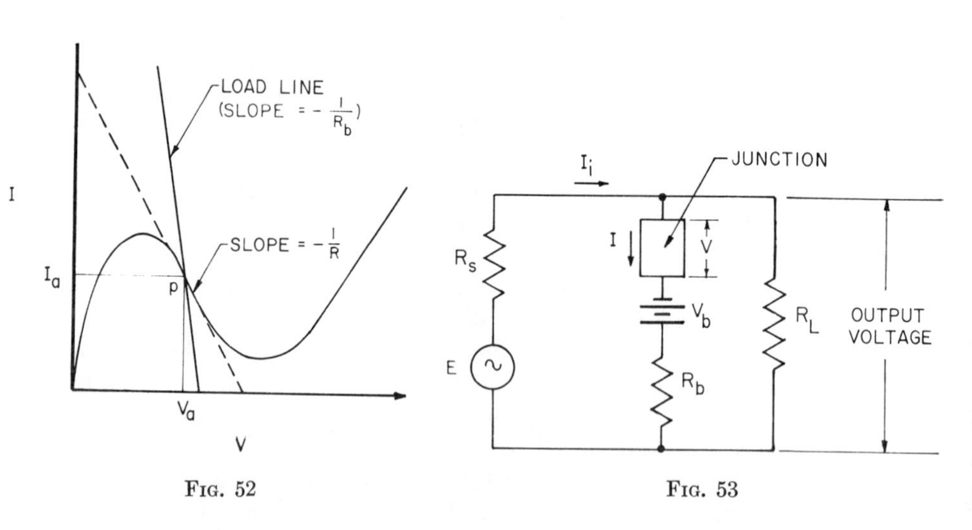

FIG. 52   FIG. 53

FIG. 52. I–V curve with negative resistance region.
FIG. 53. Negative resistance amplifier circuit diagram.

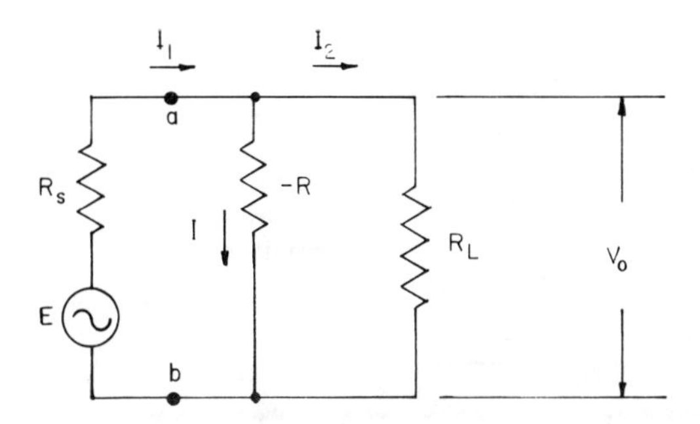

FIG. 54. Negative resistance amplifier equivalent circuit.

negative resistance region, and that the bias voltage V_b be adjusted to give operation near the inflection point, point p in Fig. 52, for zero input signal. One arrangement for biasing the device is shown in Fig. 53 in which R_s is the source resistance and R_L the load resistance. The steep load line in Fig. 52 corresponds to a very small resistance R_b in the bias voltage supply, i.e., R_b must be less than the absolute value R of the slope of the negative resistance region. The bias voltage required to establish the operating point p is given by

$$V_b = V_a + I_a\{R_b + [R_sR_L/(R_s + R_L)]\} \tag{66}$$

The operation of this circuit as an amplifier can now be given by assuming that the negative resistance device is properly biased to point p and that its resistance is $-R$, where R is a positive number. This assumes that point p in Fig. 52 is the origin of the signal currents and voltages. The resulting equivalent circuit is shown in Fig. 54. For clarity we have assumed that the load line in Fig. 52 is vertical, i.e., R_b is zero. The effect of R_b can be added by simply replacing every R term by $(R - R_b)$ in the equations given below.

The signal current I_1 produced by the input voltage E is given by

$$I_1 = E/(R_s + R_p) \tag{67}$$

where

$$R_p = -RR_L/(R_L - R) \tag{68}$$

The output voltage V_o is given by

$$V_o = I_1R_p \tag{69}$$

The input power P_i, is then

$$P_i = V_oI_1 = [E/(R_s + R_p)]^2R_p \tag{70}$$

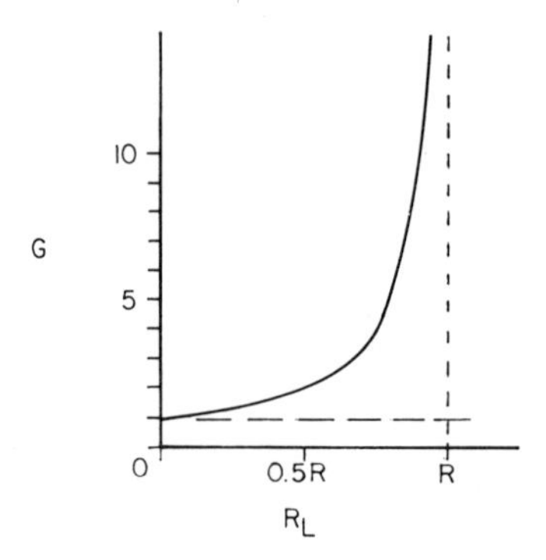

FIG. 55. Power gain as a function of load resistance for a negative resistance amplifier.

The signal output power P_o is given by

$$P_o = V_o^2/R_L = (R_p^2/R_L)[E/(R_s + R_p)]^2 \tag{71}$$

The power gain is P_o/P_i, or

$$G = R/(R - R_L) \tag{72}$$

which is plotted in Fig. 55. The current gain I_2/I_1 is equal to the power gain G. (The voltage gain is unity.) Large power gain is obtained for R_L just slightly less than R; for example, if $R_L \simeq 0.9R$, a gain of 10 is obtained. Unfortunately, for R_L close to R, the gain is very sensitive to load resistance and to R. If the value of R changes slightly because of a bias voltage drift or a temperature change, the gain will vary. It can even go to infinity at which point instability will develop. This instability is the first major problem of a negative resistance amplifier.

The maximum output power for a given source resistance R_s and a given gain G is obtained when the absolute value of the slope of the negative resistance region is

$$R = R_s/(G - 1) \tag{73}$$

Thus, for maximum output power a gain of 10^2 requires $R_s = 99R$. This rather large R_s is a problem in cryogenic amplifiers if their low-noise properties are to be used to advantage because R_s must be low to keep Johnson voltage noise low, and Eq. (73) requires that R_s be significantly greater than R.

The amplifier input resistance looking into the terminal ab (Fig. 54) is given by

$$R_i = RR_L/(R - R_L) = R_L G \tag{74}$$

Note that R_i is positive. Maximum output power is obtained from Eqs. (72)–(74) for R_i equal to R_s.

We will use these results to consider a negative resistance amplifier using

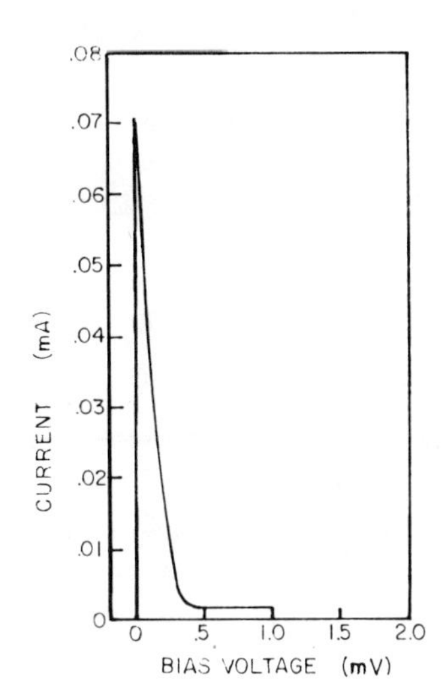

FIG. 56. Negative resistance region of a lead, lead oxide, lead Josephson junction. [After Eck *et al.*, 1965, courtesy of Plenum Publishing Corp.]

a Josephson junction. Since point contacts and weak links (Dayem bridges) do not generally have negative resistance regions, it is necessary to use thin-film dielectric barrier junctions (see Section III.B.1.b). The Josephson junction has the advantage over quasiparticle tunneling junctions of a lower bias voltage V_a (compare the I–V curves of Figs. 56 and 8) and a larger peak-to-valley ratio which will result in larger dynamic range. However, the quasiparticle junction has the important practical advantage of a somewhat larger insulation-film thickness, about 50 Å, separating the two superconducting films, compared with the insulation thickness of only about 10 Å for a Josephson junction. This means that reproducibility and construction problems should be better for quasiparticle junctions.

Eck *et al.* (1965) give the experimental negative resistance region of the lead, lead oxide, lead Josephson junction reproduced in Fig. 56. The negative resistance region of this I–V curve has an R of about 2.5 Ω at $V = V_a \simeq 0.1$ mV and $I = I_a \simeq 0.04$ mA. Assume a gain of 10^2 is desired. This requires $R_L = 2.48$ Ω. Optimum coupling to a source resistance would be obtained for the value of R_s given by Eq. (72), i.e., $R_s = 246$ Ω. This would result in a Johnson-noise voltage of 4×10^{-10} V for $T = 4°$K and a bandwidth of 1 Hz. The input resistance R_i would also be 248 Ω. The required bias voltage, according to Eq. (66), is about 10 mV assuming that R_b of Fig. 53 is much less than 2.5 Ω.

Quasiparticle and Josephson junctions have intrinsic switching times in the subnanosecond region. Thus, operation should be possible in the microwave region. However, almost no information has been published to date on these types of amplifiers, and therefore actual performance data are not available.

It appears that useful negative resistance amplifiers could be made using tunneling junctions; however, it is too early to predict how they would compete in performance with other superconducting amplifiers.

C. Inductance Modulation Amplifiers (Parametric)

The amplifiers discussed in Section III.B obtain gain by having a small signal current control the real or resistive part of an impedance and thereby also control a load current. The amplifiers discussed in this section achieve gain by modulation of the inductive part of an impedance, using properties of the superconductive state. The inductance can be either self or mutual, and it can be made to vary by applying either the signal current (i.e., a nonlinear inductance) or by an ac drive or pump current. These circuits are compared in Fig. 57. Generally, the output is tuned for impedance matching to the output amplifier. The ac power delivered to the output comes from the pump source. The input signal merely controls the flow of the ac power.

Superconductivity provides ideal properties for building a modulated inductance circuit. Johnson noise is greatly reduced (because of the low temperature and zero resistance), very high-Q coils and transformers can be obtained, and several unique properties of the superconducting state

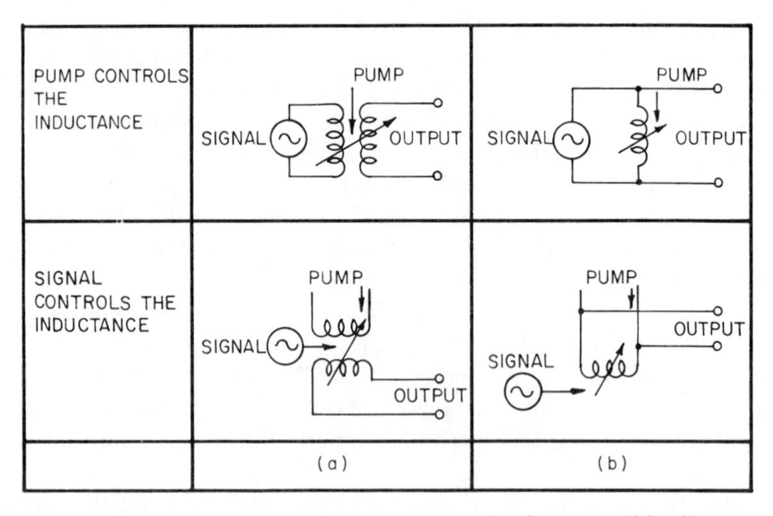

Fig. 57. Inductance modulation circuits: (a) mutual inductance; (b) self-inductance.

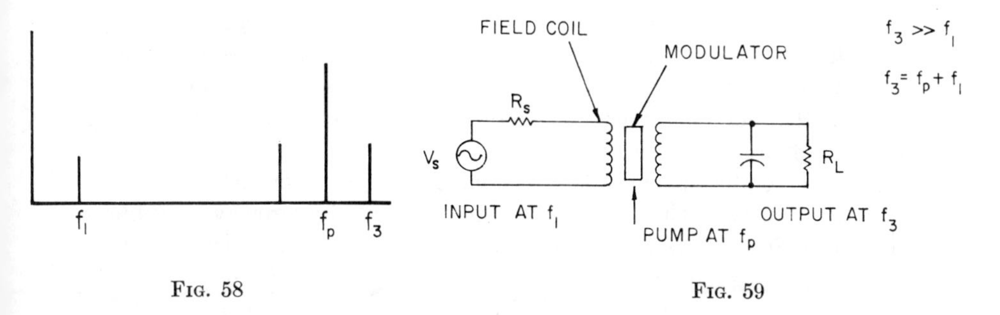

FIG. 58 FIG. 59

FIG. 58. Frequency spectrum of an up-converter.
FIG. 59. Up-converter circuit.

permit construction of the required variable inductance circuit elements. The intrinsic noise levels in superconductors are not yet fully understood, although preliminary experiments indicate that these noise levels are very low. (In fact, a good superconducting amplifier would be useful for measuring noise in superconductors themselves.) A time-varying inductance can be obtained by using one of at least five properties of the superconductive state; the Meissner effect, quantized flux, flux flow, superinductance, and the London penetration depth.

Amplifiers using reactance modulation are generally referred to as parametric amplifiers. Only the techniques of obtaining a variable inductance using the properties of superconductivity will be discussed here. A detailed discussion of parametric amplification is given by Blackwell and Kotzebue (1961).

There are two basic parametric circuits that are used for low-noise signal amplification: the negative resistance parametric amplifier, and the up-converter (Blackwell and Kotzebue, 1961). In the negative resistance amplifier the signal input and output frequencies are the same, although another frequency, called the idler, is generated internally. The power gain can be made arbitrarily large; in fact, oscillation can occur. Typically, the circuit is operated close to the point of oscillation to achieve significant gain. This amplifier thus has the disadvantage of possible instability or oscillation. Another disadvantage is the necessity for selecting only a sideband (upper or lower) generated by the mixing of the signal and pump frequencies, i.e., the pump frequency must be rejected.

The up-converter has an output signal frequency f_3 which is higher than the input signal frequency f_1. The pump frequency f_p is equal to $f_3 - f_1$, as shown in Fig. 58. The theoretical maximum power gain is f_3/f_1. This gain can be approached if $(f_1/f_3)(\alpha Q)^2 \gg 1$, where Q is the Q of the variable reactance and α is the fractional modulation of the reactance.

The disadvantage of the up-converter is that the output frequency must be considerably greater than the input frequency if large gain is to be obtained. If f_1 is dc or a very low frequency, the power gain is limited by αQ rather than by frequency.

The superconducting parametric circuits discussed here are of the up-converter type, with the exception of the distributed microwave amplifiers discussed in Section III.C.5. In addition, most of the circuits were designed and built as magnetometers rather than as amplifiers; however, one only need add an input coil to a magnetometer to make it into an amplifier. The input signal current then produces a magnetic field on the magnetometer sensor, and this field is sensed and amplified by the magnetometer. The circuit diagram of a simple up-converter based on modulated mutual inductance is shown in Fig. 59.

1. *Mechanical motion amplifiers*

a. Buchhold circuit. The simplest modulated-inductance circuit, in concept at least, is the segmented disk amplifier of Buchhold (1963) (Fig. 60). The inductance is modulated by changing the coupling between the signal coil and output coil; i.e., the segmented disk rotates between two planar coils, thereby varying the mutual inductance at a rate proportional to the number of segments and the rotation rate of the disk. The amplitude of the fractional inductance modulation is a function of the coil size and the separation between the disk and the coil. The gain is then a function of the circuit dimensions, the modulation amplitude, the frequency of modulation (ω), the signal frequency, and the circuit Q. No performance data are available for this circuit.

b. Vibrating coils. The mutual inductance between two coils can also be modulated by changing the position of one of the coils relative to the

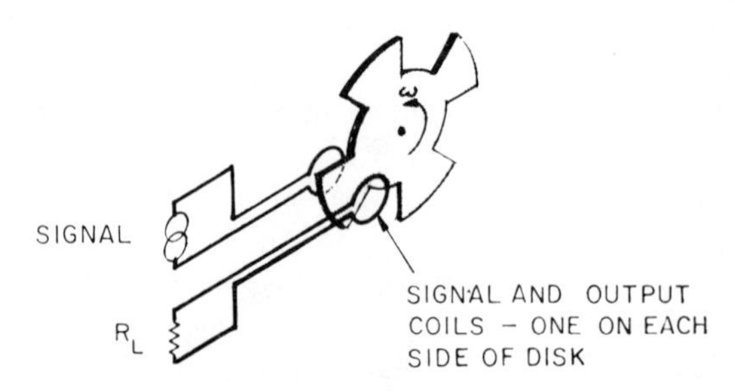

FIG. 60. Buchhold modulated inductance circuit.

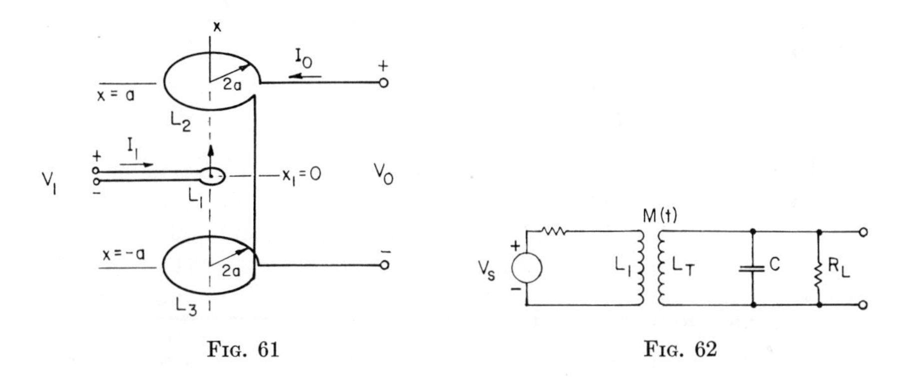

FIG. 61

FIG. 62

FIG. 61. Ries–Satterthwaite parametric modulator.
FIG. 62. Schematic of Ries–Satterthwaite parametric amplifier.

other. The Ries–Satterthwaite (1967) parametric amplifier shown in Fig. 61 is based on this principle. This is also the same principle employed in the vibrating sample magnetometer (Foner, 1959).

Output coils L_2 and L_3 (Fig. 61) are wound in opposite directions to cancel signals due to external field variations. The signal coil L_1 is mounted coaxially with and centered between the output coils. The signal coil is vibrated along the axis of the coils by connecting the coil to a loudspeaker, or tuning fork, driven by a piezoelectric bimorph. The tuning-fork resonant frequency is 1200 Hz and the Q is greater than 10^4. Amplitudes of about 0.25 mm can be obtained with this system with a drive power of less than 1 mW. The equivalent circuit of this amplifier is shown in Fig. 62.

The signal coil L_1 is vibrated relative to the Helmholtz output coils L_2 and L_3 (represented by L_T) thereby varying the mutual inductance $M(t)$. The signal voltage, resistance, and current are V_s, R_s, and I_{is}, respectively. If the signal coil is vibrated with simple harmonic motion of maximum amplitude α and frequency ω along the axis between the output coils, the mutual inductance will be given by

$$M(t) = \left.\frac{\partial M}{\partial x}\right|_{x=0} \sin \omega t \tag{75}$$

The most sensitive circuit reported by Ries–Satterthwaite (1967) has an input inductance L_1 of 5×10^{-7} H. With a signal resistance of $4 \times 10^{-7}\ \Omega$, the noise level is about 4×10^{-14} V for a 1.3-sec time constant (about 10 times the Johnson-noise voltage). The response of the amplifier is linear, and it is not subject to damage or performance change under severe overloading; i.e., as large as 10^{10} times the minimum detectable signal.

The major disadvantage of this amplifier, other than its large size, is the low operating frequency, which reduces the maximum parametric gain and the frequency response. Also, problems are encountered with spurious vibrations which cause noise and zero offset.

c. Vibrating ground plane. A technique for modulating the inductance of a coil has been developed at Stanford University (Opfer, 1970). The basic circuit is shown in Fig. 63. The superconducting ground plan is a film of tin or niobium evaporated onto the surface of a quartz crystal, which is driven electrically at frequencies from 0.1 to 1.0 MHz. A thin-film inductor of niobium is sputter deposited on a separate substrate in 500 strips 10 μ wide spaced about 10 μ apart. The area covered by the inductor is 1×1 cm and the total strip length is about 5 m. The separation between the ground

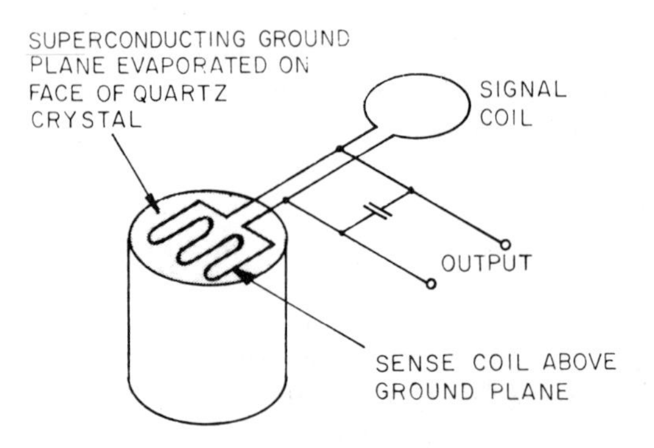

FIG. 63. Vibrating ground–plane modulator.

plane and inductor is about 1 μ. Ground-plane amplitudes of 0.1 μ are obtained giving 10% change in the inductance. The inductance is 10^{-6} H without the ground plane. With the ground plane about 1 μ from the inductor, the geometrical inductance is reduced to 6×10^{-7} H, which is close to the value of the kinetic inductance of the niobium film at 4.2°K.

Significant features of the circuit are: (1) None of the superconducting components is switched out of the completely superconducting state; thus noise due to eddy current damping, flux flow, and temperature fluctuations should be eliminated. (2) The response is linear with signal current. (3) The response should be free from zero drift.

Although, at present, a complete modulator circuit has not been assembled, the fabrication techniques, using photoetch masking and sputtering, have been perfected, and numerous circuit components have been tested. No performance data have been published.

2. *Meissner-effect amplifiers*

A circuit similar to the mechanical motion circuits was reported by Deaver and Goree (1967) in which the Meissner–Oschenfeld effect is used to provide the time-varying inductance. In this circuit, a solid superconducting post is inserted in the bore of two coaxial coils, as shown in Fig. 64. A resistive heater element is thermally bonded to one end of the post, and the other end is attached to a copper post that is fixed at the helium-bath temperature. The circuit is generally run in a sealed enclosure to prevent helium from being in direct contact with the post since helium boiling (bubbles) produces very large noise. This is similar to the thermal noise observed with cryotrons (Johnson and Chirlian, 1966).

The modulators can be switched between the superconductive and normal states at frequencies as high as 10 kHz, although the percent modulation decreases rapidly at frequencies above a few hundred hertz from a maximum of about 60 to less than 5%. Initially, it was felt that this decrease in modulation with increasing frequency was caused by the thermal time constant of the modulator. A new, hollow modulator (Fig. 65) was designed to minimize the thermal path.

The theoretical thermal response for the hollow Meissner modulator, when the device is immersed directly in liquid helium, exceeds 100 kHz. The modulation is still a strong function of frequency, falling rapidly at frequencies above 500 Hz. Signal-to-noise ratio and magnetic field resolution are relatively independent of frequency up to about 20 kHz. Furthermore, the noise was found to be a function of the magnetic field transverse to the modulator axis. The source of the noise is probably the eddy-current damping of the motion of the transverse flux during thermal cycles of the modulator.

The slow switching speed and large noise observed with Meissner

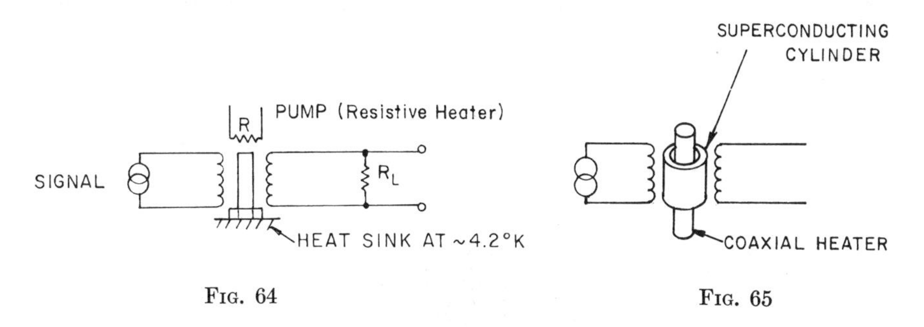

FIG. 64. Meissner-effect modulator.
FIG. 65. Hollow Meissner modulator (1-cm long, 1-mm o.d., 0.5-mm i.d.).

modulators should be dramatically improved by constructing the modulator in a manner to minimize pinning of the magnetic flux and the eddy-current restraint of flux motion. Multiple layers of superconductor, dielectric, heater, etc., should give sufficient volume of material for large signals yet minimize the speed and noise restrictions. An advantage of the Meissner-type modulator is that the response is linear with the absolute value of the applied field (signal).

3. *Quantized-flux amplifiers*

A modulator that reduces the eddy currents and, therefore, allows faster switching speeds utilizes the quantized-magnetic-flux property of superconductors to vary the inductance of a coil. A circuit based on this property is shown in Fig. 66 (Deaver and Goree, 1967).

London (1950) predicted that the magnetic fluxoid is quantized in a multiply connected superconductor. The fluxoid is given by

$$\oint (2m\mathbf{v} + (2e/c)\mathbf{A}) \cdot \mathbf{d}l = nh \tag{76}$$

where m, e, and $\mathbf{v}$ are the electron mass, charge, and velocity, respectively; $\mathbf{A}$ is the magnetic vector potential; c is the velocity of light; h is Planck's constant; and n is an integer.

If we consider a cylinder whose wall thickness is many penetration depths, then an integration path may be chosen in which the electron velocity $\mathbf{v}$ is zero around the complete path of integration. We then have the flux quantization condition

$$\oint \mathbf{A} \cdot \mathbf{d}l = nhc/2e \tag{77}$$

where $\oint \mathbf{A} \cdot \mathbf{d}l$ is the total magnetic flux linking the path of integration and $hc/2e$ is the flux quantum, $\phi_0 = 2 \times 10^{-7}$ G cm^2. This quantization condi-

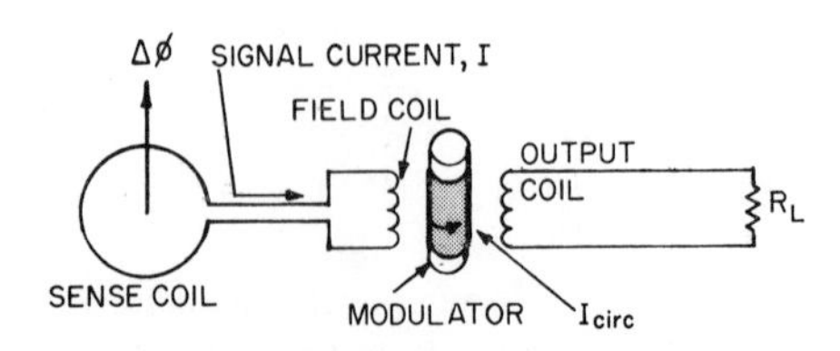

FIG. 66. Quantized-flux modulator.

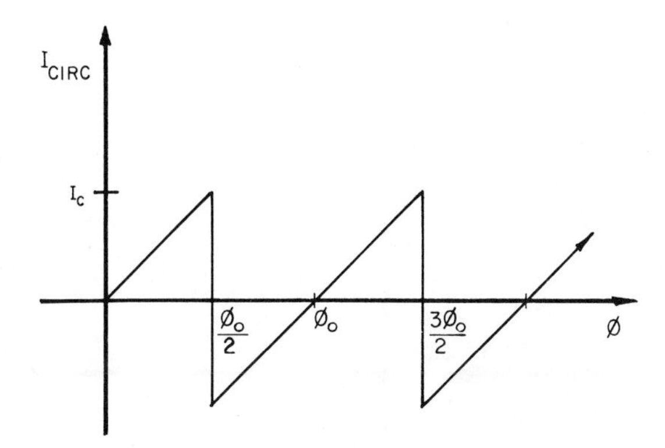

FIG. 67. Circulating current versus ambient flux for a thin cylinder.

tion means that, regardless of the ambient field, the total flux linking a superconducting cylinder will be integral multiples of the flux quantum ϕ_0.

In operation, a thin cylinder is placed inside field and output coils, as shown in Fig. 66. Assume that all the flux linking the cylinder is produced by current in the signal coil. If the cylinder is cooled below T_c in a field producing less magnetic flux than $\phi_0/2$ linking the cylinder, then persistent currents are induced in the cylinder to reduce the flux to zero, the lowest available energy state. A plot of the circulating current in the cylinder versus the flux in which the cylinder is cooled below T_c is shown in Fig. 67. As a variable inductance element, the modulator is switched between the superconductive and normal states, thereby changing the flux linking the output coil. The output, for a linearly increasing flux will be that shown in Fig. 68. The response shown in Fig. 68 is identical to the circulating current versus applied flux shown in Fig. 67. If the cylinder walls are thicker than a few London penetration depths, the output voltage will also contain a linear Meissner-effect contribution and the response will be the sum of the Meissner-effect and quantized-flux signals, as shown in Fig. 68. The output versus signal current is periodic in the flux quantum.

Meissner-effect and quantized-flux modulators have been used in magnetometer circuits in which the field coil is connected to a superconducting sense coil, as shown in Fig. 25. A change in magnetic field in the sense coil induces a current I in the circuit which is detected by means of the field it produces on the modulator through the field coil. The sense-field coil assembly is generally used so that the modulator can be isolated from ambient field changes transverse to the sensitive axis. Furthermore, the sense-field coil circuit is a dc transformer giving amplification of the field at the modulator as discussed in Section III.A.9.

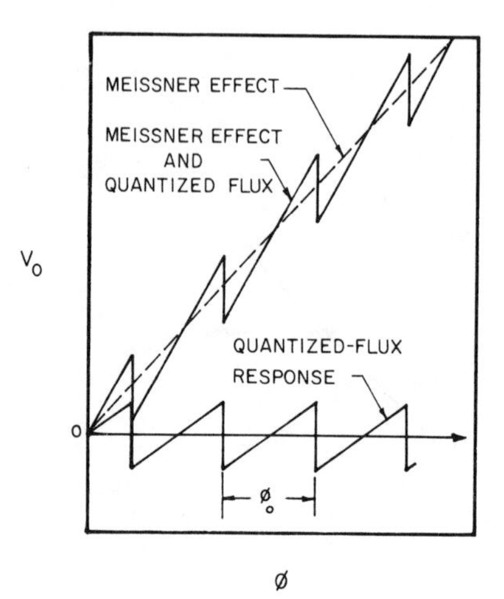

FIG. 68. Response of Meissner-effect and quantized-flux modulators.

4. *RF-driven tunneling amplifiers* (*loop with a single junction*)

This type of device consists of a closed superconducting loop containing a single weak-link junction, as shown in Fig. 69. Detailed discussions of the device are given by Silver and Zimmerman (1967), Zimmerman *et al.* (1970), and Giffard *et al.* (1972), and also in Chapter 1.

In operation this loop is driven by an rf magnetic field. This rf field induces a circulating current in the superconducting loop and the entire current must pass through the junction. For low-rf-drive levels no net magnetic flux is allowed to pass through the loop because the cylinder is essentially a closed superconducting ring. As the rf-drive level is increased, the critical current I_c of the junction is reached, and one quantum of flux ϕ_0 enters the loop via the junction. At a later time in the rf cycle, one ϕ_0 leaves the loop. This is illustrated in Fig. 70 in which the supercurrent I

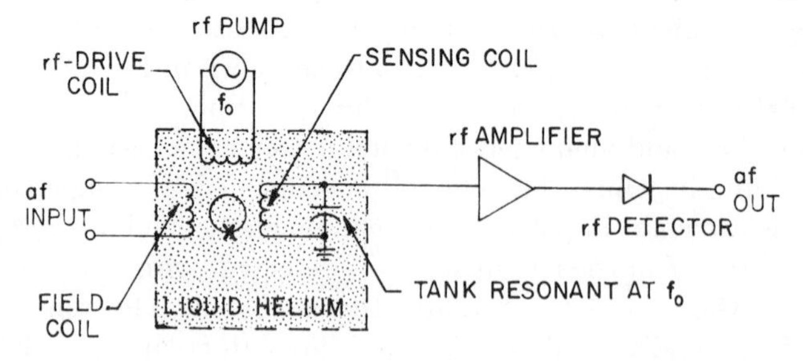

FIG. 69. Parametric amplifier with a superconducting loop containing a single junction.

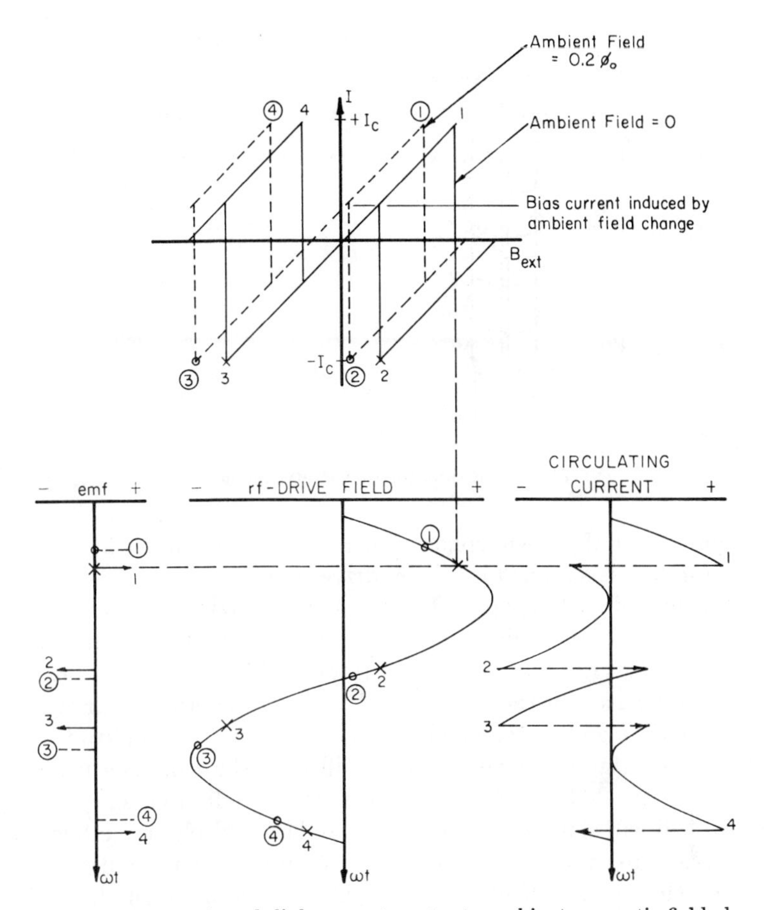

FIG. 70. Response of weak-link magnetometer to ambient magnetic field changes.

induced in the loop is shown for the arbitrary case, $I_c = 0.8\phi_0/L$, where L is the geometrical inductance of the loop. Each time a fluxoid enters or leaves the loop a step change is made in the circulating current I equal to ϕ_0/L and an emf spike is induced in the sense coil. Actually this current change will not be perfectly abrupt as shown in Fig. 70 because of the finite contribution of the kinetic inductance (or superinductance) term in Eq. (81). Exact calculations have not been made, but rough approximations indicate that the kinetic inductance term has a small contribution in most geometries that have been used. This kinetic inductance is discussed further in Section III.C.7 and in Section III.B.1.b. Note that the emf consists of a pair of spikes of alternate polarity. This emf induces a current in the sense coil tank circuit whose amplitude is related to the fundamental

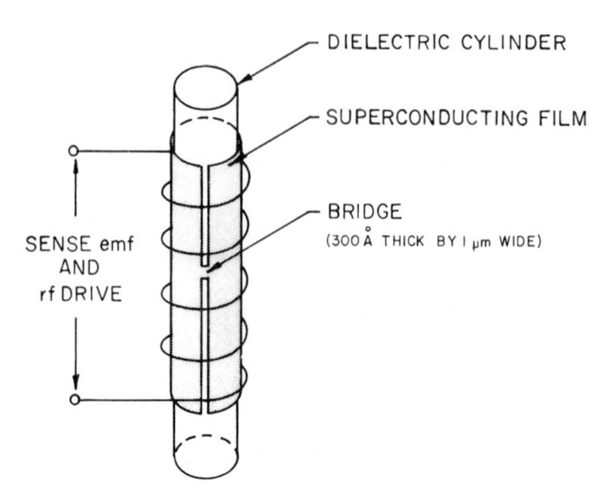

FIG. 71. Weak-link magnetometer element.

Fourier component of the emf spikes. The power thus available at the input of the rf amplifier is supplied by the rf drive source.

Now if a small dc field is applied to the loop, a persistent current will be induced in the loop and will be added to or subtracted from the current wave form of Fig. 70. This will cause the critical current to be reached earlier on one polarity of the rf cycle and later on the other polarity. Consequently the timing of the emf spikes will change and the amplitude of the Fourier fundamental will decrease. This will then show up as a change in the dc level of the output of the rf detector. An audio-frequency current in the field coil will similarly amplitude modulate and phase modulate the rf tank current and thereby be amplified and appear at the output of the rf detector. Hence, the effective mutual inductance between the rf-drive coil and the sensing coil has been made to change by the af input signal. Actually, this is a very unusual type of inductance because of the quantized nature of the flux change. The device is reactive since the emf is proportional to dI/dt rather than to I. However, strictly speaking, the af field does not change the mutual-inductance M in the relation emf $= M\, dI/dt$, but rather changes the timing of the dI/dt spikes. From a terminal viewpoint the effect appears to be an inductance change. The device does make use of energy storage and the af signal controls this energy. The device is, therefore, parametric and will obey the Manley–Rowe equations, which are applicable to parametric circuits (Manley and Rowe, 1959).

The geometry of the superconducting loop and junction of Fig. 69 can take a variety of forms. Two of these forms are (1) a cylindrical thin film with a bridge, and (2) a machined ring with a point contact.

The cylindrical film type of device which was first developed by Mercereau

(1970) and Nisenoff (1970) is made by vacuum depositing a thin superconducting film on a cylindrical substrate. A typical substrate is a quartz rod 1 mm in diam by 0.5 cm long. Typically, the film has been about 400 Å thick. The film is then scratched or etched parallel to its cylindrical axis from end to end, except for one small region which forms the Dayem bridge (Anderson and Dayem, 1964) as shown in Fig. 71. The bridge is typically about 1 μ wide by $\simeq 2$ μ long. It is not necessary that the bridge have rectangular shape, provided that a small enough region is obtained.

The machined ring geometry has been extensively studied by Zimmerman and Silver (1968). It consists of a "C" machined out of a block of superconductor, such as niobium, with a point contact made across the open end by means of a pointed screw as shown in Fig. 72. Very fine threads and small radius points are required. Differential screws can be used to obtain finer adjustment. Mechanical links have been used in many experiments to make adjustment from the outside of the liquid-helium Dewar.

Zimmerman *et al.* (1970) have developed a symmetric single-point-contact structure that minimizes thermal strains. This device can be preset at room temperature and gives satisfactory performance at liquid-helium temperature, even after repeated thermal cycles. This symmetric point contact has also been described by Giffard *et al.* (1972), and in Chapter 1.

Another geometry, which in effect uses fractional turn loops, has been used by Zimmerman (1971) to reduce the inductance of the circuit containing the point contact. By using 12 loops in parallel across the point contact, instead of 2 loops, the inductance was decreased and signal amplitude was increased by a factor of 2.5.

A logical extension of this concept is that of the toroid geometry in which a toroidal cavity in a superconducting body surrounds a point contact (Rorden and Deaver, 1971). The current through the point contact spreads out and closes on all sides around the point. This is like having an infinite number of parallel loops connected across the point contact. A toroidal rf

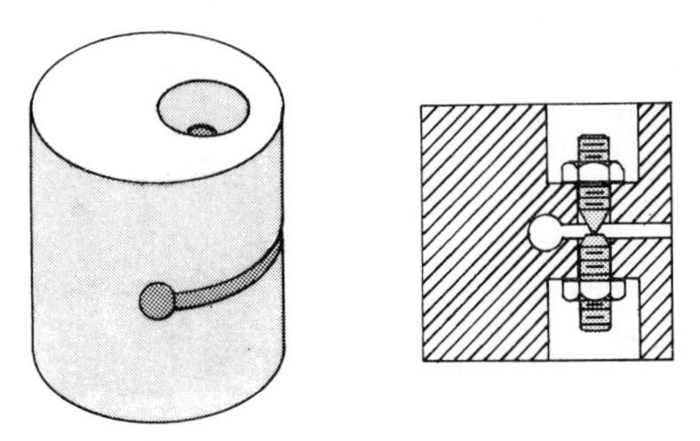

FIG. 72. Point-contact magnetometer. [After Zimmerman, 1971.]

coil is then placed in the toroidal cavity. Deaver (1971) achieved a resolution of one part in 16,000 of a flux quantum with such a device, using a 3-sec integration time.

It should be pointed out that planar film (superconductor–dielectric–superconductor) Josephson junctions are not suitable for this type of amplifier or magnetometer because of the large hysteresis in its I–V characteristic. This results from the large junction capacity, as discussed in Section II.A.2.b and in Chapter 1.

Magnetometers which have been built using this flux-flow principle have generally used rf pump frequencies in the range of 10–30 MHz. Magnetic field resolution of about 10^{-9} G rms has been obtained for a dc to 1-Hz bandpass without the use of field amplifier coils as described in Section III.A.9. Zimmerman and Frederick (1971) have operated a single-point-contact device at 300 MHz and obtained a signal-to-noise (measured from the rf-detected output) increase of about 10 relative to 30-MHz operation. They did not report actual magnetic field sensitivity data, although it is expected that similar order of magnitude improvement would be observed.

Kamper and Simmonds (1972) have pushed the operating frequency up to 9 GHz and shown that the device is sensitive to variations in magnetic field at frequencies from dc to 1 GHz. No data on field sensitivity were reported.

The input current resolution which could be expected for this type of amplifier depends upon the choice of field coil. There is a trade-off between low current resolution for many turns and low input inductance for a few turns on the field coil.

The authors achieved a current resolution using a thin-film cylindrical device pumped at 30 MHz, of 4×10^{-10} A Hz$^{-1/2}$ with a 10-turn field coil using 0.05-mm-diam niobium wire and a field coil diameter of 2.5 mm at a temperature of 4.2°K. This field coil when mounted inside a superconducting shield, had an inductance of about 200 nH. In the machined ring–point-contact-type device the field coil can consist of a solenoid placed inside the loop. Sullivan (1972) has achieved 5×10^{-12} A Hz$^{-1/2}$ current resolution at 2.0°K with a 600-turn four-layer solenoid of 0.09-mm-diam copper wire on a 1.6-mm-diam form. In this case the inductance of the field coil was about 6 mH. Sullivan used an 8-hole point-contact SQUID pumped at 30 MHz for this measurement. Note that the change in energy detected, i.e., $\frac{1}{2}L(\delta i)^2$ is a factor of about 75 smaller for the thin-film device.

The rf-detected output (see Fig. 69) varies periodically with current in the field coil. Therefore, most applications make use of servo feedback to obtain a linear response and large dynamic range. The feedback signal can be superimposed on the rf-drive coil. In fact the rf drive, sensing, and servo feedback can all make use of the same coil. Typically, an af signal is also superimposed on this same coil and then phase-sensitive detection can be used at the rf-detector output. This provides the output signal and also the required servo-feedback signal.

The intrinsic speed of this flux-flow device has not been determined; however, point contacts have been used up to x band. Thus, in most circuit applications the frequency response will be limited by either the pump frequency or the servo-feedback electronics. These flux-flow devices are generally very sensitive to stray rf fields and careful shielding of the superconducting junction is necessary.

It should be emphasized that the noise performance of these types of amplifiers has generally been measured and quoted with input signals at dc to 1-Hz frequencies.

5. *Fluxoid motion amplifiers at microwave frequencies*

The fluxoid was defined in Eq. (76) as the closed-line integral of the sum of two terms: the momentum $m\mathbf{v}$ of the electron pairs and the magnetic vector potential $\mathbf{A}$. The momentum term is only significant when the cross-sectional area in which the current flows is very small (e.g., 2×10^{-10} cm^2) so that high electron pair velocities are obtained. In larger size paths only the magnetic term remains, and the integral of it gives the magnetic flux enclosed. The integral of both terms is the London fluxoid. A fluxoid can be made to penetrate into a Type-II superconductor, or a thin film of a Type-I superconductor. Control of the motion of these fluxoids in appropriate geometries can result in a variable inductance.

Two methods of obtaining parametric amplification by means of fluxoid motion in a microwave cavity or waveguide will be considered here. The first involves the motion of fluxoids in a superconducting thin film with the magnetic field applied perpendicular to the film. The second involves the Josephson effect between agglomerated particles in a very thin superconducting film with a parallel field.

The first case consists of a geometry having radial fins placed in a circular waveguide so that the circumferential magnetic field is normal to the fins. At low microwave drive levels the microwave flux will pass around the outside of the fins without penetrating them. As the microwave power level is increased above some threshold value, the Lorentz force, $\mathbf{J} \times \mathbf{B}$, will produce a force sufficient to cause the fluxoids to move into the superconducting fins. These fluxoids will then move radially at the microwave frequency. This corresponds to an inductance change at the microwave frequency, thus making it possible to obtain parametric gain. At magnetic fields less than those required to obtain fluxoid motion, the distributed inductance of the waveguide is constant. At very high fields the films will be driven normal and the inductance will also be constant but larger in value. Inductance modulation and, therefore, parametric gain will only be obtained at intermediate field values. Most measurements have been made by using the double degenerate mode of the parametric amplifiers because of its simplicity. This means that the pump, idler, and signal frequencies are all close together. Bura (1966) used such a radial fin geometry to achieve 27-dB parametric gain at 2 GHz with a 4000-Å tin film at 2.3°K. The pump power was 2 W. Likewise, 27-dB gain was obtained in 1000-Å lead films at 4.2°K and 2.8 GHz with 6.3 W of pump power.

D'Aiello and Freedman (1966) performed experiments (discussed in the following section) with thin superconducting films parallel to the microwave magnetic field. They concluded that their amplification was achieved by means of fluxoid motion in the films rather than by means of London-penetration-depth effects for which their experiments were designed. They achieved gain only when the films were slotted.

Zimmer (1967) has conducted experiments to verify that Josephson effects can be detected in a single thin tin film in a microwave resonator similar to that used by Clorfeine (1965) (see Section III.C.6). The fact that Josephson effects can be obtained raises further questions concerning the actual gain mechanisms occurring in microwave experiments on thin films. One possible way in which Josephson junctions can result in a single thin film is by agglomeration or thermal grooving. This can occur when very thin films are deposited on noncooled substrates. Under the right conditions, islands are obtained that are connected by very thin sections. These thin sections can then act as superconducting weak links that exhibit Josephson effects.

Detailed understanding is lacking of the physical mechanisms responsible for the microwave properties which have been observed, and very little is known about the noise performance that can be obtained with these microwave amplifiers. However, workable amplifiers have been built and their performance seems to warrant further development.

6. *London-penetration-depth amplifiers*

The penetration of magnetic fields H, and electric fields E, into the surface of a superconducting body is described by the London equations:

$$\mathbf{H} = (4\pi\lambda^2/c)\nabla \times \mathbf{J} \tag{78}$$

and

$$\mathbf{E} = (4\pi\lambda^2/c^2)\dot{\mathbf{J}} \tag{79}$$

where J is the current density, and λ is the penetration depth. For a superconducting sheet with a parallel magnetic field, the currents will flow only near the surface and the current density will fall off exponentially in a London penetration depth λ. The magnetic field will likewise decrease exponentially from the surface in the same depth λ.

The value of λ is between 300 and 2000 Å for various superconducting materials at temperatures well below the transition temperature T_c. The value of λ depends upon temperature (London, 1950) according to

$$\lambda = \lambda_0[1 - (T/T_c)^4]^{-1/2} \quad (80)$$

where λ_0 is the penetration depth at $T = 0$ and is a constant for each material, e.g., $\lambda_0 = 520$ Å for tin and $\lambda_0 = 430$ Å for mercury. The value of λ_0 depends critically on surface condition, metallurgical factors, and the type of material.

Experiments have shown that λ also varies with magnetic field H. This means that the inductance of a superconductor of appropriate geometry can be nonlinear and thus parametric amplification may be possible. Pippard (1950) made measurements of λ versus H at a frequency of 9.4 GHz in fine tin wires. The value of λ increased by about 2.5% as H increased from zero to H_c for temperatures close to T_c (3.72°K). As T decreased from T_c the change in λ with H decreased to about 0.2% at $T = 3.1$°K and then increased again to about 2% for T below 2.2°K. The dependence of λ was nearly quadratic in H.

Other measurements at low frequencies have also shown that λ increases quadratically with H. At 2 MHz, Sharwin and Gantmakher (1961) obtained a 6% increase in λ for a single-crystal tin sample. For other samples λ had a complicated but strong field dependence.

Connell (1963) made measurements at 700 kHz on pure tin and indium and alloys of In–Sn, In–Tl, and In–Pb. Some of these alloys exhibited field effects on λ of over 10% as H increased from zero to H_c.

In the frequency range between 170 MHz and 3 GHz the field change in λ may be positive or negative (Glosser, 1967) depending upon the temperature. For planar samples the field effects also depend upon the orientation of the superconducting surface relative to the applied rf field. The largest effect is for a parallel arrangement (Glosser, 1967). Experiments at 8.8 GHz on flat tantalum wafers (Glosser, 1967) gave results similar to measurements by others on tin (Dresselhus *et al.*, 1963; Richards, 1962). In many of these measurements the imaginary (reactive) part X_S of the surface impedance is measured instead of λ; however, X_S and λ are closely related and the general behavior of one can be judged from the behavior of the other.

Attempts to explain the field effects of λ have met with some success

(Ginsburg and Landau, 1950; Maki, 1965). However, some of the more complicated effects are not yet understood, especially the changes in the sign of the field dependence with temperature in the frequency ranges from 2 to 174 MHz and 3 to 8 GHz. The theories are based on the redistribution of the normal and superconducting electrons as functions of current.

The field effects on λ are large enough to provide parametric amplification at microwave frequencies. Clorfeine (1964) was able to achieve oscillation and also parametric gain of 11 dB at 6.06 GHz for 250 Å tin films operated at 2°K. The tin films were deposited on a rutile (TiO_2) dielectric resonator. The resonator was placed on a quartz crystal, and this structure was placed in a stainless-steel waveguide. The resonator was operated in the doubly degenerate mode, i.e., the signal, pump, and idler frequencies were in the same resonance band and spaced in arithmetic progression. Amplification was achieved with pump power levels of only about 0.2 μW. It is not clear to what extent the amplification was influenced by thermal grooving in the very thin tin films, i.e., the tin forms islands instead of a continuous smooth film. Zimmer (1967) has been able to verify that Josephson effects can be obtained between the islands in such films thereby indicating the possibility that the parametric amplification of Clorfeine might be partly due to Josephson effects or at least altered by such effects (Clorfeine, 1965).

D'Aiello and Freedman (1966) also were able to obtain parametric amplification in a system similar to that of Clorfeine. However, they could only obtain gain when the tin films had slots cut in from the edges. Since the tin films were vacuum deposited at 80°K, it is unlikely that thermal grooving occurred. D'Aiello and Freedman account for the amplification as being produced by the interaction between the motion of flux vortices or fluxoids in the tin film and the microwave fields rather than being due to changes in λ. The use of fluxoid motion to achieve inductance changes is discussed in Section III.C.4. We conclude that the mechanisms contributing to the gain in the microwave amplifiers of Clorfeine and D'Aiello and Freedman are not well delineated at the time of this writing. However, in principle, it is certainly possible to obtain parametric amplification by means of changes in λ with magnetic field.

7. *Superinductance amplifiers*

The inductance of a circuit is comprised of two components; a magnetic part that is a function of the magnetic field produced by a current in the circuit and a kinetic part that is a function of the velocity of the charge carriers that produce the current. Under certain conditions, described below, the kinetic component of the inductance can be equal to or larger than the magnetic component; and, it can be modulated with applied field, temperature, or current.

Consider the energy E, associated with a current flowing in a conductor:

$$E = (B^2/8\pi)\, dV + n(\tfrac{1}{2}mv^2)\, dV \tag{81}$$

where B is the magnetic field intensity, V is the volume, n is the density of electrons, m is the electron mass, and v is electron velocity. The magnetic term $\int B^2/8\pi\, dV$ can also be written as a function of the geometric inductance L of the conductor

$$E_{\text{mag}} = \tfrac{1}{2}LI^2 \tag{82}$$

where I is the net transport current. The remaining term is a function of the kinetic energy of the charges and can be expressed in terms of a kinetic inductance such that

$$\tfrac{1}{2}L_{\text{k}}I^2 = \int_V ne(\tfrac{1}{2}mv^2)\, dV \tag{83}$$

The current $I = nevA$, where A is the cross-sectional area of the conductor and v is the electron velocity. Thus,

$$\tfrac{1}{2}L_{\text{k}}(nevA)^2 = ne(\tfrac{1}{2}mv^2)\,V \tag{84}$$

if the density of electrons n is constant. There, the kinetic inductance is

$$L_{\text{k}} = (m/ne^2)\,l/A \tag{85}$$

where l is the length and A is the cross-sectional area of the conductor. It is of interest to note that the ratio of the kinetic inductance to the geometric inductance increases as the dimensions of the film decrease. If we are dealing with a superconductor, n will be the density of superelectrons, and the kinetic inductance can be written in terms of the well-known penetration depth:

$$\lambda = m/\mu_0 ne^2 \tag{86}$$

where μ_0 is the permeability of free space. From Eqs. (85) and (86), the kinetic inductance is

$$L_{\text{k}} = \mu_0\lambda^2(l/A) \tag{87}$$

Since the penetration depth is a function of temperature, field, and current (Section III.C.6) a circuit can be constructed whose kinetic inductance can be modulated by applied field, temperature, or current. The question remains: Is the inductance change large enough to have any physical utility? A film 100 Å thick by 1 cm long by 1 cm wide with a penetration depth of 1000 Å may have a kinetic inductance (Little, 1967) of about 10^{-8} H. The geometric inductance for this same film is also about 10^{-8} H. Thus, to obtain values for the kinetic inductance that are significantly higher than the geometric inductance, the inductor must be operated very near the superconducting transition temperature T_c, where the penetration depth, λ, is large. As we have seen in Section III.C.6, Eq. (80), the penetration depth is a function of temperature as given by

$$\lambda^2 = \lambda_0^2[1 - (T/T_c)^4]^{-1} \tag{88}$$

At $T \to T_c$

$$\lambda \simeq \lambda_0(T_c - T)^{-1/2} \tag{89}$$

Thus, the penetration depth becomes very large near T_c, and since the kinetic inductance varies as λ^2 we may also obtain large L_k.

Little (1967) has studied the kinetic inductance of aluminum films and found that L_k increases by more than an order of magnitude at temperatures within 0.1°K of T_c. For the films reported $T_c \simeq 1.9$°K and L_k varied from 0.1 to $\simeq 1$ μH with temperatures from 1.7 to 1.8°K.

Thin films may be thermally switched very rapidly at helium temperatures and therefore, the kinetic inductance may be modulated by factors of 10 or more. Magnetic fields also affect L_k. Little (1967) has shown that a field change from 0 to 15 G may change L_k by factors of 10. This is a rather large field change, however, and magnetic field does not appear to be as suitable as temperature for modulating L_k.

The penetration depth, and L_k, will also be changed by variations in the transport current. In discussing this type of modulation in a magnetometer circuit, Goodkind and Stolfa (1970) show that changes in the transport current from 0 to I_c will cause an increase in the kinetic inductance of up to 50%. Current changes are easy to implement and although the magnitude of the inductance change is relatively small, current modulation of kinetic inductance has been the only type used in a superconducting device. This work was reported by Goodkind and Stolfa (1970) in which the variations in kinetic inductance of a bridge-type magnetometer were used to sense changes in magnetic field. The bridge was in a superconducting ring inductively coupled to an rf-drive coil. This circuit was discussed in Section III.C.4.

In general, superinductance is a nonlinear energy-storage system and fits the conditions for the active element in a parametric circuit. Thus, superinductance circuits should find application in numerous types of parametric amplifiers, especially since inductance modulation by as much as a factor of about 10 is possible.

Acknowledgments

The authors gratefully acknowledge the technical advice received from Mr. Louis Rorden and Mr. Leonard Orsak, and the drafting of the figures by Ms. Patricia Heaton.

References

Anderson, P. W., and Dayem, A. H. (1964). *Phys. Rev. Lett.* **13**, 195.
Beasley, M. R., Labusch, R., and Webb, W. W. (1969). *Phys. Rev.* **181**, 682.
Blackwell, L. A., and Kotzebue, K. L. (1961). "Semiconductor Diode Parametric Amplifiers." Prentice-Hall, Englewood Cliffs, New Jersey.
Bremer, J. W. (1962). "Superconducting Devices." McGraw-Hill, New York.
Brenneman, A. E. (1963). *Proc. IEEE* **51**, 442.
Buchhold, T. A. (1963). Cryogenic DC to AC Amplifier, U.S. Patent No. 3,098,189.
Buck, D. A. (1956). *Proc. IRE* **44**, 482.
Bura, P. (1966). *Appl. Phys. Lett.* **8**, 155.
Clarke, J. (1966a). *Phil. Mag.* **13**, 115.
Clarke, J. (1966b). *New Sci.*, 611.
Clarke, J. (1969). *Proc. Roy. Soc. London* **A308**, 447.
Clarke, J. (1971). *Phys. Rev.* **4**, 2963.
Clarke, J., and Paterson, J. L. (1971). *Appl. Phys. Lett.* **19**, 469.
Clarke, J., Tennant, W. E., and Woody, D. (1971). *J. Appl. Phys.* **42**, 3859.
Clorfeine, A. S. (1964). *Proc. IEEE* **52**, 844.
Clorfeine, A. S. (1965). *Proc. IEEE* **53**, 388.
Connell, R. A. (1963). *Phys. Rev.* **129**, 1952.
Dahm, A. J., Denenstein, A., Langenberg, D. N., Parker, W. H., Rogovin, D., and Scalapino, D. J. (1969). *Phys. Rev. Lett.* **22**, 1416.
D'Aiello, R. V., and Freedman, S. J. (1966). *Appl. Phys. Lett.* **9**, 323.
Deaver, B. S., Jr. (1971). Private communication.
Deaver, B. S., Jr., and Goree, W. S. (1967). *Rev. Sci. Instr.* **38**, 311.
Dresselhaus, M. S., Douglass, D. H., and Kyhl, R. L. (1963). *In* "Proc. Int. Conf. Low Temp. Phys. 8th" (R. O. Davies, ed.), pp. 328. Butterworth, London.
Devroomen A. R., and Van Baarle, C. (1957). *Physica* **23**, 785.
Eck, R. E., Scalapino, D. J., and Taylor, B. N. (1965). *In* "Proc. Int. Conf. on Low Temp. Phys. 9th (J. G. Daunt, *et al.*, eds.), pp. 415. Plenum, New York.
Fiske, M. D., and Giaever, I. (1964). *Proc. IEEE* **52**, 1155.
Foner, S. (1959). *Rev. Sci. Instr.* **30**, 548.
Forgacs, R., and Warnick, A. (1967). *Rev. Sci. Instr.* **38**, 214.
Fulton, T. A. (1971). *Appl. Phys. Lett.* **19**, 311.
Gange, R. A. (1964). *Proc. IEEE* **52**, 1216.
Giaever, I. (1960). *Phys. Rev. Lett.* **5**, 147.
Giaever, I., and Megerle, K. (1962). *IRE Trans. Electron Devices* **9**, 459.
Giffard, R. P., Webb, R. A., and Wheatley, J. C. (1972). *J. Low Temp. Phys.* **6**, 533.
Ginzburg, V. L., and Landau, L. D. (1950). *Zh. Eksp. Teor. Fiz.* **20**, 1064.
Glosser, R. (1967). *Phys. Rev.* **156**, 500.

Goodkind, J. M., and Stolfa, D. L. (1970). *Rev. Sci. Instr.* **41**, 799.
Goodman, W. L., Hesterman, V. W., Rorden, L. H., and Goree, W. S. (1973). *Superconducting Instrument Systems, IEEE Special Issue on Superconductivity* (T. Van Duzer, ed.).
Goree, W. S. (1972). In "Proc. Appl. Superconductivity Conf., Annapolis, Md.," pp. 640, *IEEE*, New York.
Gygax, S. (1961). *Angew. Math. Phys.* **12**, 289.
Huebner, R. P., and Govednik, R. E. (1966). *Rev. Sci. Instr.* **27**, 1675.
Ittner, W. B., III, and Kraus, C. J. (1961). *Sci. Amer.* **205**, 124.
IRE Subcommittee on Noise (1960). (H. A. Haus, Chairman of Subcommittee), *Proc. IRE* **48**, 69.
Johnson, A. K., and Chirlian, P. M. (1966). *IEEE Trans. Magn.* **MAG-2**, 390.
Josephson, B. D. (1962). *Phys. Lett.* **1**, 251.
Kachinskii, V. N. (1965). *Cryogenics* **5**, 34.
Kamper, R. A., and Simmonds, M. B. (1972). *Appl. Phys. Lett.* **20**, 270.
Langenberg, D. N., Scalapino, D. J., and Taylor, B. N. (1966). *Proc. IEEE* **54**, 560.
Little, W. A. (1967). *In* "Proc. Symp. on Phys. of Superconducting Devices," (B. S. Deaver, Jr., and W. S. Goree, eds.), pp. S-1. Univ. of Virginia, Charlottesville, Virginia.
London, F. (1950). "Superfluids," Vol. 1. Wiley, New York.
Lukens, J. E., Warburton, R. J., and Webb, W. W. (1971). *J. Appl. Phys.* **42**, 27.
McCumber, D. E. (1968). *J. Appl. Phys.* **39**, 3113.
McWane, J. W., Neighbor, J. E., and Newbower, R. S. (1966). *Rev. Sci. Instr.* **37**, 1602.
Maki, K. (1965). *Phys. Rev. Lett.* **14**, 98.
Manley, J. M., and Rowe, H. E. (1959). *Proc. IRE* **47**, 2115.
Matisoo, J. (1967). *Proc. IEEE* **55**, 172.
Meissner, H. W. (1963). Superconducting Contact Device, U. S. Patent No. 3,096,421 issued 7-2-63.
Mercereau, J. E. (1970). *Rev. Phys. Appl.* **5**, 13.
Meyerhoff, A. J., Cassidy, C. R., Hebeler, C. B., and Huang, C. C. (1964). *Adv. Cryog. Eng.* **9**, 360.
Miller, J. C., Wine, C. M., and Cosentine, L. S. (1964). *Proc. IEEE* **52**, 1223.
Motchenbacher, C. D., and Fitchen, F. C. (1973). "Low-Noise Electronic Design." Wiley, New York.
Newhouse, V. L. (1964). "Applied Superconductivity." Wiley, New York.
Newhouse, V. L., and Bremer, J. W. (1959). *J. Appl. Phys.* **30**, 1458.
Newhouse, V. L., Bremer, J. W., and Edwards, H. H. (1960). *Proc. IRE* **48**, 1395.
Newhouse, V. L., and Edwards, H. H. (1964). *Proc. IEEE* **52**, 1191.
Newhouse, V. L., and Edwards, H. H. (1965). *In* "Amplifier and Memory Devices" (N. Prywes, ed.), Chaps. 18 and 20. McGraw-Hill, New York.
Newhouse, V. L., Mundy, J. L., Joynson, R. E., and Meiklejohn, W. H. (1967). *Rev. Sci. Instr.* **38**, 798.
Newhouse, V. L. (1969). *In* "Treatise on Superconductivity" (R. Parks, ed.). Dekker, New York.
Nisenoff, M. (1970). *Rev. Phys. Appl.* **5**, 21.
Opfer, J. (1970). *Rev. Phys. Appl.* **5**, 37.
Pippard, A. B. (1950). *Proc. Roy. Soc. London* **A203**, 210.
Pritchard, J. P., Jr., and Schroen, W. H. (1968). *IEEE Trans. Magn.* **MAG-4**, 320.
Radhakrishnan, V., and Newhouse, V. L. (1971). *Rev. Sci. Instr.* **42**, 129.
Richards, P. L. (1962). *Phys. Rev.* **126**, 912.

Ries, R. P., and Satterthwaite, C. B. (1967). *Rev. Sci. Instr.* **38**, 1203.
Roberts, B. W. (1966). Nat. Bur. Stand. Tech. Note No. 408.
Roberts, B. W. (1969). Nat. Bur. Stand. Tech. Note No. 482.
Rorden, L. H. (1965). Final Rep. on U. S. Naval Research Laboratories Contract No. Nonr 3249(00)X with Standford Research Inst., Menlo Park, California, AD 660 050.
Rorden, L. H., and Deaver, B. S., Jr. (1971). Private communication.
Schiller, C. K., and Bulow, H. (1969a). *J. Phys. Chem. Solids* **30**, 1977.
Schiller, C. K., and Bulow, H. (1969b). *J. Appl. Phys.* **40**, 4179.
Sharvin, Y. V., and Gantmakher, V. F. (1961). *Sov. Phys.—JETP* **12**, 866.
Siegwarth, J. D., and Sullivan, D. B. (1971). *Rev. Sci. Instr.* **43**, 153.
Silver, A. H., and Zimmerman, J. E. (1967). *Phys. Rev.* **157**, 317.
Slade, A. E., and McMahon, H. O. (1958). *Proc. West Joint Compt. Conf.* **13**, 103.
Stewart, W. C. (1968). *Appl. Phys. Lett.* **12**, 277.
Sullivan, D. B. (1972). *Rev. Sci. Instr.* **43**, 499.
Templeton, I. M. (1955). *J. Sci. Instr.* **32**, 314.
Tinkham, M. (1963). *Phys. Rev.* **129**, 2413.
Young, D. R. (1959). *In* "Progress in Cryogenics" (K. Mendelssohn, ed.), Vol. 1, pp. 1–33. Academic Press, New York.
Zimmer, H. (1967). *Appl. Phys. Lett.* **10**, 193.
Zimmerman, J. E., and Silver, A. H. (1968). *J. Appl. Phys.* **39**, 2679.
Zimmerman, J. E., Thiene, P., and Harding, J. T. (1970). *J. Appl. Phys.* **41**, 1572.
Zimmerman, J. E. (1971). *J. Appl. Phys.* **42**, 4483.
Zimmerman, J. E., and Frederick, N. V. (1971). *Appl. Phys. Lett.* **19**, 16.

Chapter 3

Computer Memory

A. R. SASS†

Quasar Electronics Corporation
Franklin Park, Illinois

W. C. STEWART and L. S. COSENTINO

RCA Laboratories
Princeton, New Jersey

† Formerly of IBM System Products Division, East Fishkill, New York.

I. Introduction

In the past few years a number of excellent texts have been written which have discussed the computer applications of superconductivity (Newhouse, 1964; Bremer, 1962). These texts have more than adequately described the foundations of this field and pointed out its salient features.

This chapter will be devoted to one facet of the field, namely, superconductive random-access memories. This facet appears to be the most likely initial candidate for the computer marketplace. Superconductive associative memories (memories which can perform logic) combine the properties of memory discussed in this chapter and logic discussed in Chapter 2, Volume I. Owing to this natural extension, a study of superconductive associative memories offers relatively no new information concerning superconductivity *per se*. Thus for reasons of clarity and in order to focus on salient points, associative memories will not be discussed in this chapter (an excellent review of the subject is given by Hanlon, 1966).

The discussion here will be self-contained; however, the reader is referred to the above-mentioned texts and will be referred to general technical literature for a broader view and more detail, respectively.

In a computer memory there are two basic properties that must be present for proper operation. First, there must exist storage units each of which is capable of storing binary ("1" or "0") information. This implies a physical phenomena exhibiting two stable states. A superconducting loop carrying current or zero current (or clockwise or counterclockwise current) satisfies this requirement.

The second requirement is the ability to select one storage unit in order to read it or write into it. The selection is as follows. The storage unit has a gate which is normally closed, but which during reading or writing must be opened. Just as a gate in a fence is only efficient if it can be opened only by exceeding a nonzero pressure threshold, the gate of the storage unit must also have a threshold. A superconductor, switchable to the normal state when its critical field is exceeded, satisfies this requirement.

The combination of a superconducting loop, capable of storing a persistent current, with a portion of the loop (the gate) switchable into the normal state in order to change the current, forms the basic element of a superconductive memory element. An array of these elements, appropriately interconnected, constitutes the memory itself. The array must be interconnected and accessible so that selected elements can be "written" or "read" when required. Specific interconnection schemes referred to as "organizations" will be discussed. Some of these organizations are similar to those used in nonsuperconductive memories; it will be shown that each organization places different (in character and degree of achievability) requirements on superconductive material specifications.

Although the above characteristics are central to the concept of a superconductive memory, they are not the characteristics which make such a memory interesting from a competitive viewpoint.

It was recognized rather early in the planning of superconductive memories that its chief virtue was one of amenability to batch fabrication; namely, large arrays of memory cells could be made simultaneously. The cost impact of such a product is apparent. These arrays, superconductive memory cells, and associated wiring are in the form of thin films. The array fabrication consists of alternate vacuum deposition of metal and insulating layers. During early development stages, the memory cells and associated wiring were defined in the metal layers by means of masking during metal evaporation; later chemical-etch techniques, having higher resolution for high-density elements were adopted. It should be recognized that these concepts of batch fabrication, dating back to the late 1950s, were quite new and exciting for their day and still represent a high order of device integration.

Added to the idea of batch fabrication of high-density cells were some of the associated virtues of superconductive elements and interconnections. These virtues were low-power dissipation of the memory cells during reading and writing and zero dissipation of the current carrying lines in the memory arrays themselves. These properties have a significant impact on the cost of driving and sensing electronics.

While the need for a liquid-helium refrigerator for the memory imposed a specific economic overhead on this memory technology, the virtues of batch fabrication and low-memory-power dissipation were great enough so as to offset this overhead for an interesting range of applications.

This chapter will be devoted to expanding upon the subjects described in this Introduction section as well as to speculating upon the future course of superconductive memories.

II. Theory of Superconducting Memory Structures

Since analyses of the circuit operation of cryoelectric devices can be no more accurate than the equivalent circuit representations of the devices, it is desirable to develop relations by which the geometrical structures of the devices can be characterized by lumped parameters for inductance, capacitance, and resistance, or if necessary, by simple distributed networks. For computer applications, the circuit topology consists primarily of strip

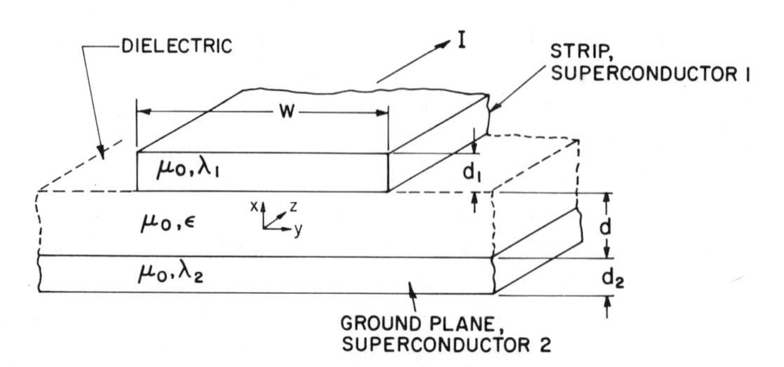

FIG. 1. Superconducting strip line.

line, in which a thin sheet of superconductor lies over a single, large, insulated superconducting ground plane; the strip thickness and that of the insulating layer is much smaller than the strip width, which is in turn much smaller than the length of the line. The primary influence of the superconducting properties of the circuit elements is the exhibition of zero electrical resistance in those portions which experience temperature and magnetic fields below the critical values. Second only to the absence of resistance is the ability of the superconductors to screen out dc magnetic fields from their interior, with the exception of a thin layer near the surface. As a result, inductance relations for a configuration of superconductors will generally differ from those of the same configuration comprised of normal conductors.

An approximate approach, which is often adequate for many purposes, is to consider the superconductors as perfectly diamagnetic and ignore the effects of a finite magnetic penetration depth. There results in this approximation an exact analog between electrostatic and magnetostatic field solutions for a given geometry. In the case of the strip line carrying a total current I shown in Fig. 1, and for which d, $d_1 \ll W$, the problem is essentially one dimensional. An essentially uniformly distributed diamagnetic screening current of $-I$ exists in the ground-plane surface just beneath the strip to give

$$H = I/W \tag{1}$$

in the dielectric region between strip and ground plane. On the basis either of magnetostatic energy storage or magnetic flux per unit of current, the inductance per unit length of this configuration is then computed to be

$$L/l = \mu_0 d/W \quad \mathrm{H\ m^{-1}} \tag{2}$$

An identical expression holds to the same degree of approximation for microstrip lines of normal conductors at high frequencies. For superconductors, however, the frequency range includes dc. The electrostatic analog (Newhouse *et al.*, 1960) shows for $W/d \geq 10$, that the magnetic field at the ground-plane surface beneath the middle of the strip is always greater than 80% of the value given by (1); the resulting inductance values are even closer to that given by (2) since inductance in this case may be calculated from the average field between strip and ground plane at the center. A further result for strips of finite width shown by this analysis is the tendency for the current to peak at the edges instead of assuming a uniform distribution.

A. The London Equations

The penetration of a magnetic field less than the critical value into a small region near the surface of a superconductor can be treated by use of the phenomenological London equations (London, 1961) for the superfluid component of the two-fluid model

$$\mathrm{curl}(\mu_0 \lambda^2 \mathbf{J}) = -\mathbf{B} \tag{3}$$

and

$$\mathbf{E} = -\frac{\partial}{\partial t}(\mu_0 \lambda^2 \mathbf{J}) \tag{4}$$

Combining these relations with Maxwell's equations gives the following partial differential equations governing the spatial distribution of H and J in a superconductor:

$$\nabla^2 \mathbf{H} = \mathbf{H}/\lambda^2 \tag{5a}$$

$$\nabla^2 \mathbf{J} = \mathbf{J}/\lambda^2 \tag{5b}$$

It is a general property of the solutions to Eqs. (5) to decrease very rapidly toward the interior of the superconductor. The penetration depth is the distance from the surface of a semiinfinite superconducting slab at which H and J have fallen to e^{-1} times the surface values. Swihart (1961) and Meyer (1961) consider in detail the applicability of (3)–(5) to superconducting shielded strip-line structures. On the basis of a one-dimensional field analysis (i.e., a strip of infinite width in which the current density at any depth is uniform across the width), they conclude that the static London equations provide a simple characterization of such structures for frequencies up to 1 GHz, with temperatures $\leq 0.95T_c$, and with an effective penetration depth (Ittner, 1960) consistent with modern nonlocal theories used for λ. Subject to these constraints, and with the assumption of a perfectly lossless dielectric, lossless and dispersionless wave propagation is obtained in a form equivalent to a TEM transmission line mode for which equivalent values of per-unit-length series inductance and shunt capacitance can be obtained.

The validity of the one-dimensional field analysis for structures of finite size rests upon the assumptions that (a) the lengths of the strips are much longer than the widths, (b) the current distribution in the ground plane is essentially a mirror image of that in the strip, and (c) the current is essentially uniformly distributed across the width of the strip. Condition (a) is easily achieved in practice and will not be considered further. Sass (1964) has shown that for d large compared with λ_2, the distribution of current per unit width $j(y)$ at the surface of the ground plane for a specified current distribution in a strip of finite width is the same as the charge distribution obtained in the analogous electrostatic problem by the method of images. The current density within the ground plane resulting from the penetration of the field is obtained by multiplying $j(y)$ by a penetration function

$$J_2(x,y) = j(y) \cosh[(x' + d_2)/\lambda_2]/\lambda_2 \sinh(d_2/\lambda_2) \qquad (6)$$

where $x' = (x - d/2)$, and the other quantities are defined in Fig. 1. Thus for W large compared to d, d_1, and d_2, the shielding current in the ground plane is concentrated under the strip and essentially mirrors the width distribution of the strip current, satisfying assumption (b) above.

The question of uniformity of current distribution across the width of the strip mentioned in (c) above is more difficult to resolve since no general solution of this problem for the finite geometry of Fig. 1 has been given. Some peaking of the current density at the edges of the strip is to be expected if d_1 is comparable to or larger than λ_1, the penetration depth of the strip. As a result, the surface current density at the facing surfaces of the strip and the ground plane, and the tangential magnetic field between the strip and the ground plane in the central region will both be somewhat lower than if the current were uniformly distributed. The amount of this reduction will be approximately proportional to the amount of current in the strip edges in excess of the fraction of the current which is uniformly distributed. As will be discussed subsequently, the per-unit-length strip inductance depends directly on the quantity of magnetic flux linking a perpendicular plane between the center of the strip and the ground plane plus contributions from the surface current density at the boundary of the plane. Thus, edge peaking gives a slightly lower inductance than a uniform current distribution in the London model.

The justification for considering edge peaking to be a slight effect in a shielded geometry, where W is large compared to film thicknesses and separation distances, comes from the following considerations. It has been shown, for a completely *unshielded* superconducting strip of finite width and a thickness comparable with λ (Rhoderick and Wilson, 1962), that the edge peaking reduces the surface current density and tangential magnetic field at the center of the strip only by a factor of $\pi/2$ from the case of uniformly distributed current. Sass and Skurnick (1965) have shown that bringing a second strip of equal width and thickness facing the first within a distance less than or equal to $6\lambda_1$ restricts the edge peaking of current density to a distance of approximately $10(d_1 + d)/W$ from the edge. For typical strip-line dimensions, the result will be a decrease in inductance from the uniform current density case on the order of 2%. The additional width of the ground plane beyond the edges of the strip in Fig. 1 should serve to further even out current variations in the strip. Thus, to a reasonably high degree of accuracy, the assumption of uniform current density in the one-dimensional field analysis can be used for deriving specific inductance relations. The approximation becomes less valid as the ratio W/d is reduced to smaller values.

B. Capacitance and Inductance

Since the electrostatic charges which establish the transverse electric field across the dielectric layer reside on the conductor surfaces in superconductors, as in good normal conductors, the capacitance per unit length for the structure of Fig. 1 is given simply by

$$C/l = \epsilon W/d \quad \text{F m}^{-1} \tag{7}$$

when $d \ll W$, where ϵ is the permittivity of the dielectric and the edge fringing is neglected.

The standard approach for calculating self-inductance consists of equating the magnetostatic free energy K from the field solution to $LI^2/2$ and solving for L. In normal conductors and insulators K is simply $\mu_0 H^2/2$. For London superconductors, however,

$$K = \mu_0 H^2/2 + \mu_0 \lambda^2 J^2/2 \tag{8}$$

The second term is the kinetic energy of the superelectrons which is furnished by a momentary electric field to establish current. Since this energy is not lost to the ionic lattice of the superconductor as in normal conductors, no further energy from an electric field is required to maintain a steady current, as indicated by (4). An opposing electric field is actually required to halt the supercurrent, whereupon the superelectron kinetic energy is recovered by the source, along with the energy initially required to establish the H field.

It has been shown (Sass and Stewart, 1968) that the above procedure leads to the following definitions for mutual and self-inductance which are applicable to the perfectly general case of an arbitrary three-dimensional configuration of London superconductors carrying N independent mesh currents:

$$L_i = \partial\Phi_i/\partial I_i \quad \text{henrys} \tag{9a}$$

$$M_{jk} = M_{kj} = \partial\Phi_j/\partial I_k \quad \text{henrys} \tag{9b}$$

where

$$\Phi_i = \iint_{s_i} \mathbf{B}\cdot d\mathbf{s} + \oint_{C_i} \mu_0\lambda^2\mathbf{J}\cdot d\mathbf{l} \tag{10}$$

is the London fluxoid of the ith mesh contributed by the total fields of all current sources. The surface s_i in (10) is bounded by the curve C_i which lies totally within the superconductors bounding the ith mesh (although C_i is denoted as a closed curve, it does not include the external voltage or current source to which the ith mesh may be connected; the curve must be considered as closed along a suitably chosen interface between the source and the loop whose inductance is being defined). For one-dimensional geometries that are considered to be of infinite length, Φ may be evaluated on a per-unit-length basis.

For the strip-line geometry of Fig. 1, the field solutions for a z-directed current I in the strip are

Strip:

$$\mathbf{H} = [I/W][\sinh(x_1/\lambda_1)/\sinh(d_1/\lambda_1)]\hat{\jmath} \tag{11a}$$

$$\mathbf{J} = [I/\lambda_1 W][\cosh(x_1/\lambda_1)/\sinh(d_1/\lambda_1)]\hat{k} \tag{11b}$$

Dielectric:

$$\mathbf{H} = -[I/W]\hat{\jmath} \tag{11c}$$

Ground plane:

$$\mathbf{H} = -[I/W][\sinh(x_2/\lambda_2)/\sinh(d_2/\lambda_2)]\hat{\jmath} \tag{11d}$$

$$\mathbf{J} = -[I/\lambda_2 W][\cosh(x_2/\lambda_2)/\sinh(d_2/\lambda_2]\hat{k} \tag{11e}$$

where $\hat{\jmath}$ and $\hat{k}$ are unit vectors in the y and z directions, respectively; $x_1 = x - (d/2 + d_1)$, and $x_2 = x + (d/2 + d_2)$. A mesh current may be defined by assuming that the strip current I is returned through the ground plane. The fluxoid is easily evaluated for a surface lying in the x–z plane, for which

$$\Phi/l = (\mu_0 I/W)[d + \lambda_1 \coth(d_1/\lambda_1) + \lambda_2 \coth(d_2/\lambda_2)] \quad \text{Wb m}^{-1} \tag{12}$$

and

$$L/l = (\mu_0/W)[d + \lambda_1 \coth(d_1/\lambda_1) + \lambda_2 \coth(d_2/\lambda_2)] \quad \text{H m}^{-1} \tag{13}$$

Note that expressions for H within the superconductors are not explicitly used in evaluating Φ and L. The first term on the right-hand side of (13) is contributed by the magnetic flux in the dielectric and is the same as the value obtained in (2) previously. The last two terms in (13) are the combined contributions of magnetic flux penetrating the superconducting strip and ground plane, plus the superelectron kinetic-energy terms.

If the strip current is not returned through the ground plane by means of a physical connection, (13) will still be applicable if the strip is arranged to return such that the image currents under the strip, which are confined near the upper surface of the ground plane, can form a closed path under the strip. Otherwise, inductance terms (which may be large) associated with the closure of the image currents must be added to (13). The case of superposed parallel strips over a ground plane separated by insulating layers can be treated by applying (7) to obtain direct interelectrode capacitances and (9) to obtain self- and mutual-inductance values.

C. Resistance

When the strip of Fig. 1 consists of a thin normal conductor over a superconducting ground plane, Meyers (1961) has shown that the skin effect is negligible over the range of frequencies for which (3) and (4) are applicable. Thus the equivalent series impedance of the line contains a resistive part equal to $\rho/Wd \ \Omega \ \mathrm{m}^{-1}$, where ρ is the resistivity of the strip. Although the fluxoid concept is no longer applicable, consideration of the field energy as originally outlined (Meyers, 1961) or, equivalently, consideration of the low-frequency internal impedance (cf. Ramo and Whinnery, 1953) shows that $\lambda_1 \coth(d_1/\lambda_1)$ in (13) is replaced by $d_1/3$, and the remaining terms are unchanged.

D. Lumped Equivalent Circuit

When a short length of the strip line forming a cryotron gate is switched into the normal state, the resistance introduced is typically 10^{-3} times the characteristic impedance of the strip line. The resulting transients which propagate along the line give an effective time constant (Sass *et al.*, 1967c) equal to the total inductance of the loop containing the gate divided by the total gate resistance; capacitive effects do not enter unless the transients are to be propagated over lengths greater than several feet. The change of inductance which occurs when an in-line cryotron gate is switched from superconducting to normal and back again can have a significant influence on circuit behavior (Meyers, 1962) since the gate occupies a substantial fraction of the total length of its circuit loop. This effect is much smaller in crossed-film cryotron circuitry and is usually neglected in equivalent circuit analysis. The low-frequency lumped equivalent circuit for shielded cryotron devices, therefore, consists of inductances and time-varying resistances.

For a circuit of N meshes, in which the ith mesh consists of a totally superconducting loop with no voltage sources, but in which there exists inductive coupling to other meshes, Kirchhoff's voltage law gives

$$O = M_{i1}\frac{dI_1}{dt} + M_{i2}\frac{dI_2}{dt} + \cdots + L_i\frac{dI_i}{d2} + \cdots + M_{iN}\frac{di_N}{dt} \tag{14}$$

Replacing L_i and the Ms by their definitions (9a) and (9b) we obtain

$$O = \frac{\partial\Phi_i}{\partial I_1}\frac{dI_1}{dt} + \frac{\partial\Phi_i}{\partial I_2}\frac{dI_2}{dt} + \cdots + \frac{\partial\Phi_i}{\partial I_i}\frac{dI_i}{dt} + \cdots + \frac{\partial\Phi_i}{\partial I_N}\frac{dI_N}{dt} = \frac{d\Phi_i}{dt} \tag{15}$$

This result, which shows the invariance with time of the fluxoid in a closed superconducting loop, demonstrates the consistency of our equivalent circuit definitions with the fluxoid conservation principle of the London theory. Although the quantization of the fluxoid plays a central role in the understanding of many superconducting phenomena, the current levels and inductance values found in practical cryoelectric computer circuits involve fluxoid values of many thousands of fundamental quantum units. It is within these circumstances that Φ may be considered a smooth, well-behaved linear function of circuit mesh currents, as implied by the field equations (3) and (4), and that the inductance parameters have well-defined values.

Matisoo (1967) has demonstrated that thin-film tunnel junctions with barriers sufficiently thin to exhibit the Josephson supercurrent, are suitable as cryotron gates. The value of I_{max}, the maximum supercurrent which can be supported, is modulated by a controlling magnetic field as in ordinary cryotrons. These devices switch from the zero-voltage state to a finite-voltage state (typically 1–3 mV) in subnanosecond times. The detailed mechanism of this operation is described elsewhere in this book. Once the gate has switched, the current–voltage characteristic exhibits hysteresis; the nearly constant voltage quasiparticle tunneling curve is followed for decreasing gate current. As the current falls below a value typically on the order of 10–20% of I_{max}, the junction usually reverts to the zero-voltage state. Since the superconducting films which comprise the tunnel junction do not enter the normal state, there is no inductance change during switching.

We conclude this section by noting that the lumped-linear-equivalent-circuit concepts reviewed here have successfully provided an excellent characterization of the superconducting memory devices to be described. The notable exception is the continuous film memory cell to be discussed in Section III.A; this device consists of an inherently continuous structure which has not been satisfactorily modeled by a lumped linear circuit. A further exception, alluded to previously, occurs for cells in which loop inductances and critical currents are sufficiently small that fluxoid quantization effects become pronounced.

III. Superconductive Memory Cells

A. Early Device Research

1. *Introduction*

Von Neumann is credited by Buckingham (1958) as being the first person to suggest, before World War II, that superconductors might be used as memory elements. Many memory cells based on superconductive principles have been invented since that time. Buck (1956) developed the cryotron and showed how it could be used as a superconductive computer component. An element (latching flip–flop) having two stable states and comprised of six cryotrons was described. The state of storage (1 or 0) was determined by the path through which current flowed. This type of memory element was little used because a second method with cryotrons, namely, persistent current storage, proved to be much simpler to implement and to have superior operating properties. All the cells to be subsequently described utilize persistent current storage in a superconducting path, part of which can be made resistive by means of either a magnetic control field or current-induced switching. A review by Rhoderick (1959) also describes some of the first cryoelectric storage cells.

2. *Early persistent current cryotron cells*

The early persistent current cryotron cells (Hayes, 1960; Newhouse *et al.*, 1960) are exemplified by the elements of Fig. 2. For a more complete description of a "crossed film" cryotron, see chapter 2.

An input control can switch one branch of the storage loop resistive. This branch is made to have a low inductance L_L relative to the other branch of the loop so that if the control is not activated, most of the loop input current I_0 will flow in this branch. If however, the input control is activated, then loop current will eventually flow in the high-inductance

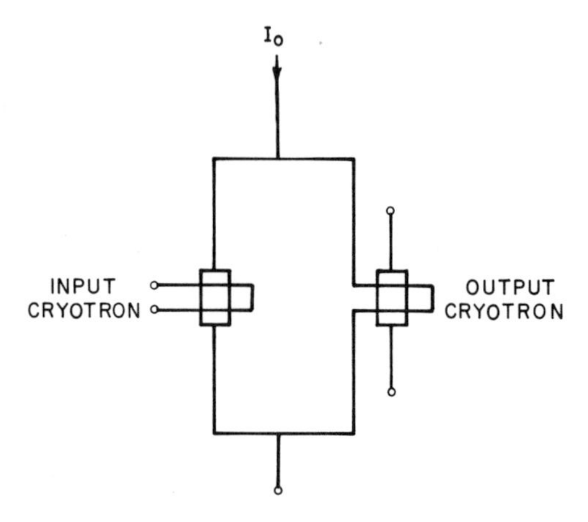

FIG. 2. Persistent current cryotron storage cell.

L_H branch only. The control can then be deactivated and the low-inductance branch will return to the superconducting state. If I_0 is then switched off, a circulating persistent current of magnitude $I_0L_H/(L_L + L_H)$ will be stored in a clockwise direction around the loop. This current can be removed by activating the input cryotron control when I_0 is not present. The output or sensing cryotron gate is either resistive or superconducting as determined by the current in its control which is part of the storage loop. Several thin-film memory arrays were produced using the basic cell (Ittner and Kraus, 1961; Bremer, 1962), and modified versions were proposed as building blocks for memory systems and other computer functions (Haynes, 1960).

3. *Early noncryotron-type storage loops*

In addition to the cryotron-type storage cells, persistent current loops were devised in many other forms, one of the earliest being that of Casimir-Jonker and DeHaas (1935). Three cells which appeared almost simultaneously in 1957 are the persistatron (Buckingham, 1958), the persistor (Crittenden, 1958), and the Crowe cell (Crowe, 1958). The common feature of these cells is that switching is accomplished in a part of the storage loop by current in the film rather than by an externally controlled magnetic field.

The persistatron consists of a ring or loop of superconductive material with two segments of unequal inductance connected in parallel. The inductances of the branches are arranged to be different by geometrical or material design. Current injected into the loop divides between the two segments in inverse proportion to the inductance of each branch. The side with the lower inductance will carry most of the current until its critical current is reached. Any further increase in current is shunted to the other branch while the first branch current remains at the critical value. If the drive current is then removed, both branch currents decrease (again in inverse proportion to the inductance of each segment). Thus the low-inductance path returns to the superconducting state and a persistent current will be stored. If the inductance of the branch which is switched is much smaller than that of the other branch, essentially all of the drive current will be stored. A succeeding drive pulse in the same direction as the first will put current into the switchable branch in a direction opposite to that of the stored current so that no switching occurs and there is no change in the value of the persistent current after this pulse is over. On the other hand, a subsequent pulse of drive current in the opposite direction than the original pulse will add to the stored current and switch the low-inductance-branch resistive. When the pulse is over, a persistent current will flow in the ring in the opposite direction than the first stored current. An output voltage will be produced when the low-inductance branch becomes resistive, and energy will also be dissipated during that time. Sensing is accomplished by viewing the voltage across the cell directly or by means of another ring inductively coupled to the storage ring. Vail *et al.* (1960) demonstrated persistatron operation with 1-μsec pulses at a repetition frequency of 16 kHz while Stewart *et al.* (1961) observed complete switching of a persistatron in less than 10 nsec.

The persistor is a device similar to that just described, being comprised of a superconducting inductor in parallel with a superconductive element which can be switched to the normal state by a current above the critical value. Crittenden *et al.* (1960) have operated these devices at a 15-MHz repetition rate using 30-nsec pulses although elements have been operated with read or write pulses of 15-nsec duration. By arranging drive coils so as to inductively couple pulses to the cell, coincident-current operation could be obtained with the persistor elements arranged in the usual two-dimensional X–Y array. A sense line would wind serially through the matrix connecting all the outputs of the individual cells.

The Crowe cell (Crowe, 1957; Garwin, 1957) also makes use of stored persistent current in a superconducting path. See Fig. 3. A physical hole in a "hard" superconductive material is bridged by a "soft" superconductive material which can be switched resistive when the current induced in it by the drive lines exceeds the critical value. At this point, flux can link the bridge and thread the hole. As the drive pulse is turning off, the induced current in the crossbar is also decreasing, reaching zero value and then increasing in the opposite direction. When the drive current returns to zero

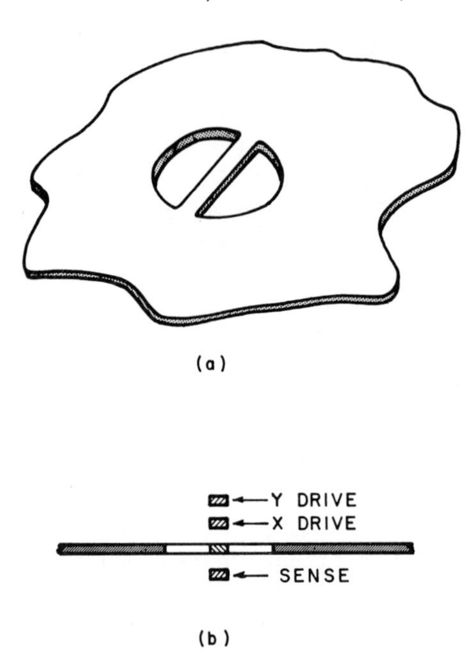

FIG. 3. Crowe cell: (a) Storage film with hole and bridge of cell; (b) cross section of cell including sense and drive lines.

value, there is then flux trapped in the hole, and a persistent current around the hole maintains the trapped flux. The output can be inductively sensed by a line beneath the crossbar on the other side of the hole.

Thus far isothermal operation of the above storage devices has been assumed so that heating effects are negligible. In each case, the switched segment must return to the superconducting state before the drive current is removed in order to store the maximum value of current in the loop. Otherwise any circulating current would partially or completely decay toward zero because of the resistance present. When heating of the film takes place rapidly, then Joule heating effects cause a thermal runaway situation in which the film temperature quickly rises causing the critical current of the film to decrease faster than the decay of persistent current in the loop. This has also been called adiabatic operation. If the thermal recovery time of the film is greater than the electrical time constant (L/R) of the loop, then the circulating current will decay to zero before the film becomes superconducting again and no current will be stored in the cell. Both isothermal and adiabatic operation of the Crowe cell have been observed and analyzed in some detail (Crowe, 1957; Edwards and Newhouse, 1965). The rise time of the drive pulse determined the type of operation, with fast rise times (~5 μsec) resulting in adiabatic operation and slow rise times (~1 sec) resulting in isothermal operation. Since no persistent current is established in the cell with adiabatic operation, it is necessary to apply a "priming" pulse at the very beginning of the information storage cycle. The pulse is so shaped that the leading portion causes the film to go normal while the trailing portion is reduced in magnitude and held on long enough to cause a predetermined current of a value below the critical current to be established in the loop when the pulse terminates after the film has cooled to the ambient temperature. Thereafter read and write pulses of lower magnitude than the priming pulse are applied which can cause the film to switch only if the addition of the induced current and the circulating current exceed the critical current of the film. Each drive pulse must be of sufficient duration that the film cools before the pulse ends so that current of one polarity or the other always circulates in the cell between application of pulses. In this manner, memory operation in the adiabatic mode can be effected.

If heating is appreciable during cell operation and drive pulses are not maintained long enough for thermal recovery to be complete before drive current is removed, then the amount of current stored in the cell (if any) will be dependent on such factors as the relative thermal and electrical time constants, drive-pulse rise and fall times, drive-pulse amplitude and duration, and the actual temperature change in the switched film.

4. *The continuous film memory*

In addition to the cell just described with physical holes in a medium to trap flux, Crowe (1957) also proposed a memory using a continuous sheet of superconductive material. See Fig. 4. Drive-line current induced an image current in the sheet which, after reaching a critical value, switched the film and allowed flux to "punch through" and link a portion of the film. It was found that flux was trapped in the sheet after removal of the drive current, as had also been seen by Alers (1957). Drive wires were visualized as running at right angles to each other over the sheet while a sense line beneath the sheet intersected both wires at a 45° angle. It was felt that many problems of the Crowe cell were associated with the edges of the holes which caused large variations in the field needed to switch the bridge which lead to lack of uniformity and poor reproducibility (Burns *et al.*, 1960). The continuous film memory (CFM) was expected to overcome these difficulties besides eliminating the need for accurate registration of the bridge with the drive lines over holes. The geometrical simplicity of the CFM and the ease with which such a structure might be batch-fabricated with extremely high-packing densities account for the enthusiasm generated

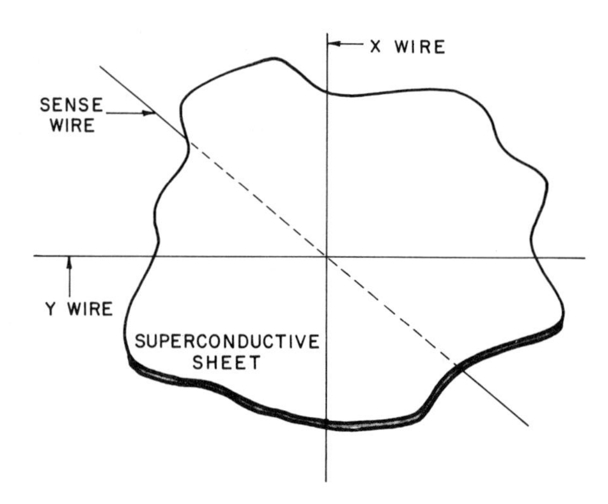

FIG. 4. Continuous film memory cell.

by the concept and the great efforts made to produce large arrays of CFM cells.

Burns *et al.* (1960, 1961) reported work on small arrays of CFM cells (6×1 and later 10×10) which further enhanced the outlook for a practical cryoelectric memory. Small groups of cells were found to operate with excellent reproducibility and with sense signals of several millivolts. This sense voltage was deemed adequate since the almost perfect shielding properties of superconductors provided such a quiet background that the signal-to-noise ratio was impressive and better than that of existing magnetic memories. Also, if operation was kept within the proper range of drive currents, a half-select pulse would produce no sense signal whatsoever as opposed to the magnetic core memory where "δ-noise" resulted from the small differences in the sense voltages caused by unselected cores on a driven line. The CFM cell was also found to be quite fast since write-in was accomplished with a pulse as short as 3 nsec as evidenced by readout obtained with a slower pulse (Burns *et al.*, 1961). By comparison, Crowe cells had a thermal recovery time of about 100 nsec (Crowe, 1957). It was estimated that capacities of CFM memories could be as high as $10^6/\text{in.}^3$ by stacking together 1000 planes, each 10 mils thick with drive lines spaced on 10-mil centers. Cryotron decoding trees could be put down at the same time as the memory thus effecting economies in interconnections.

Unfortunately, initial expectations were never realized and the problems encountered were so serious that today (1969) there is no longer a serious effort to utilize the CFM as a building block for a large cryoelectric memory system. It will be instructive to determine why the promises of the CFM were never fulfilled.

One fundamental shortcoming is the lack of a *workable* model of CFM cell operation. A simple model does suffice to explain, qualitatively, memory action in a cell. Time-coincident equal-amplitude current pulses applied to the X and Y drive lines cause the tin memory film beneath the X–Y line intersection to switch to the normal state, thereby permitting flux to penetrate the film. When the drive currents are removed, this flux becomes trapped in the "magnetic holes" and is maintained by currents flowing in the plane of the tin film. The distribution and amplitude of current in the plane are problems which do not lend themselves to exact solution and are virtually impossible to calculate. When currents of opposite polarity than the original pulses are later applied to the same drive lines, the field of the stored persistent current adds to the field of the drive currents causing the tin to switch again but this time storing current in the opposite direction. If drive currents of the same polarity follow the original pulses, the drive and stored current fields subtract and no switching occurs. When the film is switched to normal, flux lines will cut across the sense line producing a voltage across that line designating a stored "1". Absence of voltage designates a "0".

As will be described in Section IV, for a cell to operate properly, a drive pulse on only one line (half-select) must not switch the tin to normal. Coincident X–Y drive pulses of the opposite polarity than the previous pulses must switch the tin to normal while coincident pulses of the same polarity as the previous pulses must not switch the tin to normal. Assuming equal currents on the two lines, these requirements determine the largest and smallest current amplitudes that can be used to select the given cell.

One interpretation of this model is illustrated in Fig. 5 (after Burns *et al.*, 1960) and shows how the current induced in the superconducting film images the drive current until the drive amplitude exceeds I_c, the critical current of the film, after which the film current no longer increases and flux penetrates the film, links the sense line, and produces an output voltage. When the drive current is reduced below I_c the bridge returns to the superconducting state and flux is trapped in the tin sheet supported by rings of current around the normal regions.

There are many difficulties with the above model. To begin with, the impossibility of determining the stored current distribution has already been noted. Another complication results from the heating that takes place at the cell which further affects the stored current. In the model of Barnard *et al.* (1964), for example, if heating effects are assumed negligible (isothermal mode), the stored current is proportional to the drive current while for an optimum heating mode, the stored current is independent of the

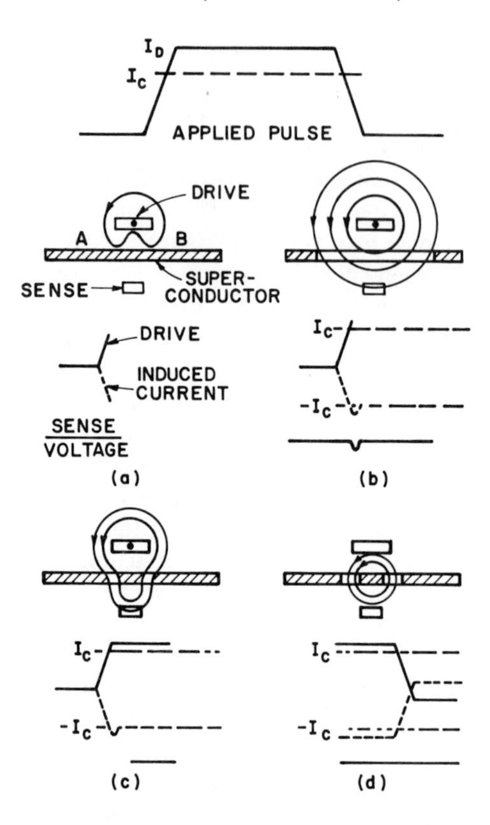

FIG. 5. CFM cell operation. [Burns *et al.*, 1961.]

drive current. To determine the actual mode of operation, electrical and thermal time constants of the cell would have to be known, and these are difficult to calculate or measure. Another problem arises in determining how the drive currents interact with each other and with the stored current. Most early analyses (Burns, 1960; Barnard *et al.*, 1964) assumed vector addition of the drive currents while some later models (Edwards and Newhouse, 1965) favor algebraic addition of these currents. Most analyses also assume algebraic addition of the resultant drive current and the stored current. In addition, there are always the unknown effects of physical defects and impurities in the storage structure on the cell properties. These items demonstrate the lack of a simple model to explain CFM operation. Much more complicated models have been proposed (e.g., the vortex theory of Pearl, 1965) but none has been successful in predicting cell-operating properties with accuracy.

Another problem with the CFM involves tolerances or ranges over which the drive currents may vary without spurious results from the memory. As was mentioned above, proper cell operation results when either X or Y current does not affect the cell, and coincidence of X and Y current may affect the cell state. The minimum current, which when applied in time coincidence to the X and Y lines is sufficient to switch the tin and produce a stable sense signal, is designated as the threshold current of operation I_{T} for a cell. The minimum current, which when applied to either the X or Y line alone produces switching of the tin film, is designated as the disturb current I_{D}. The fractional operating margin of a cell M is defined in terms of permissible variation from the mid-range value of current at which the cell will operate or $M = \pm(I_D - I_{\mathrm{T}})/(I_D + I_{\mathrm{T}})$.

As will be discussed in Section IV, for an array of cells to function properly, it is most important that there be sufficient uniformity of threshold currents over the plane and large margins for each cell so that the entire array will work at the same current with some tolerance about that current. If the memory is to consist of many planes stacked together, the same constraints on uniformity and margins also apply from plane to plane as well as from cell to cell.

Analyses of drive-current tolerances for a CFM cell have been done by several authors (Burns, 1964; Barnard *et al.*, 1964; Edwards and Newhouse, 1965; Gamby and Maller, 1965; and Atherton and Tyler, 1966) with predictions varying widely depending on the various assumptions made concerning summation of drive currents, dependence of stored current on the drive current, and details of the disturb mechanism. Again, the need for a model is obvious. The computed tolerances in most cases are maximum values from which must be subtracted any effects caused by fabrication nonuniformities, variations in the helium-bath temperature, or differences in the drive-current wave shape.

The operating margin of a plane M_{P} is defined in terms of the permissible variation from the mid-range value of current at which the array of cells will operate or $M_{\mathrm{P}} = \pm(I_{\mathrm{D}}' - I_{\mathrm{T}}')/(I_{\mathrm{D}}' + I_{\mathrm{T}}')$, where I_{T}' is the highest cell threshold current and I_{D}' is the lowest disturb current of any cell on the plane. It is clear that although individual cells on a plane may have operating margins, unless these margins overlap, the array cannot be operated at a common current. Little success has been achieved in making CFM cell arrays with overlapping working ranges.

Another factor acting to further reduce working tolerances is the sensitivity of the CFM cell to various patterns of half-select pulses or disturbs (Burns, 1964; Barnard *et al.*, 1964; Edwards and Newhouse, 1965; and Gamby and Maller, 1965). The major effect of disturb pattern testing is to lower the highest permissible cell disturb current, which correspondingly decreases practical cell and plane tolerances, since a cell would be subjected to such patterns of half-select pulses as neighboring cells are cycled under realistic operating conditions.

The disturb mechanism is another phenomenon which is imperfectly understood but which probably involves the lack of strong flux pinning sites at a cell. Barnard *et al.* (1964) have suggested a particular type of memory film structure, having physical microscopic voids to act as flux pinning sites, which might make the CFM less sensitive to disturbs. Experimentally it has been found that disturb effects are cumulative. Large numbers of half-select pulses disturb a cell much more than a single pulse, the effect of which may not even be seen. Burns (1964) attributes disturb effects to small grains and impurities at the storage locations. Barnard *et al.* (1964) invoke the possibility of a superconducting- to normal-state phase transition with a toe on the curve. A small amount of resistance is introduced into the cell during each half-select pulse until the stored current is reduced to a value such that a half-select pulse will no longer exceed the toe of the curve. Feissel *et al.* (1966) describe two distinct types of disturbs. One is sensitive to the drive-pulse rise time and results in a decrease of stored flux, while the other changes the distribution of stored current without altering the stored flux and is insensitive to drive-pulse rise time. Pearl (1965) accounts for disturb effects by the production and migration of vortices at film irregularities and defects at low values of currents which annihilate some of the stored flux.

Gamby and Maller (1965) reported that CFM cells having in-line drives, i.e., with X and Y lines running parallel and one above the other over a storage location, operate with better cell tolerances (up to $\pm 20\%$) and have a lower sensitivity to disturbs. However, all the other problems with uniformity and reproducibility still must be overcome in order to get good array tolerances. In addition, the in-line geometry necessitates a lower packing density and longer drive lines.

It is generally agreed that the detailed structure of the thin storage film greatly influences the operating features of the CFM cell. Such variables as evaporation rate, grain size, substrate condition and temperature, film stresses, crystal orientation, contaminants in the vacuum system, and others all may play some part in determining film properties. Many of these items are either not well understood or are incapable of being precisely controlled.

In summarizing the great amount of effort expended on the CFM, it may be concluded that the geometrical simplicity of the device proved to be its downfall. The severe requirements placed on microscopic structure at a storage location which must be kept uniform over a large substrate and from plane to plane cannot be met with state-of-the-art technology. The distribution of currents stored in a metal sheet does not lend itself to analysis, making a working model impossible. The dependence of the bit-selection process on a current previously stored in a cell increases further the burden of uniformity at the storage sites. The lessons learned from the CFM investigations point the way to improved cells, some of which are described in subsequent sections.

B. Device Evolution

1. *Modified CFM cells*

Improvements in operating margins of memory cells that retain the feature of trapping flux in a continuous film have been obtained by modifying the drive and sense-line configurations. Goser and Kirchner (1967) added a third drive line as shown in Fig. 6a to the in-line cell of Gamby and Maller (1965). A matrix array of these cells is driven by three independent decoders as shown in Fig. 6b. The drive lines are threaded so that activation of one drive line by each decoder gives three coincident drive currents at only one storage cell in the matrix. No more than one drive

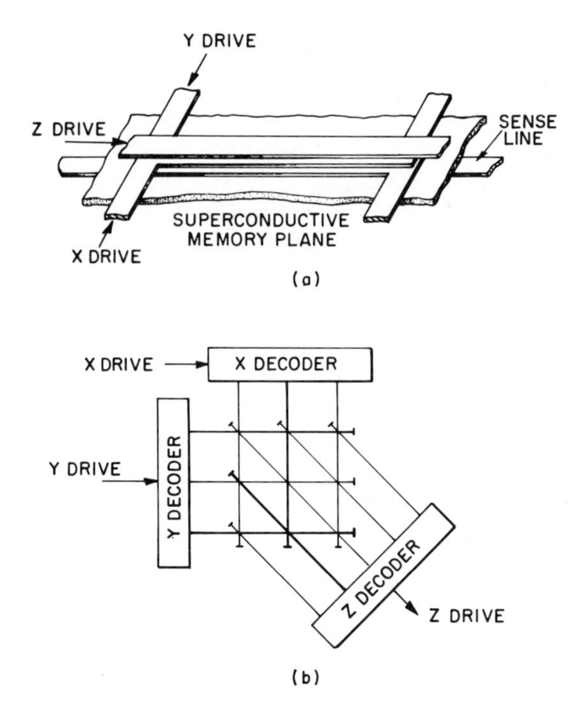

FIG. 6. Alternate CFMs: (a) Triple coincidence CFM. [Goser and Kirchner, 1967.] (b) Three-dimensional access technique.

current appears at any unselected cell location. Thus the disturb current level is only one-third of the full select value rather than one-half. Goser and Kirchner report that an experimental cell which exhibited a ±12% margin with double coincidence rose to ±20% with triple coincidence.

A modification of the double-coincident in-line CFM cell places the two drive lines on opposite sides of the continuous memory film and dispenses with the separate sense line (Goser, 1967). This cell is applicable only to a word-organized memory, which is to be discussed subsequently. Cell operation requires the coincidence of pulses of one polarity on both drive lines to trap flux in one direction, but a pulse of opposite polarity is required only on the word drive line to reverse the trapped flux while the digit line is used for sensing. As a result, alternating-polarity disturb pulses occur on only one drive line rather than on both. Since the critical field for thin films is lower when the field is applied to both sides, the threshold for full switching with both lines should be lower than the threshold for disturb effects caused by digit currents in unselected words. Measured margins of an experimental cell were ±25 and ±18% for the two drive lines. The addition of a low-transition-temperature sense line to this cell has also been proposed (Kadereit and Goser, 1968) for performing nondestructive matching operations for a content-addressed memory.

The increased cell margins obtained in the modified CFM structures are achieved at the expense of additional drive lines and decoder circuitry in one case, and the necessity for word organization in the other. Both of these factors dictate requirements on the overall memory system whose implications are discussed in Section IV. In addition, the cell-to-cell uniformity characteristics of the continuous film remain unchanged by modifications in drive-line geometry.

2. *Bridge cell*

One of the most interesting aspects of the bridge cell (Fig. 7) was that it was discovered at a point in time when continuous film types of memory arrays had passed their zenith in interest (Ahrons, 1965). The bridge cell with a structured geometry was a throwback to the early days of superconductive cell development.

There was, however, one novel feature of the bridge-cell design, namely, the fact that the storage loop was in the "third dimension." Upon application of a digit current I_d to the cell, the entire current flows through the gate. This phenomena can be understood by either employing a "minimum stored energy" argument or by studying the equivalent circuit of the loop (Fig. 8). The existence of the mutual inductance, and its particular magnitude, are responsible for the effect.

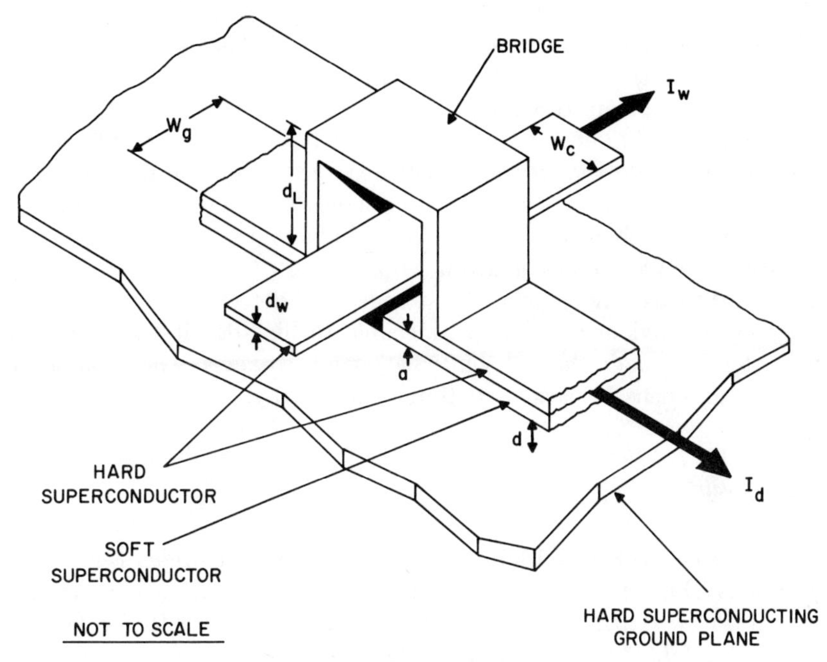

FIG. 7. Bridge cell.

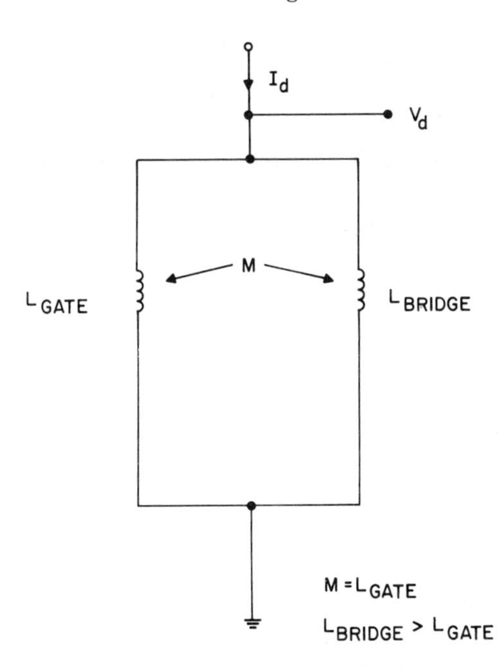

FIG. 8. Bridge-cell equivalent circuit.

Upon application of the write current I_w, the gate becomes resistive, and I_d is diverted to the bridge. When I_w and I_d are removed (in that order), I_d is stored in the loop. It is interesting to note (again consider the equivalent circuit), that *all* of I_d is stored. In order to detect the stored current, I_w is turned on and the stored current decays through the resistive gate; the resultant voltage V_d is sensed at the indicated terminals. This voltage is given by $L\dot{I}_d$, where L is the *total* loop inductance, $L_{bridge}-L_{gate}$; this is to be contrasted to the persistatron cell.

In addition to having efficient storage and detection capability, the bridge cell is obviously (Fig. 7) amenable to high-density-array layout. This latter fact made the cell somewhat attractive, even when comparing it to the more densely packed continuous film array.

3. *Loop cell*

The loop cell (Gange, 1967) can be described as a bridge cell in two dimensions (Fig. 9). It does not have the favorable mutual inductance between the two parts of the superconducting loop, as does the bridge cell, thus the total I_d is not stored, and total possible sense voltage cannot be

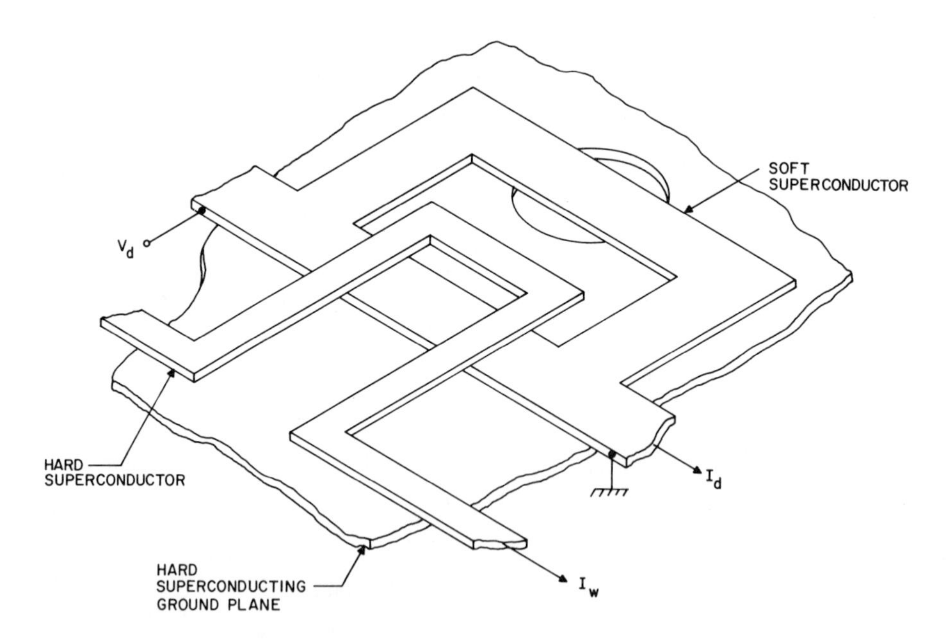

FIG. 9. Loop cell. [Gange, 1967.]

detected. (The reader is encouraged to verify this by examining Fig. 8 when $M = 0$.) However, the hole in the ground plane increases the inductance of the nongate side of the loop cell by more than an order of magnitude above the gate side of the loop. Thus it can be shown from Fig. 8 when $M = 0$ that more than 90% of I_d is stored and more than 80% of the total possible sense voltage per unit I_d can be detected.

Thus the basic tradeoff between the bridge and loop cell is the higher density capability of the former versus the easier manufacturability of the latter. The loop cell can be fabricated with fewer metal layer depositions. Furthermore, the superconductive loop is made in one step as opposed to that of the bridge cell where superconductive contacts must be made. Regarding the higher density capability of the bridge cell, it should be noted that the total flux sensed during detection is less than that of the loop cell when the two-cell line widths are adjusted for equal density, (Gange, personal communication).

4. *Multiaperture cell*

A cell recently developed at General Electric (Mundy and Newhouse, 1968) is depicted in Fig. 10 along with its equivalent electrical circuit. The device consists of an insulated Nb ground plane, a storage line and loop structure of Sn, and an insulated Pb access cryotron control line. Inductance L_1 is made much smaller than L_2 and L_3 so that none of the drive current

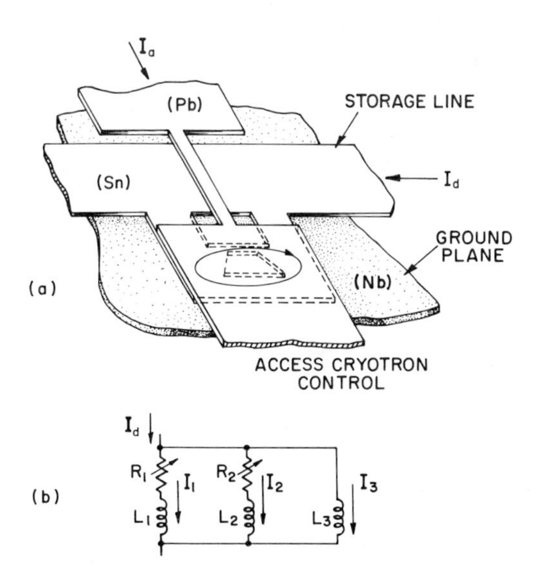

FIG. 10. Multiaperture cell: (a) Structure; (b) equivalent circuit.

I_d enters branches 2 and 3 when the access current I_a, and hence R_1, are zero. The application of I_a in coincidence with I_d switches in the cryotron gate resistance R_1 and diverts I_d to branches 2 and 3 where it initially divides inversely to the inductance ratio of the two branches. When the critical current of branch 2 is exceeded, any additional current goes to branch 3. Branch 1 thus comprises an access switch to the persistatron–persistor–type cell formed by branches 2 and 3.

The inclusion of an access switch at each cell greatly reduces the sensitivity of the cell to disturb pulses; there is no inductive coupling of the cryotron control line to the persistent current loop, and the value of L_1 nearly eliminates the interaction of I_d with stored current when the cryotron is not switched. In the ideal case, the only restriction on current I_a is that it produce a control field sufficient to switch the access cryotron gate. Similarly, the current I_d must be sufficiently high that the critical current of branch 2 is exceeded but not high enough for the critical current of branch 3 to be reached. The operating current ranges therefore depend ideally on the magnitudes of critical current levels and inductances in the cell which are governed by the cell geometry, and can be made large if space is available. Additional factors such as finite access cryotron resistance, nonzero critical current of branch 1 during operation, and the influence of the cryotron control line field on the critical current values for branches 2 and 3, however, arise in practice to further limit the range of operation. Operating margins between ± 37 and $\pm 47\%$ in I_d have been measured experimentally. The direct sense signal resulting from time varying flux linking the outer aperture during switching is too small for direct detection in small, densely packed cells. Integrating techniques, such as current stretch sensing (to be discussed in the sensing-techniques subsection), are therefore required.

C. Three-Wire Cell

The concept of the three-wire cell involves operating principles that were different from those employed, at the time, in continuous film memories. It is useful to review these concepts before proceeding to examples of actual cells and later to arrays and systems (Sass *et al.* 1966).

A schematic of the cell is shown in Fig. 11. The purpose of X and Y lines will be more fully explained in a later section; here let it suffice to say that in an X–Y array of cells, a cell is "selected" for reading or writing by activating the particular X and Y lines, which intersect at the cell location, and switching the cell's gate resistance. The digit line is used to enter information during writing and to sense information during reading.

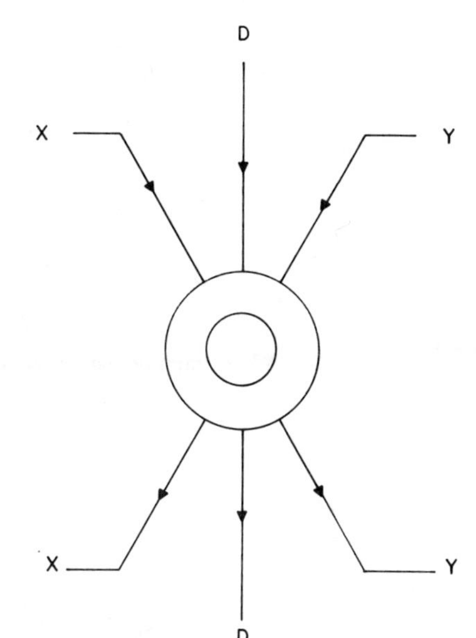

FIG. 11. Three-wire-cell schematic [Sass *et al.*, 1966].

The following operational rules are considered ideal:

1. The X and Y currents do not contribute flux (current) to the stored information, and levels of X–Y switching current are not affected by stored information.
2. The digit current does not contribute to the selection. This rule also applies to digit current during reading when it is actually in the form of stored current.
3. Film thicknesses are such that small variations in thickness do not greatly affect selection current levels. Thicknesses larger than the coherence length are considered acceptable.

It is useful to compare these rules to those governing operation of the continuous film cell. In the latter, the absence of a digit current line immediately invalidates rule 2 and places the burden of selection *and* storage on the X and Y lines (in contrast to rule 1). During readout of the continuous film cell, logical detection is based on interaction between X and Y currents and stored current (in contrast to rule 1). Similarly rule 3 is not in effect in the continuous film cell since in that case, storage film thickness is about 1000–3000 Å.

The rationale for the three-wire cell rules will now be discussed. Owing to the X–Y selection requirement (for both three-wire and continuous film cells), the X or Y current alone must not "select" *any* cell in an array while

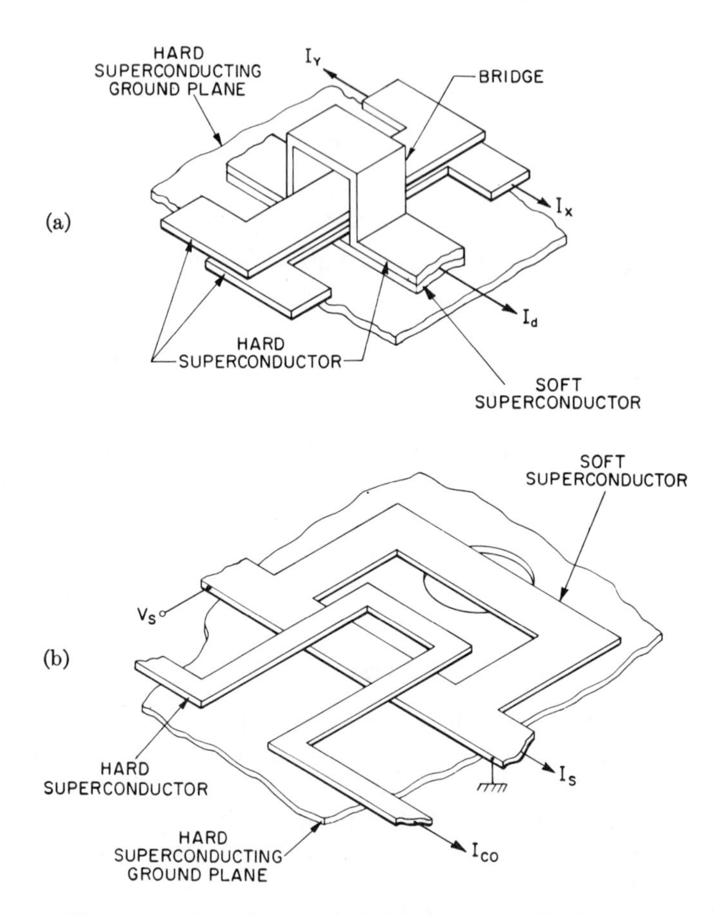

FIG. 12. Three-wire cells: (a) Bridge cell [Sass *et al.*, 1966]; (b) Loop cell [Gange, 1967].

the X *and* Y current must "select" *every* cell at some time during memory operation. In a later section, the equations describing this logical situation and the resulting tolerances on X and Y current levels, based on variations in effective critical fields of the gates, is derived. It will be shown that for excessive variations there is a zero operating range for X and Y currents to perform logical X–Y selection—resulting in improper memory operation.

Here it can be seen that for the three-wire cell, rule 1 guarantees that variations in stored current levels will not contribute to the effective variations in critical fields of the memory array gates—as it does in the continuous film cell array. In any event, rule 2 ensures uniform stored currents

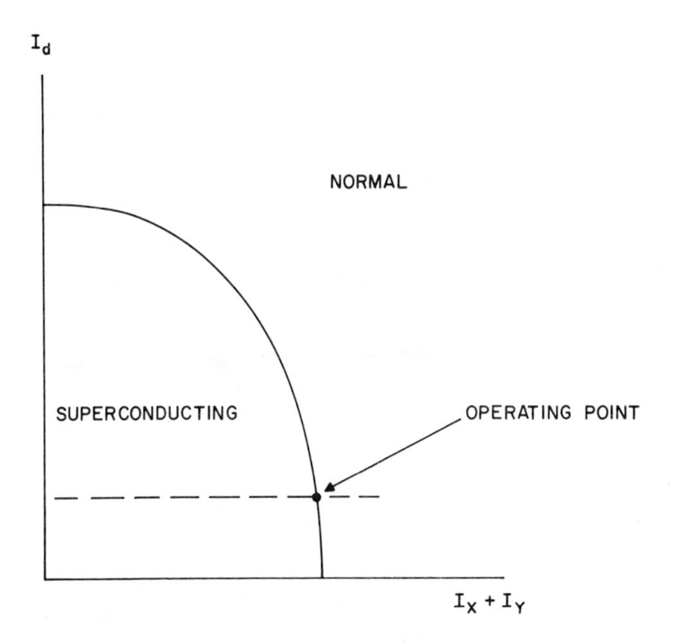

FIG. 13. Memory cell gate—phase diagram.

in three-wire arrays. It can be said that rules 1 and 2 reinforce each other in this respect. In the continuous film cell array, variations in stored currents result in variations in X–Y current levels which further increase variation in stored currents. This is basically a runaway situation, exaggerated by the thin-film nature (in opposition to rule 3) of this device.

Two three-wire cells are shown in Figs. 12a and 12b; these are basically the bridge and loop cell described previously. The first part of rule 1 is satisfied by the placement of the X–Y lines relative to the cell loops. The operating curve of the cells' gates are shown in Fig. 13. The operating point indicates agreement with the remainder of rule 1 and with rule 2. Finally gate thicknesses are of the order of 10,000 Å.

D. SENSING TECHNIQUES

The introduction indicated that this chapter would not be all inclusive insofar as memory cell design is concerned.† However, all known detection techniques can be lumped into five major categories and thus will be treated in turn.

† Associative memory cells are not considered.

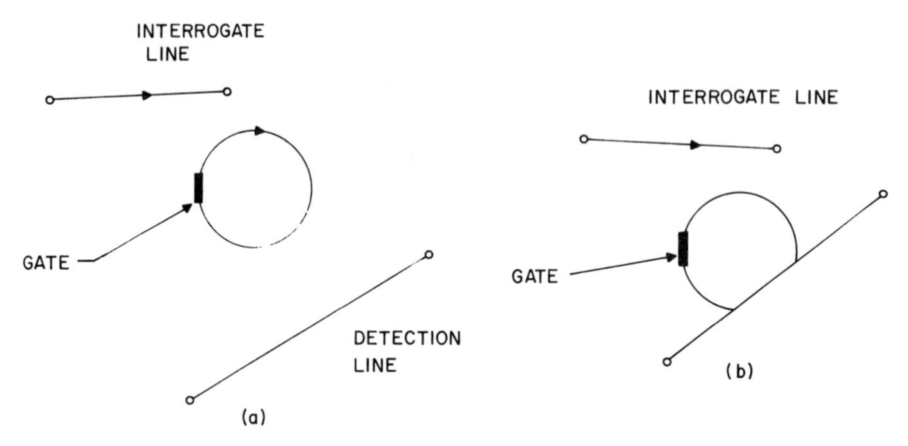

FIG. 14. Inductive detection: (a) Nongalvanic; (b) galvanic.

Each technique is, of course, a method of determining the presence–absence or sense of persistent current in a loop. Some feeling for cell designs not specifically treated in this chapter can be gained from this discussion.

1. *Inductive sensing*

This method is schematically shown in Fig. 14. A persistent current is destroyed by the interrogate line (which switches a portion of the loop resistive by cryotron action), and the time rate of decay of flux is detected. Memory cells such as the persistatron, bridge cell, and loop cell are read in this manner. Of course this is a destructive readout (DRO) so that the information must be rewritten if it is to be preserved.

2. *Resistive sensing*

This method is shown in Fig. 15. Neither the persistent current nor the interrogating current (nor the gate current) can separately switch the gate resistive; however, a combination of interrogating current and persistent current can do so. In fact, by proper arrangement of the interrogating and persistent current, the gate will switch resistive only if the persistent current is parallel (as opposed to antiparallel) to the interrogating current. Thus the presence–absence or sense of the persistent current can be detected. Note that this method of sensing is a nondestructive readout (NDRO) since the persistent current is not destroyed during its detection.

The reason for the existence of the interrogating current is the same as for inductive sensing and will be more apparent after reading Section IV on memory organization. Here it is sufficient to say that usually a single

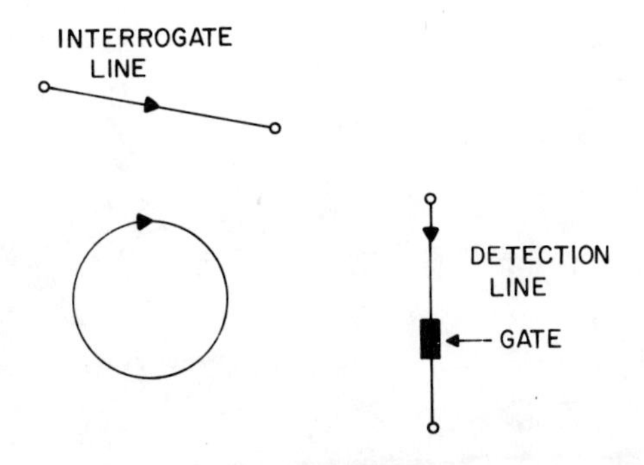

FIG. 15. Resistive detection.

detection line contains the gates of many memory cells. In order to read one cell along the line (and detect the resistance of one gate by only connecting to the ends of the detection line), the interrogate current (one per cell along a single detection line) basically "selects" the cell to be detected.

3. *Flux shuttling*

A more recent form of NDRO detection, called flux shuttling, is shown in Fig. 16 (Sass and Nagle, 1967). Here the interrogate current is the control of a variable inductance which increases as the interrogate current

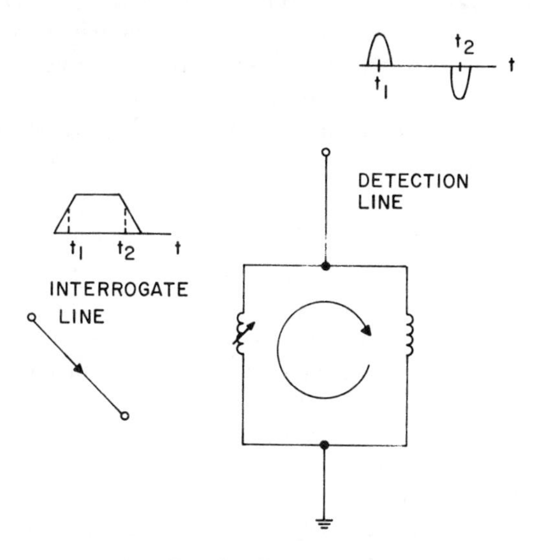

FIG. 16. Flux-shuttling equivalent circuit.

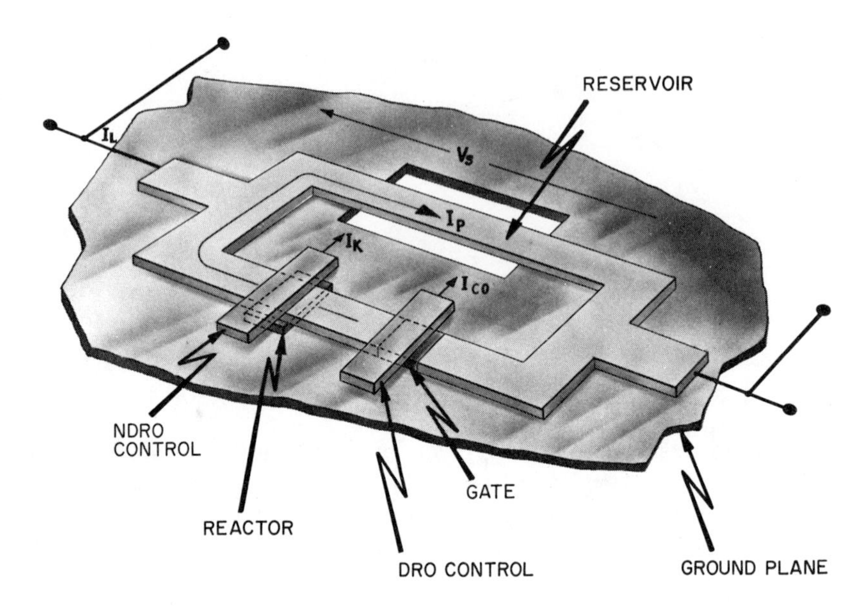

FIG. 17. Thin-film superconductive NDRO memory cell: flux-shuttling form. [Sass and Nagle, 1967.]

is turned on. Since the flux in the loop is constant as long as the loop is superconducting, the persistent current decreases. As shown in the figure, this causes a positive voltage on the detection line. Similarly, when the control is turned off, the persistent current returns to its initial value which causes a negative voltage on the detection line. A physical realization of this cell is shown in Fig. 17, where the variable inductance is achieved by the existence of the reactor; in the superconducting state the reactor excludes flux due to the Meissner effect, while in the normal state, it does not (hence the name "flux shuttling").

4. *Current-stretch sensing*

A more recent form of DRO sensing is called current-stretch sensing (Newhouse and Edwards, 1967) and is depicted in Fig. 18. As the gate of a memory cell is switched resistive by an interrogate current, its stored flux is completely transferred to the large superconducting loop. This sets up a persistent current in this loop whose magnitude with respect to the original persistent current is in the same proportion as the inductance of the cell loop is to the large loop.

The current in the large loop, which is static, can be sensed by the superconducting amplifier. One of the basic advantages of this method is that,

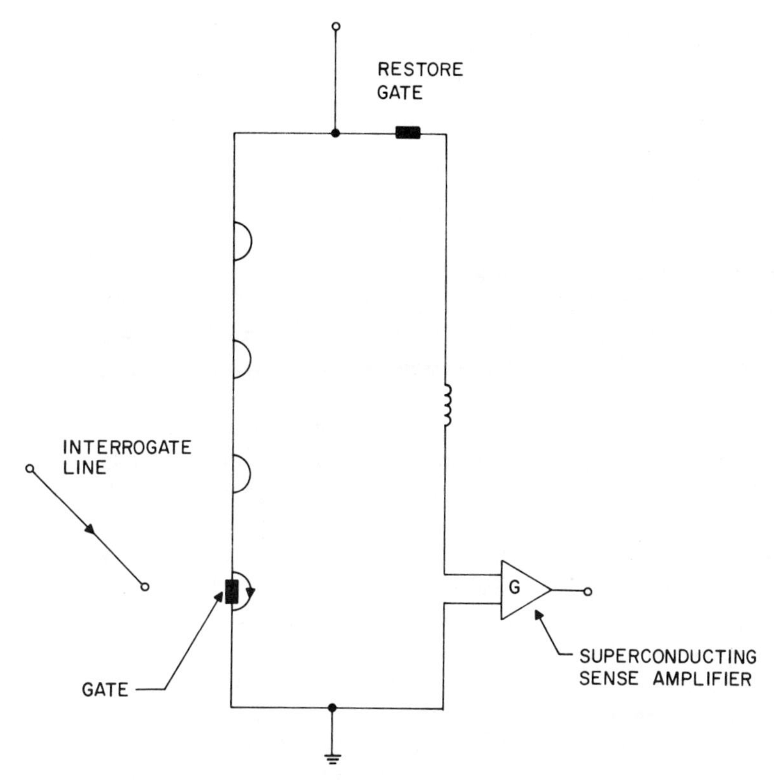

FIG. 18. Current stretch sensing.

while inductive sensing is most efficient at a bandwidth corresponding to the memory-cell time constant, in this case the current in the large loop can be sensed at a lower bandwidth (since it is static).†

IV. Memory Organization

The cost of a superconducting storage cell (or magnetic core) is considerably lower than a current driver or sense amplifier. Thus memory organizations have evolved which, in effect, balance out this inequality. The interesting fact is that the organizations used for magnetic memories apply almost directly to superconductive memories. As will be shown, these organizations are based on the inherent nonlinear current-field switching characteristic of a superconductive cell; this characteristic resembles that of a magnetic core. As was indicated in the last section, the reversal of

† See note added in proof: "5. Continuous sheet sensing," p. 266.

circulating current (or flux) in a superconductive loop is based on exceeding a critical field; similarly the reversal of magnetization in a core is based on exceeding a coercive force.

A. Word Organization

Probably the simplest organization to understand is "word organization," which is pictured in Fig. 19a. In this organization a memory cell is "selected" by activating the two wires that intersect at the cell location; "selection" is defined as the process of reading or writing the cell. This type of selection is known as "current coincident selection."

By activating one word line and certain digit lines, a pattern of cells is selected for writing along the word line. This pattern corresponds to the pattern of activated digit lines referred to as a "word." "1" is written into a selected cell, "0" into an unselected cell. Note that for this write operation, the function of the word line is only that of defining the row of cells while the digit line is used to actually write the information. In terms of the critical field of the memory cells at digit- and word-line intersections, the digit current must be smaller than that which will generate the critical field. This is because cells on other word lines must not be "disturbed" during the writing of the selected word.

Similarly, the word current must be larger than the critical field since every cell along a word line must be selected during writing, regardless of whether a "1" or "0" is to be written. During reading a word line is pulsed, and the voltages along the digit lines indicate the information in the stored word.

A bridge cell, word-organized memory array is shown in Fig. 19b. A current-steering arrangement is shown in the word dimension. Such decoders will be discussed in Section IV.C. Note that this memory satisfies the above criteria for writing. Digit-line current only contributes to stored flux; digit current is less than that needed to switch the gate resistive in the absence of word current. Word-line current does not contribute to stored flux; its value must exceed that needed to switch the gate resistive.

In order to read the bridge-cell memory, a word current is generated, and stored flux is destroyed along every cell (containing a "1") on the line. By detecting the voltage across the terminals of each digit line, the word can be detected; the number of sense amplifiers equals the number of digit lines. Note that this sensing arrangement is a destructive readout (DRO) so that during memory operation each read is followed by a write where either the old or new information is stored.

The basic advantage of a word-organized memory is the minimal demand

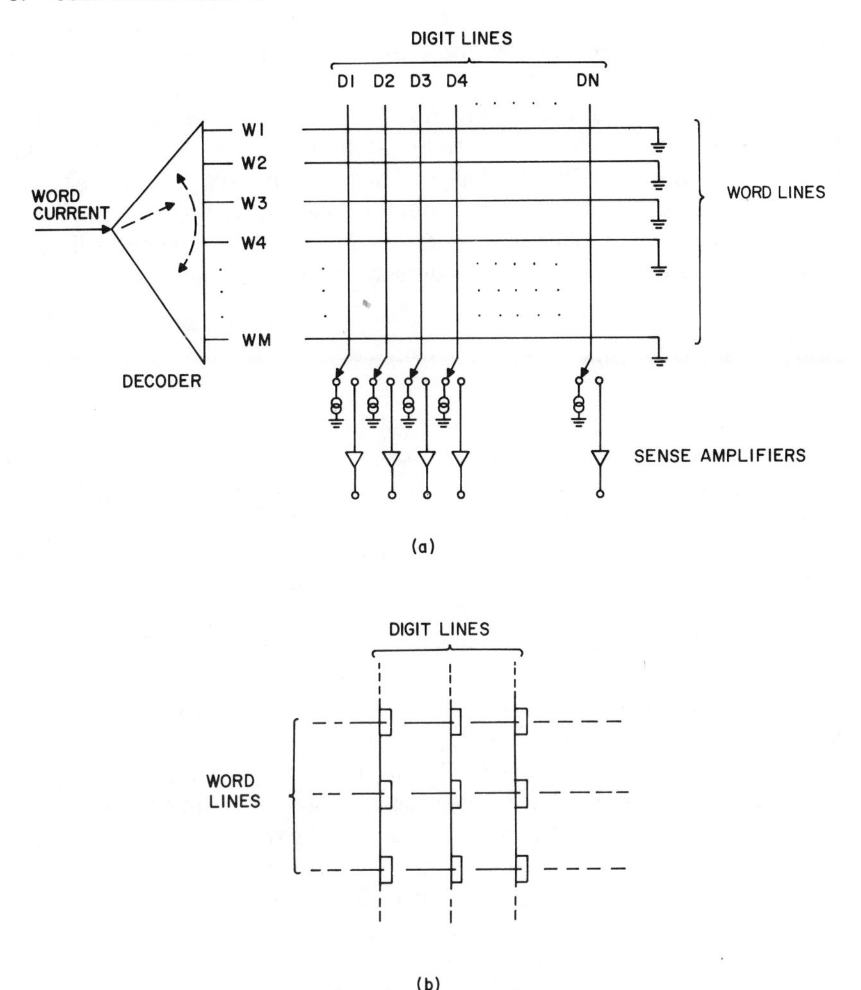

FIG. 19. Word organization: (a) Schematic; (b) Bridge-cell array.

placed on the control of material parameters in the cell array. Experience indicates that the effective critical fields of the gates of superconductive cells vary from cell to cell. This is due both to imperfect tracking of material constants and to variations in word- and digit-line widths (remember the word or digit current needed to switch a gate is proportional to line width). However, for a word-organized array, it is only necessary that the word current be larger than that needed to switch the gate in the memory having the highest effective critical field. Similarly the digit current must only be lower than that needed to switch the gate having the lowest effective critical field. Thus the degree of materials and geometrical control is indeed minimal.

The disadvantage of a word-organized array is that for large memories the word-dimension size places extreme requirements on the decoder. This will be discussed later in Section VI. Superconductive memory sizes of about 16 million bits are most meaningful. With memory word lengths of about 64 bits, this implies a requirement of about 250,000 word lines; a requirement not easily met by decoders as will be discussed below.

B. Bit Organization

A bit-organized memory minimizes the demands on the decorder which was discussed in the last section. Shown in Figs. 20a and 20b is such an organization, which is sometimes referred to as a three-dimensional organization thus contrasting it to word organization which is basically two-dimensional.

A cell in a digit plane is selected by energizing the pair of lines which intersect at the cell location. A word is selected by energizing the corresponding lines on each digit plane, as shown in the figure. Thus, corresponding selection lines on different digit planes can be serially connected as in Fig. 20c.

Considering the 16-million-bit (64-digit-word) memory discussed in the last section, two decoders each capable of handling about 500 lines are needed. ($W^{1/2} = 500$, $W = 250{,}000$). This is in sharp contrast to the word-organized memory requiring one decoder capable of handling 250,000 lines.

The disadvantage of bit organization is that the requirements on material and geometry control are more stringent than word organization. The current in a single selection line must be less than that needed to exceed the critical field of the memory cell having the lowest effective critical field in the memory (since, as discussed above, a cell is selected by a coincidence of two energized lines). Similarly the sum of the currents in two coincident selection lines must generate a field which exceeds the highest critical field cell in the memory. Since a decoder merely steers current, ideally the current in the selection lines (in a dimension) must be the same. Thus the above-stated requirements can be expressed as follows:

$$I_X < C_1 H_c\text{, lowest;} \qquad I_Y < C_2 H_c\text{, lowest;}$$
$$(I_X/C_1) + (I_Y/C_2) > H_c\text{, highest} \tag{16}$$

C_1 and C_2 are, of course, geometrical constants. For a symmetrical situation, $C_1 = C_2$. Thus

$$C_1 H_c\text{, lowest} > I_X = I_Y > \tfrac{1}{2}(C_1 H_c\text{, highest}) \tag{17}$$

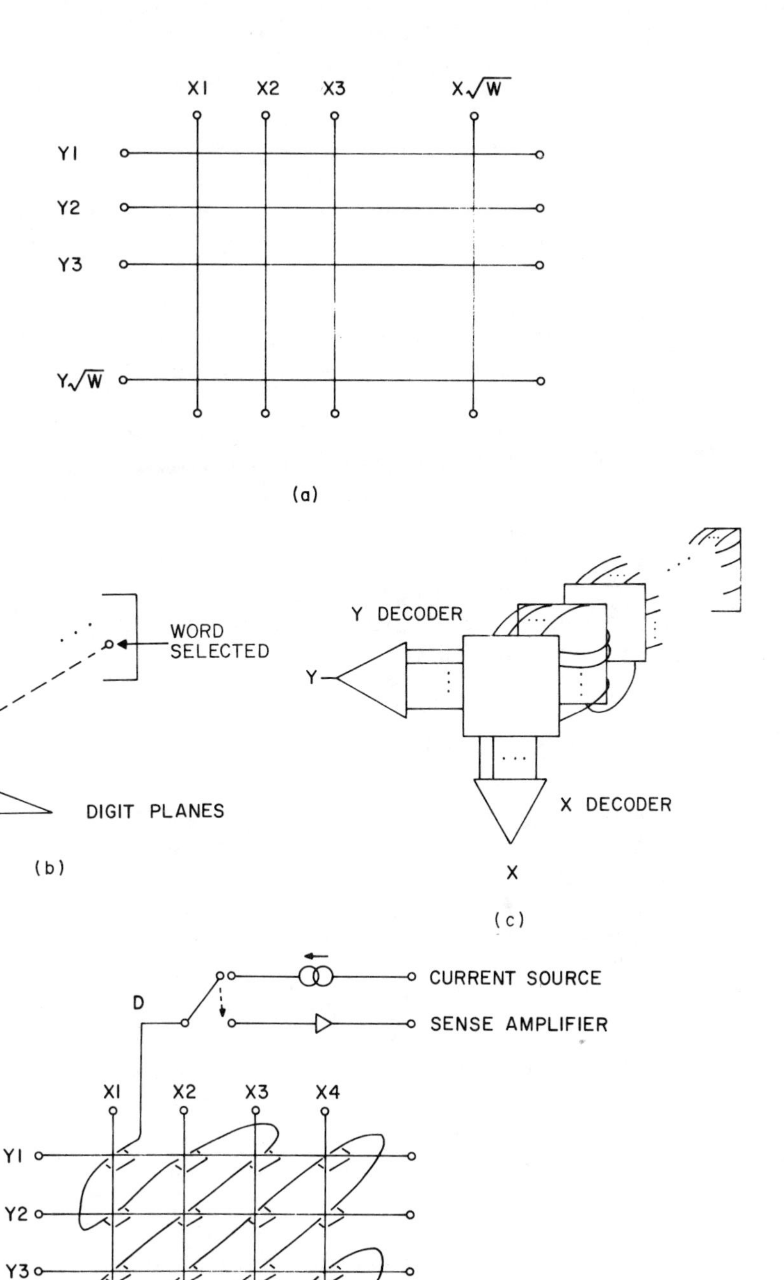

FIG. 20. Bit organization: (a) Digit plane; (b) memory; (c) memory connections; (d) loop-cell digit plane.

As contrasted to the word-organized memory, where the word current only has a lower bound for proper memory operation, in this case the selection currents have an "operating margin." This has implications regarding the control of current-supply stability and control of the decoder so that it does deliver approximately the same current to every line in memory.

The best situation that can be expected is perfect control of material parameters and line geometry so that H_c, lowest $= H_c$, highest. In this situation, Eqs. (16) and (17) indicate that I_X and I_Y can be set in the middle of the operating margin and can be allowed to drift $\pm 33\frac{1}{3}\%$ of this value without interfering with memory operation.

In Fig. 20d is a three-wire-loop cell-memory digit plane indicating not only the X and Y selection lines but a "digit line." This line, as explained in a previous section, is used to carry the current stored in the selected memory cell during writing and to detect the voltage across the cell during reading. Thus the digit plane in Fig. 20d has one sense amplifier (not shown) associated with it.

C. Cryotron Decoder

One of the early selling points of a bit-organized superconductive memory was the possibility of utilizing a cryotron decoder. Since such decoders can be fabricated in the same manner as the memory-cells, it can be integrated with the memory arrays. This would then contribute to the low cost of the memory itself. As will be shown, the number of cryotrons in a decoder is far less than the number of cryotrons in the memory array. Thus, the decoder cost is a small fraction of the memory cost.

A cryotron "tree" decoder is shown in Fig. 21. The tree current is less than that which would switch a tree gate. By activating one address line in each level of the decoder, only one nonresistive path exists between the apex of the tree and the set of selection lines. Upon application of a current step to the tree apex, current first splits equally among all selection lines. After five time constants τ, 90% of the current is flowing in the "selected" selection line. A digit plane of a bit-organized array with two decoders is shown in Fig. 22a. The combination of activated address lines defines the "address" of the selected cell. Note that a nine-level decoder is needed to steer current to one of the 500 X lines mentioned earlier.

In Fig. 22, the memory with integrated decoders is shown. Note that, consistent with the bit organization described in the last section, the corresponding decoder address lines in each digit plane can be serially connected. Rather than having only a pair of decoders for the entire memory,

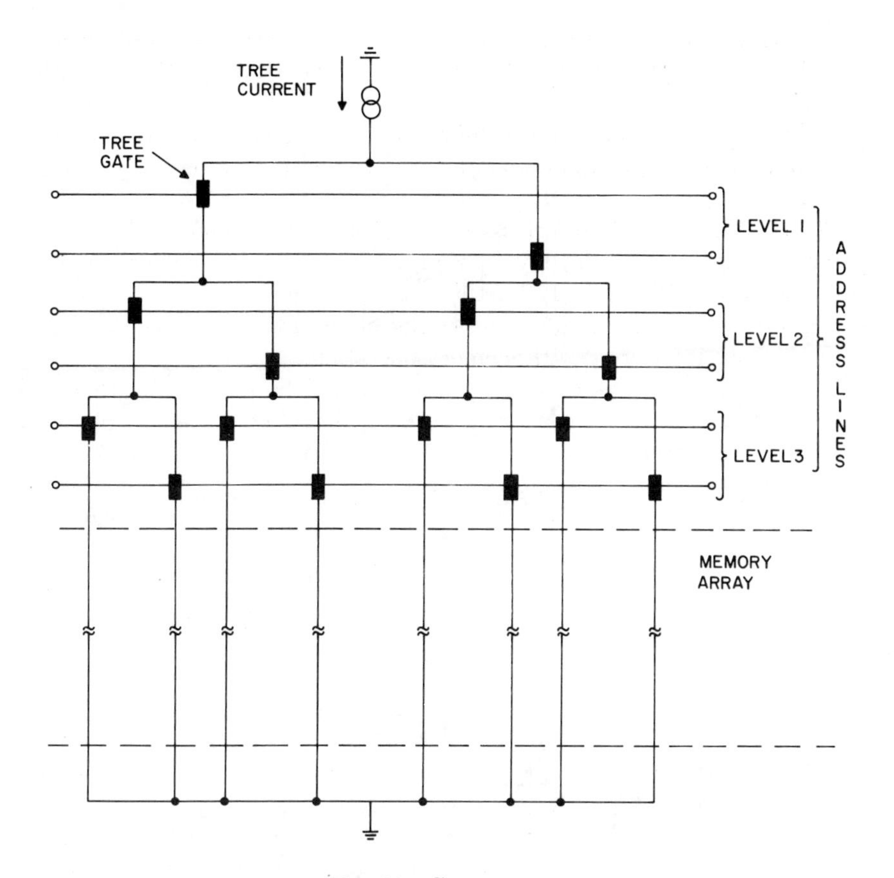

FIG. 21. Cryotron tree.

as shown in Fig. 20c, each digit plane has a pair of decoders. This arrangement is logically redundant but results in a substantial reduction of the number of connections to each digit plane. For example, in a 16-million-bit memory containing 64 digit words (a 250,000-bit digit plane), the number of connections to a digit plane is as follows (see Fig. 22): 2 tree inputs; 18 address lines; 9 per tree; and 1 input to the digit line of the three-wire array (not shown in the figure). Note that each of the above numbers are to be multiplied by 2 to account for current return lines. Thus, the entire memory stack is inputed by 18 address lines, 2 tree lines, and 64 digit lines—a small number of inputs for a 16-million-bit memory. This result should be compared to Fig. 20c, where there are more than 1000 connections per plane and to the memory stack.

The above-mentioned advantages of cryotron decoders must be compared

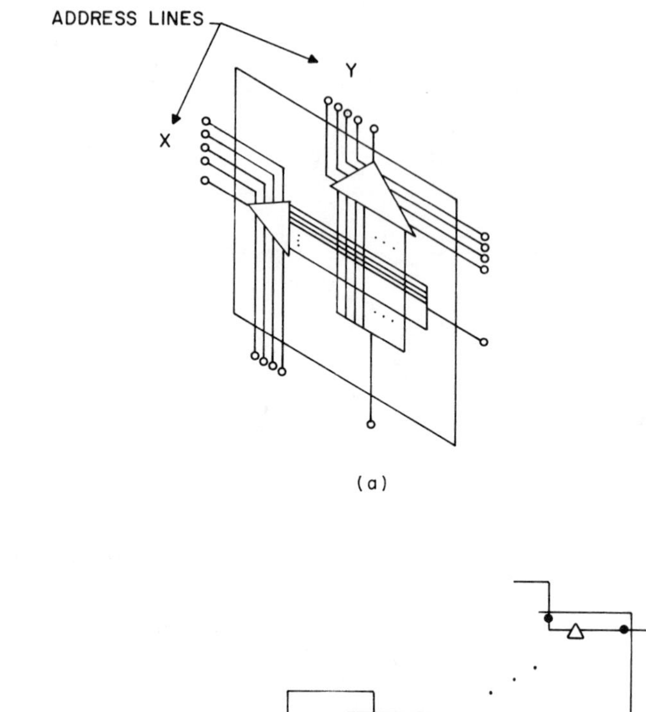

FIG. 22. Cryotron decoders: (a) Digit plane; (b) memory.

to their two disadvantages. First, the time constant of the decoder itself must be examined. An equivalent circuit of a cryotron tree is shown in Fig. 23. Rough "order of magnitude" reasoning indicates that

$$\tau = nL/R \tag{18}$$

where L is the inductance of a selection line, R is the resistance of a gate in the cryotron tree, and n is the number of levels in the tree.

As was mentioned, a 250,000-bit digit plane requires decoders such that

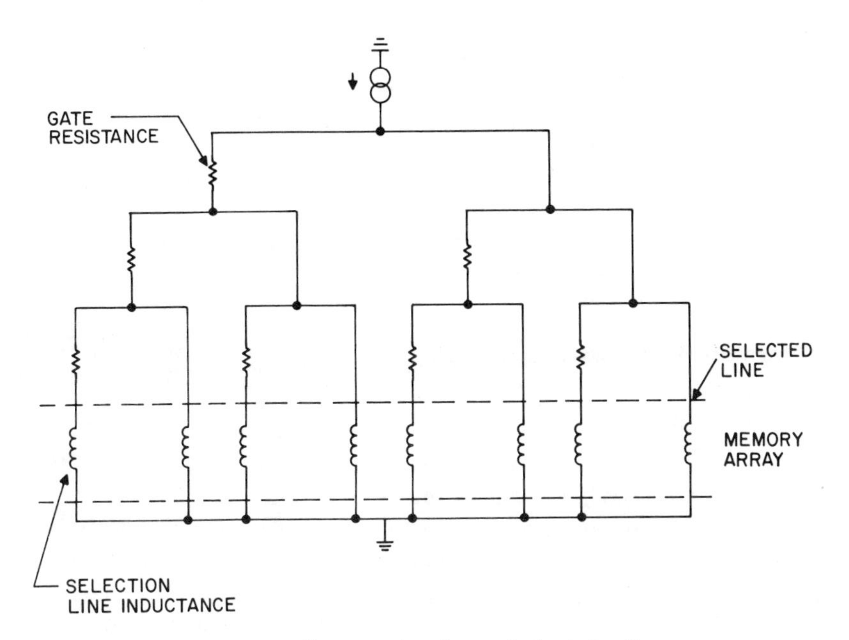

FIG. 23. Cryotron decoder equivalent circuit.

$n = 9$. For a memory cell occupying an 8- $\times$ 8-mil square, a selection line is 500×8 mil $= 4$ in. long. The inductance of this (1-mil-wide) line is 10^{-12} H/square $\times$ 4000 mils $= 4$ nH. For crossed film cryotrons, $R = 2 \times 10^{-3}\ \Omega$. Thus $5\tau = 22.5\ \mu$sec. This time constant is excessive since a complete read or write cycle corresponds to current steered into and out of a desired selection line. By increasing the gate resistance R by an order of magnitude, memory cycle time can be reduced to an acceptable level. This can only be done successfully by using longer cryotron gates; however, this requires additional space for the tree.

An alternative is to use cryotrons with Josephson tunnel junction gates. Anacker (1969) has considered a bit-organized array with two tunnel cryotrons per memory cell and a third tunnel cryotron for NDRO sensing. For a 64×10^3 cell array driven by a tunnel cryotron decoder, total reading and writing cycle times are estimated to be 46 and 30 nsec, respectively.

A second and not as easily solvable disadvantage of the superconductive tree is not a consequence of the tree *per se* or even superconductivity itself. The very advantages, discussed above, gained by integrating the decoder and array, result in a serious intolerance of the digit plane to failures in the decoder. Failures near the tree apex result in large fractions of unusable memory cells. Efficient redundancy techniques to alleviate this problem have not yet been developed, particularly for the above-mentioned "long cryotron gate" trees where real estate is at a premium. The use of only two cryotron decoders, connected as shown in Fig. 20c is not acceptable owing to the resulting large inductance of each selection line and its effect on decoder time constant.

In Section IV.D, a superconductive memory system having room-temperature decoders, and connected similarly to Fig. 20c, will be described. The aforementioned advantages of cryotron decoders are lost; however, a more practical system evolves.

D. Hybrid (Word-Bit) System Organization and Characteristics

1. *System description*

To tailor a memory system specifically to the unique properties of the three-wire cryoelectric cell, the hybrid AB system was devised (Gange, 1967, 1968). The organization is word bit in character rather than being purely bit organized or word organized. Advantages claimed for this organization include digit strip lengths that are shorter than those of a bit-organized system and the feasibility of testing individual planes rather than stacks of planes. Another important feature of this design involves the use of room-temperature decoders rather than decoders deposited directly on the substrate. Because of the special properties of a cryoelectric memory system to be discussed in subsequent sections, only two such decoders are required for the complete system. In addition, many other benefits result from the use of external decoders. Fabrication and evaluation of memory planes are greatly simplified increasing yield; commercially available decoders can be used at room temperature; many more memory cells can be put on one substrate; cycle times are no longer limited by cryotron decoding trees; thermal loading is reduced; and, finally, perfect planes are no longer absolutely necessary. On the last point, extra elements can be included on the plane and only the good portions would actually be used. Without a cryotron decoder, the number of input wires and interconnections needed is greatly increased but still falls within the capability of present technology.

The hybrid AB system employs three-wire cryoelectric memory cells as basic elements. Thus a write operation requires current coincidence of an A drive line, a B drive line, and a D digit line, while a read operation requires only AB current coincidence.

The A lines provide the word dimension and thread from plane to plane through a memory stack. Let A_T be the total number of such lines. The B lines are segmented into d parts and only B_T/d lines appear on each

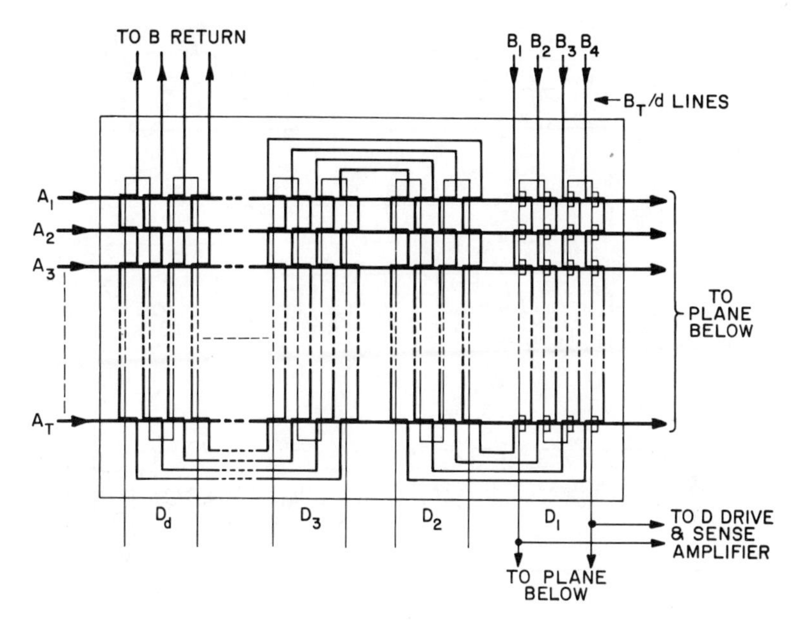

FIG. 24. AB organized array. [Sass *et al.*, 1967b.]

plane where B_T is the total number of B lines in the memory. (See Fig. 24.) The B lines on each plane thread through a number D_P of digit lines on a plane that is equal to the number of digits in a word. Each D line includes a number of digit strips equal to B_T/d which are serially connected together as shown in Fig. 25. Each digit strip is orthogonal to the A lines and contains a number of cells equal to A_T. Each B line threads through one strip of a D line, then turns and threads through one strip of the next D line, and on through the plane. Thus the number of words on a plane, W_P, will be equal to $A_T \times B_T/d$. The total number of words in the system will be $W_T = P \times A_T \times B_T/d$, where P is the total number of planes. Similarly the number of bits on a plane, N_P, will be equal to the number of bits on a strip, A_T, multiplied by the number of strips on a D line, B_T/d, multiplied by the number of D lines on a plane, d. Thus $N_P = A_T \times B_T$. The total number of bits in the memory will then be $N_T = P \times A_T \times B_T$.

Since each D line on a plane is associated with one digit of all words on that plane corresponding digits on different planes can be serially connected. Write current is then applied serially to all D lines in one group while output sense signal will come from only one member of the group. The number of D lines which can be so grouped will be limited by loading effects and speed considerations.

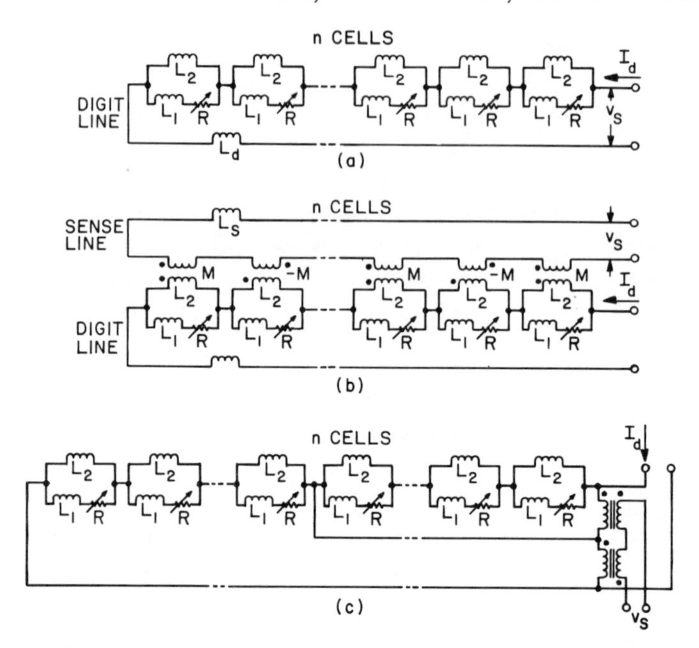

FIG. 25. Memory array equivalent circuit: (a) Digit line; (b) digit line and separate sense line for write-noise cancellation; (c) center-tapped digit line for write-noise cancellation. [Sass *et al.*, 1967b.]

Extra A, B, and D lines can be included on a substrate and the good portions selected after a preliminary checkout. Thus redundancy is an important feature of the hybrid AB system and 100% defect-free planes are no longer required.

Assuming an A-line decoder of a levels, then $A_T = 2^a$. Similarly, $B_T = 2^b$ with a B-line decoder of b levels. Then the capacity or total number of bits in the memory is $P \times 2^{a+b}$ with a total number of words $W_T = P \times 2^{a+b}/d$. The number of wires into the system is minimized when $P = d$, the situation implied before, and for this case $a = b = m$, and $A_T = B_T$. Then $N_T = P \times 2^{2m}$ and $W_T = 2^{2m}$. For any memory, the capacity and number of digits in a word will then determine the system structure. A proposed 10^8-bit system is described in detail (Gange, 1967, 1968) which requires 8704 wires into the memory and occupies about 1 ft^3 in liquid helium.

An actual system containing about 14,000 active loop cells was operated at a 4 μsec repetition rate in a closed-loop routine (Gange, 1967, 1968). Single planes containing 262,144 bits have also been produced and partially tested with packing densities on the order of 13,000 bits/in.^{2} (Gange, 1967, 1968, 1969). These planes will serve as building blocks for multimillion-bit cryoelectric memory systems.

2. *General electrical characteristics*

The results of Section II can be applied to the hybrid AB structure to determine system electrical characteristics (Sass *et al.*, 1967a, b, 1969). The periodic capacitive and inductive loading effects, including those of interconnections, can be considered as effective distributed inductances and capacitances equal to the low-frequency equivalent values. For typical dimensions and materials, characteristic impedances are real and low and pulse degradation is negligible. All loading acts to decrease the propagation velocity, which, together with the length of the lines, will determine the minimum cycle time for the system. Sass *et al.* (1969) estimate that for a 250-nsec one-way propagation time, one transistor can drive or sense roughly 50,000 bits. The absence of half-select noise in loop cells helps make the sensing of so many cells on one line possible. The low characteristic impedance and resultant small back emf of the lines result in modest driver requirements of about 200 mA and 5 V. No other electronic limitations on system performance are envisioned for capacities up to 10^8 bits.

3. *Noise considerations*

a. Random noise. Large capacity cryoelectric memories contain closely packed cells which produce sense signals smaller than those of conventional magnetic core memories, ranging from 40 μV to hundreds of microvolts at the input to the sense amplifier (Sass *et al.*, 1967a, b). Therefore, the effects of random noise on system performance must be carefully assessed. Blatt (1964) has analyzed magnetic film memory systems, and his results were extended to cryoelectric memories by Sass *et al.* (1967b) and will be briefly reviewed here.

The analysis assumes that the noise contribution from normal cryotron gates in cells at 3.5°K is negligible and that the major source of random noise is in the first stage of the sense amplifier. The output of the amplifier goes to a threshold element biased at $V_m/2$, where V_m is the expected peak sense-signal amplitude at the output of the amplifier. This biasing arrangement provides equal probability for a "0" or a "1" error. The signal-to-noise ratio at the input of the threshold detector determines the mean time between errors (MTBE) as given by†

$$\text{MTBE} = \frac{1}{f}\frac{1}{m}\left[\frac{2}{1 - \operatorname{erf}(\alpha/2\sqrt{2})}\right] \tag{19}$$

† Note that in the reference, Sass *et al.* (1967b), a factor of 2 was omitted in the denominator of the argument of the error function.

where f is the number of readouts per second, m is the number of digits interrogated in parallel, α^2 is the signal-to-noise ratio, and MTBE is in seconds. For a MTBE of 10,000 h (over a year) or better, α should be equal to or greater than 16 in a memory with a 1-MHz repetition rate with 100 bits read out in parallel. With an amplifier frequency response limited to the reciprocal of the sense-signal duration, it turns out that the peak sense-signal amplitude must be about 20 times the rms noise voltage of the amplifier for the above conditions to be met.

b. Deterministic noise. The digit line of a three-wire cell is utilized both to deliver current to be stored during a write operation and as a digit (sense) line during a read operation. (See Fig. 25a.) A pulse of digit current produces a transient ("write noise") which is considerably larger than the sense signal and may overload the sense amplifier. A recovery time must be allotted between a write and a read pulse which slows down total memory cycle time.

Several solutions to this problem have been employed. One scheme involves adding a separate sense line arranged so that coupling of adjacent cells (or groups of cells) on the digit line is of opposing polarity as shown in Fig. 25b. This scheme has been found to be quite effective by Sass *et al.* (1967b) reducing write noise by a factor of at least 10^3. Another method provides for write noise-cancellation in the coupling transformers at the terminals of the digit line (Sass *et al.*, 1966). The digit line is divided into an even number of groups containing equal numbers of cells, and the transformer secondaries are connected in series opposition as shown in Fig. 25c. Well-matched transformer cores are required for this case. Other similar cancellation techniques are possible.

When drive lines are pulsed during the read portion of the memory cycle, some of the energy can couple to the sense line of an array causing unwanted voltages to appear across it ("read noise") which may mask the true sense signal. Some of the coupling mechanisms and techniques for noise minimization will be described, following the analysis of Keneman and Cosentino (1969).

To reduce inductive coupling, drive and sense lines should cross at right angles wherever possible and the maximum distance possible should be maintained between them. While the ground plane of a cryoelectric array helps greatly to decouple drive and sense lines, further precautions are necessary to keep actual coupling to negligible values. The worst case mutual inductance between a drive line and a sense line is calculated to be

$$M = (\mu_0 h_D/\pi W)\{\tan^{-1}[(W + s)/h] - \tan^{-1}(s/h)\} \qquad (20)$$

where M is the mutual inductance per unit length, h_D is the height of the

D line above the ground plane, s is the separation between the drive line and the digit line, W is the width of the drive and digit lines, and h is the height of the drive line above the ground plane. In some cases the coupled voltage which will be across the sense line (in-mode) can be appreciable.

In-mode noise can also be capacitively coupled from a drive line to a sense line. The analysis cited produces the following result for this type of noise coupled from an A drive line to a D digit line:

$$V_n = (\omega^3 L_A L_D C_D I_A/12)[1 + (3/N) + (2/N^2)] \tag{21}$$

where V_n is in volts, N is the number of crossings of the A drive line with the D digit line, L_A is the inductance of the A line from crossing 1 to crossing N, L_D is the total inductance of the D line between crossings, and C_D is the total direct capacitance between the D and the A lines. This noise can be reduced by routing the drive line in a "go and return" arrangement such that cancellation of coupled noise takes place on the sense line.

Common-mode noise is also of interest because of the possibility of its conversion to in-mode noise. A drive line may get charged to some average potential which then causes currents to flow to system ground through unbalanced circuits producing an in-mode voltage across the sense line. The use of balanced drivers reduces the magnitude of the problem but does not solve it entirely. Keneman and Cosentino (1969) show how a cryoelectric array can be modeled as a simple circuit for common-mode noise considerations to determine the effect of a selection current in a particular drive line on a particular sense line. The actual analysis and results are strongly dependent on array layout, cell geometry, and the materials used.

When stacks of memory planes are interconnected to obtain a large memory capacity, additional components of in-mode noise may be produced between planes directly or by conversion of common-mode voltages on the various planes. Some of the techniques which were found to be useful for reducing noise on a single plane (e.g., "go and return" layout) are also applicable to stacks. Again, specific details of connections and organization are important as well as the design of the stack holder.

4. *Thermal characteristics*

The use of external decoders in the hybrid system results in a large number of drive conductors in the liquid-helium environment. However, only a few wires are pulsed at a particular time so that I^2R heating is negligible. Conductor materials with low thermal conductivity and reasonable electrical resistivity are therefore required. Lead (Pb) has been found to be a useful material in this respect. A 10^8-bit memory system, utilizing lead conductors and three stages of cooling at progressively lower temperatures, was found to require ~5 W of cooling with only a few milliwatts of I^2R heating. The cost of a refrigerator to provide this cooling is estimated to be about one-half the cost of the total memory system (Gange, 1967, 1968). This point will be discussed further in Sections VI.A and VI.B.

E. Ryotron Decoder

This section would not be complete without mentioning ryotron decoders (Gange, 1964). A ryotron is a field controlled variable inductance; a decoder employing ryotron gates can be seen by replacing the gates in Fig. 21 by inductances whose value can be increased by activating the corresponding address line. Thus in the circuit in Fig. 23, the resistors are to be replaced by very large inductors (larger than the selection line inductance). Analysis of this circuit indicates that a current step applied to the apex of the tree transfers to the selected path *instantaneously*. Thus the ryotron decoder speed is as fast as the switching speed of the ryotrons themselves.

While this concept is exceedingly elegant, it is noted that for proper ryotron operation

$$L_{\text{selection line}} \ll L_{\text{Hr}} \tag{22}$$

$$L_{\text{Lr}} \ll L_{\text{Hr}} \tag{23}$$

where L_{Lr} and L_{Hr} are the low- and high-ryotron-inductance states, respectively. This condition is difficult to meet with practical components in the limited space allowable for the individual ryotron. As an example, a typical ryotron is shown in Fig. 26 where the ryotron is in its high-inductance state when its ground plane is normally conducting (Miller *et al.*, 1964).

A method for achieving high-inductance ratios by using geometries which involve field control of the kinetic energy of superelectrons has been suggested (Gange, personal communication); however, physical realization of this concept remains for the enterprising experimentalist.

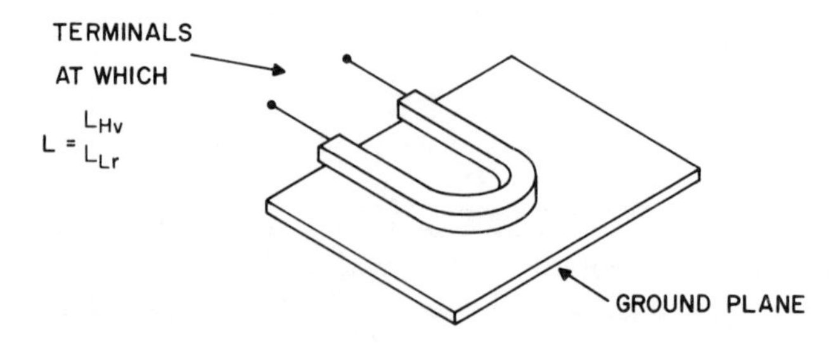

Fig. 26. Horseshoe ryotron.

V. Fabrication Technology

The emerging interest in cryoelectric circuitry for computer applications in the late 1950s and early 1960s was paralleled by, and served as a partial stimulus for, a period of active progress in thin-film and vacuum technology. Arrays of practical size required a level of circuit integration that had not been approached in other fields. While methods of preparing the artwork which ultimately defines the configuration of the circuit are quite similar among batch fabrication technologies, the requirements for good definition and positional accuracy of the smallest circuit element over large working areas in cryoelectrics rather quickly found a practical limit in the ratio of 1000:1 for maximum-to-minimum pattern dimensions in a single artwork pattern (Pritchard *et al.*, 1964). Problems associated with the fabrication of the circuits themselves, on the other hand, were peculiar to the field of cryoelectrics because of the specific materials employed.

Many of the fabrication techniques developed industrially have commercial significance; information concerning them is subject to proprietary restrictions. The published literature therefore cannot accurately reflect the present state of the art in practice, and, as a result, more data are available concerning methods which have been superseded by newer refinements. The brief treatment given here, however, is not sufficiently detailed in scope to suffer seriously from such circumstances.

A. Thin-Film Deposition

The materials most commonly employed in cryoelectric circuitry are: (1) lead or niobium as "hard" superconductors for those portions of the circuit which remain superconducting during operation, such as ground planes, drive and control lines, contacts, etc.; (2) tin or indium as the "soft" superconductors which can be switched by currents or fields; and (3) silicon monoxide or photoresist as a dielectric separating layer. Layer thicknesses typically range from 0.3 to 1 μ. All but the last of these are formed by vacuum deposition; substrates are usually glass.

Studies (Caswell, 1961a) of the influence of gases on vacuum-evaporated tin films have shown that oxygen and water vapor are the major constituents of the residual vacuum system atmosphere which affect superconducting characteristics. These gases decrease the surface mobility of the deposited tin atoms so that a finer-grain structure is obtained. In films deposited through stencil masks, which give sloping rather than sharp edges, the thin edges of fine-grained films are electrically continuous. This results in a higher and broader critical field transition for the film. Such effects are minimized by maintaining the residual partial pressures of O_2, H_2O, and CO_2 below 0.05, 0.4, and 0.8 μTorr, respectively, while depositing tin at a rate of 50 Å/sec. Films formed in this manner have a sloping edge that consists of an agglomerated island structure which is electrically discontinuous. Other means of suppressing the edge effects include mechanical trimming or etching away of the edges (Delano, 1960), and overlapping the entire film with approximately 100 Å of gold or copper to depress the critical field of the thin edges, but not of the central portion of the tin film, by means of the proximity effect (Ames and Seki, 1964). Only the last two of these remedies are practical for circuit fabrication.

The partial pressure requirements given above can be relaxed by a factor of 10 without any significant contamination of the bulk of the tin film by residual gases occurring in typical systems. Critical temperature variations can occur in tin films because of differential thermal contraction of substrate and film between room and liquid-helium temperature. Increases as high as 0.23°K have been found (Blumberg and Seraphim, 1962). Such variations would be greatly reduced if the films were formed with both diad axes of the white tin crystallites in the plane of the film. However, no techniques which accomplish this on amorphous substrates is known.

Indium films (Caswell, 1961b) are qualitatively similar to those of tin in their sensitivity to residual oxygen in the vacuum system, and are considerably less sensitive to water vapor, a major atmospheric constituent of most vacuum systems. Although the critical temperature is reproducible within several millidegrees, the indium films have approximately 60% of the normal resistivity of tin, and are less stable than tin. The low-temperature properties of lead films are also affected by deposition parameters (Caswell *et al.*, 1963). However, uniformity of critical field and critical temperature are of secondary importance for control lines and contacts. As a result, the sensitivity of lead films to contamination by residual gases is low, and precautions taken for the deposition of tin or indium films will be adequate for lead films.

The high values of critical field and critical temperature of niobium make this material attractive for use in cryoelectric circuits. A high sensitivity to residual gases shared by the transition-element superconductors make them difficult to work with in thin-film form. Recently, a successful method for fabricating Nb ground planes for cryotron arrays was reported (Joynson *et al.*, 1967). High evaporation rates were obtained by focusing a 6-kW electron beam onto a Nb block. Heating of the substrates to approximately 400°C, the gettering action of freshly deposited Nb on the vacuum system walls, and a deposition rate of 300 Å min^{-1} gave 2-μ thick films with a transition temperature essentially equal to that of the bulk material.

Anodization of the layer to give 1500 Å of oxide provided insulation as well as a reduction in microroughness.

A great deal of attention has been given to silicon monoxide since it can be vacuum evaporated by techniques compatible with those used for lead and tin, has a high sticking coefficient which minimizes surface migration, and is a reasonably good insulator. Uneven heating of the source material, because of the low thermal conductivity of silicon monoxide powder or pellets, as well as the emission of small glowing fragments which strike the substrate, can be minimized by careful consideration of source design. Unlike lead, tin, and indium films, in which residual stresses quickly anneal to an immeasurably low value at room temperature (Budo and Priest, 1963), films of silicon monoxide generally exhibit stresses which depend upon residual gas pressure P, deposition rate R, and angle of incidence upon the substrate. For low partial gas pressures and high deposition rates at normal incidence, the "natural" stress of pure silicon monoxide is tensile, changing to compressive stress as R/P drops below 5×10^6 Å $\sec^{-1}$ Torr^{-1} for water vapor, 3.5×10^6 for dry oxygen, and 1×10^6 for dry air (Hill and Hoffman, 1967). Other dry gases do not give compressive stress. Exposure of the films to the atmosphere promotes additional oxidation of the silicon monoxide by oxygen and water vapor. The resulting increase in compressive stress can cause reticulation and peeling of the films. Stable films of silicon monoxide can be deposited at 16 Å $\sec^{-1}$ by maintaining partial pressures of water vapor and oxygen below 2.5×10^{-6} and 3.5×10^{-6} Torr, respectively, by prebaking water vapor from the source material, and ensuring that the angle of incidence to the substrate is less than 15°.

Although the use of polymer dielectric films, formed by vacuum deposition of the monomer and polymerized by exposure to an electron beam (Christy, 1960) or ultraviolet radiation (Caswell and Budo, 1963), has been proposed for eliminating stress and small pinholes in the insulating layer, poor pattern definition with stencil masks and contamination of other deposition sources in the vacuum system weigh against such methods.

B. Masking Techniques

Most of the early approaches to the fabrication of cryoelectric circuitry employed the successive deposition of metal and insulating layers through stencil masks in a high vacuum system during a single evacuation cycle. The mechanization of such a process to give adequate control over geometry and material characteristics requires an elaborate system for handling multiple masks, evaporation sources, and substrates. Masks, formed by

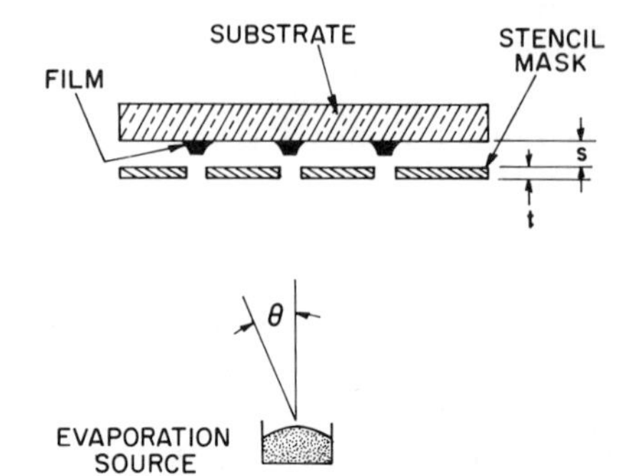

FIG. 27. Vacuum deposition: Illustration of the variation in cross sections of films simultaneously deposited through stencil masks.

machining or etching patterns in a thin rigid sheet must be placed in close proximity to the substrate between the substrate and the evaporation source. As can be seen from Fig. 27 the mask–substrate spacing must be very small to minimize tapering edges of the deposited films which result from a distributed evaporation source and multiple bounces of the evaporant molecules. Similarly, the thickness of the mask should be sufficiently smaller than the width of a slit so that deposited pattern widths will not vary significantly over small deposition angles θ. Rigidity, however, is sacrificed with thinner masks. Slit patterns as narrow as 2 mils etched in 1-mil-thick beryllium copper have insufficient uniformity for large capacity arrays (Burns, 1964). Electroforming techniques, whereby the mask is reinforced by electroplating between slits, allows the slit edges to be defined by much thinner layers and permits 1-mil-wide slits to be formed with less than 10% variation. Stencil mask patterns have been machined in $\frac{1}{32}$-in.-thick aluminum alloy sheets with a pantograph system which gives ± 1-mil registration accuracy from mask to mask (Ames *et al.*, 1962).

Masks are fastened to rigid frames which provide support along the outside edges. As masks and/or substrates are moved into position within the vacuum system, registration is typically obtained by having locating pins on the substrate holder engage holes (Caswell and Priest, 1962) or accurately machined edges (Adams *et al.*, 1966) on the mask holder. Various mechanical systems have been described in which masks and substrates can be independently positioned over fixed resistance heated evaporation sources, and in which evaporation sources and masks are moved into position under a fixed substrate. These systems typically accommodate 5 to 12 sources, 10 to 20 masks, and 1 substrate. Pumping is achieved with diffusion pumps with 6-in. or larger throat diameters in conjunction with cryogenic traps. Film thickness is monitored by special ionization gauges (Giedd and Perkins, 1960) which sense the rate of deposition, or by accumulating a portion of the evaporated vapor on an oscillating quartz crystal (Behrndt and Love, 1961) whose frequency varies with the mass of the deposit. A prototype production facility was developed at General Electric (Adams *et al.*, 1966) in which up to 20 substrates could be sequentially processed in seven isolated evaporation stations during a single pump-down cycle.

In any vacuum deposition system, the masks are exposed to heat radiating from the evaporation source as well as the hot vapor itself. A temperature rise of 10°C can increase the dimensions of a 2-in.-wide mask as much as 0.5 mil; a nonuniformly heated mask may buckle because of differential expansion and severely alter the mask–substrate spacing. In order to avoid these problems and the inherent resolution limits dictated by stencil mask techniques, Pritchard and co-workers (1964) developed a photomask–photoresist–etching process for forming the cryoelectric circuitry. With this technique, a uniform layer of metal is vacuum deposited on the substrate without stencil masking. The sample is then removed from the vacuum system and coated with photoresist. Contact printing of the desired pattern configuration is obtained by exposing the sensitized photoresist to uv illumination through a photographic transparency of the pattern. Chemical development of the exposed photoresist uncovers the portions of the metal film which are then removed by a suitable chemical etchant. The remaining protective photoresist which defined the final metallic film pattern is stripped, and the sample is ready for the deposition of the next thin-film or insulating layer. Such a process permits the use of insulating layers which need not be formed in the same vacuum environment as the metal layers. In particular, Pritchard found that polymerized photoresist itself is a satisfactory insulator for cryoelectric circuitry. When intimate contact is required between portions of two separate metal layers, the surface contamination introduced by exposing the first metal layer to the atmosphere and chemicals can be sputtered away in a low-pressure glow discharge immediately before deposition of the second layer (Pierce, 1968). The resulting ionic and electronic bombardment of the sample serves to further polymerize the photoresist insulation layers.

The photomask–photoresist process eliminates the limited resolution stencil mask step in the process of transferring the original artwork pattern to the final circuitry. Registration of the photomask with previously patterned circuit layers can be under the manual control of an operator equipped with optical aids. Between one and two orders of magnitude improvement in device packing densities is obtained at the cost of increased handling of samples during fabrication. The advantages of both techniques have been incorporated in a hybrid process (Adams *et al.*, 1966) whereby the final layer of lead in cryotron circuits is photoetched with high resolution, while the coarser tin and insulator patterns can be deposited in a single pump-down in vacuum.

VI. Present Outlook

A. Economics

The superconductive memory has an economic consideration that is unlike those of any existing electronic memory, namely, that of a refrigeration requirement. This requirement imposes a fixed capital investment that can be amortized only by spreading it over a large number of memory bits. Thus in order to compete with other memory technologies, the superconductive memory must promise large capacity.

In order to estimate this capacity the following assumptions can be made. A refrigerator cost of $20,000.00 is reasonable—not for the first few units produced, but as a good estimate of the cost-curve asymptote.

Furthermore a cost per bit of 0.2¢ for the system can be assumed. This assumption is certainly not very defensible for the following reason. It is not "price per bit" as advertised by a computer manufacturer, but rather a sum total of manufacturing cost average over the product line. Since the method of the costing can vary from one manufacturer to another, by a wide range, the 0.2¢ quoted is somewhat ambiguous. However, using the commonly advertised price of core memories (about 3¢/bit) as a reference, the 0.2¢ figure can be given limited meaning.

Taking that figure, it is seen that the refrigerator is amortized (assuming it to be 50% of the memory cost) at about a 20-million-bit memory capacity. Larger capacities yield lower bit costs. As was indicated earlier, peripheral electronics do not grow directly proportionately with memory capacity. Refrigerator cost is also relatively constant with respect to memory size. Thus, a growth to a 100-million-bit system can be expected to produce about twice the peripheral electronics cost.

The application of this type of memory has long been advertised as a replacement of electromechanical storage because of the inherently favorable relative performance capability of the superconductive memory (i.e., access times of microseconds rather than milliseconds would be feasible). While the cost of the superconductive memory would not compete with electro-mechanical storage on equal footing, some of the former's access-time advantage can be traded for decreased peripheral electronics cost. This would yield an extremely competitive product which could revolutionize large capacity storage as well be indicated in Section VI.C.

B. Refrigeration

Rather than devoting this section to a discussion of various refrigeration alternatives, a general treatment of the system requirements for the refrigerator will be presented. The rationale is that knowledge concerning the latter subject is considerably more plentiful. Furthermore, it is hoped that among the readers of this book will be refrigeration specialists who will be stimulated by these requirements.

The memory stack itself is to occupy 1 ft^3 at a temperature of 3.5°K. Surprisingly enough, the stack itself dissipates a negligible amount of power (assuming the hybrid approach mentioned above). The principal heat load of the system is the wiring between the stack and room temperature. This wiring (lumped together) can be visualized as a $\frac{1}{2}$-in.-diam metal rod, one end of which is at room temperature, and the other end is at 3.5°K.

The basic approach in minimizing the effect of this load is to minimize the sum of the electrical power dissipation and thermal power conduction. It has been shown (Gange, 1968) that, for the 16-million-bit hybrid system, the refrigerator must supply approximately 1 W of cooling power to maintain the stack at 3.5°K.

The combined requirements of a 1-ft^3 working volume at 3.5°K and a power load of 1 W are, in principle, well within the state of the art in helium refrigeration. However, at the time of the writing of this manuscript, the marriage of the memory and refrigerator had not yet been reported in the literature. This indicates a need for engineering effort to be devoted to this area.

C. Memory Systems

As mentioned in Section VI.A, the justification of superconductive memories is dependent on a need for large capacity electronic storage. A critic of this statement could easily argue that special function "logical memories" could be justifiable at higher cost and/or lower capacity. While in the Introduction it was pointed out that such memories would not be considered here, an answer to these critics is pertinent.

The computer industry has, thus far, failed to produce proved machine designs which employ "logical memories" of sufficient performance and applicability to justify their production. The failure has been at this application level since the superconductive memory cells and accessing circuitry are technically feasible.

Thus, the only well-defined plan is the production of large capacity random-access memories. The burden on the technology is great since it must achieve large capacity even for an initial entry or result in zero revenue until it is capable of this capacity. The latter plan is one which would obviously be unpopular.

The rapid advances in semiconductor memory which are not subject to these initial entry requirements will probably result in its initial capture of this market. However, the recent announcement of Josephson junction storage cells with picosecond switching speeds (Zappe, 1974) indicates that superconductive memories which are faster than semiconductor memories may soon appear (Anacker, 1969).

In conclusion, initial superconductive memories will be random access, low cost, and large capacity to be used as replacements for electromechanical drums. For this application the access time of such a memory, while slow compared to main memories, would add a speed dimension to large capacity storage. The traditional disparity between the speed of main and secondary storage would be greatly decreased, thereby enhancing system throughput.

References

Adams, C. N., Bremer, J., Oka, A., and Ummel, M. (1966). *IEEE Trans. Magn.* **MAG-2,** 385.

Ahrons, R. (1965). Proc. Intermagn. Conf., **9,** 1-1.

Alers, P. B. (1957). *Phys. Rev.* **105,** 104.

Ames, I., and Gendron, M. F., and Seai, H. (1962). *In* "Trans. 9th National Vacuum Symp.," p. 133. MacMillan, New York.

Ames, I., and Seki, H. (1964). *J. Appl. Phys.* **35,** 2066.

Anacker, W. (1969). *IEEE Trans. Magn.* **MAG-5,** 968.

Atherton, D. L., and Tyler, A. R. (1966). *Proc. IEEE (London)* **113,** 1575.

Barnard, J. D., Blumberg, R. H., and Caswell, H. L. (1964). *Proc. IEEE* **52,** 1177.

Behrndt, K. H., and Love, R. W. (1961). *In* "Trans. 7th National Vacuum Symp.," p. 87. Pergamon, London.

Blatt, H. (1964). U.S. Dept. of Commerce, Rep. No. AD-605323.

Blumberg, R. H., and Seraphim, D. P. (1962). *J. Appl. Phys.* **33,** 163.

Bremer, J. W. (1962). "Superconductive Devices," p. 88, McGraw-Hill, New York.

Buck, D. A. (1956). *Proc. IRE* **44,** 482.

Buckingham, M. J. (1958). "Proc. Int. Conf. Low Temp. Phys. Chem. 5th, 1957.," (J. R. Dillinger, ed.), pp. 229–232. Univ. of Wisconsin Press, Madison.

Budo, Y., and Priest, J. R. (1963). *Solid State Electron.* **6,** 159.

Burns, L. L. (1964). *Proc. IEEE* **52,** 1164.

Burns, L. L., Christiansen, D. A., and Gange, R. A. (1963). *AFIPS Conf. Proc.* **24,** 91.

Burns, L. L., Leck, G. W., Alphonse, G. A., and Katz, R. W. (1960). *Solid State Electron.* **1,** 343.

Burns, L. L., Alphonse, G. A., and Leck, G. W. (1961). *IRE Trans. Electron. Comput.* **EC-10,** 438.

Burns, L. L., and Sass, A. R. (1965). *J. Appl. Phys.* **36,** 1105.

Casmir-Jonker, J. M., and DeHaas, W. J. (1935). Physika **2,** 935.

Caswell, H. L. (1961a). *J. Appl. Phys.* **32,** 105.

Caswell, H. L. (1961b). *J. Appl. Phys.* **32,** 2641.

Caswell, H. L., and Budo, Y. (1965). *Solid State Electron.* **8,** 479.

Caswell, H. L., and Priest, J. R. (1962). *In* "Trans. 9th National Vacuum Symp.," p. 138. MacMillan, New York.
Caswell, H. L., Priest, J. R., and Budo, Y. (1963). *J. Appl. Phys.* **34**, 3261.
Christy, R. W. (1960). Symposium on Superconductive Techniques for Computing Systems, ONR Symposium Rep. No. ACR-50.
Crittenden, E. C., Jr. (1958). *In* "Proc. Int. Conf. Low Temp. Phys. Chem. 5th, 1957." (J. R. Dillenger, ed.), pp. 232–234. Univ. of Wisconsin Press, Madison.
Crittenden, E. C., Jr., Copper, J. N., and Schmidlin, F. W. (1960). *Proc. IRE* **48**, 1233.
Crowe, J. W. (1957). *IBM J. Res. Develop* **1**, 295.
Crowe, J. W. (1958). *In* "Proc. Int. Conf. Low Temp. Phys. Chem. 5th, 1957." (J. R. Dillenger, ed.), pp. 238–241. Univ. of Wisconsin Press, Madison.
Delano, R. B. (1960). *Solid State Electron.* **1**, 381.
Edwards, H. H., and Newhouse, V. L. (1965). Proc. Intermagn. Conf., 9.3-1.
Feissel, H., Gallet, F., and Hug, J., (1966). *IEEE Trans. Magn.* **MAG-2**, 406.
Gamby, P., and Maller, V. A. J. (1965). Proc. Intermagn. Conf., 9.2-1.
Gange, R. A. (1964). *Proc. IEEE* **52**, 1216.
Gange, R. A. (1967). *Electronics,* Apr. 17, 111.
Gange, R. A. (1968). *Proc. IEEE* **56**, 1679.
Gange, R. A. (1969). *Electronics* Mar. 17, 108.
Garwin, R. L. (1957). *IBM J. Res. Develop.* **1**, 304.
Giedd, G. R., and Perkins, M. H. (1960). *Rev. Sci. Instr.* **31**, 733.
Goser, K. (1967). *Proc. IEEE* **55**, 1493.
Goser, K., and Kirchner, H. (1967). *Proc. IEEE* **55**, 592.
Hanlon, A. G. (1966). *IEEE Trans. Electron. Comput.* **EC-15**, 509.
Haynes, M. K. (1960). *Solid State Electron.* **1**, 399.
Hill, A. E. and Hoffman, R. R. (1967). *Brit. J. Appl. Phys.* **18**, 13.
Ittner, W. B. III (1960). *Phys. Rev.* **119**, 1591.
Ittner, W. B. III, and Kraus, C. J. (1961). *Sci. Amer.* **205**, 124.
Joynson, R. E., Neugebauer, C. A., and Rairden, J. R. (1967). *J. Vacuum Sci. Tech.* **4**, 171.
Kadereit, H., and Goser, K. (1968). *Proc. IEEE* **56**, 121.
Keneman, S. A., and Cosentino, L. S. (1969). *IEEE Trans. Magn.* **MAG-5**, 412.
London, F. (1961). "Superfluids," 2nd ed., Vol. I. Dover, New York.
Matisoo, J. (1967). *Proc. IEEE* **55**, 172.
Meyers, N. H. (1961). *Proc. IEEE* **49**, 1640.
Meyers, N. H. (1962). *Proc. IRE* **50**, 2352.
Miller, J. C., Wine, C. M., and Cosentino, L. S. (1964). *Proc. IEEE* **52**, 1223.
Mundy, J., and Newhouse, V. (1968). *IEEE Trans. Magn.* **MAG-4**, 705.
Newhouse, V. L. (1964). "Applied Superconductivity." Wiley, New York.
Newhouse, V. L., and Drapeau, R. E. (1965). *IEEE Trans. Magn.* **MAG-1**, 324.
Newhouse, V. L., and Edwards, H. H. (1967). *Radio Electron. Eng.* **33**, No. 3, 161.
Newhouse, V. L., Bremer, J. N., and Edwards, H. H. (1960). *Proc. IRE* **48**, 1395.
Pearl, J. (1965). Thesis, Polytechnic Institute of Brooklyn.
Pierce, J. T. (1968). Intermagn. Conf., Washington, D. C. Paper 6-3.
Pritchard, J. P., Jr., Pierce, J. T., and Slay, G. B. (1964). *Proc. IEEE* **52**, 1207.
Ramo, S., and Whinnery, J. R. (1953). "In Fields and Waves in Modern Radio," 2nd ed., pp. 239–55. Wiley, New York.
Rhoderick, E. H. (1959). *Brit. J. Appl. Phys.* **10**, 193.
Rhoderick, E. H., and Wilson, E. M. (1962). *Nature* **194**, 1167.
Sass, A. R. (1964). *J. Appl. Phys.* **35**, 516.

Sass, A. R. and Nagle, E. M. (1967). *IEEE Trans. Magn.* **MAG-3,** 268.
Sass, A. R., Nagle, E. M., and Burns, L. L. (1966). *IEEE Trans. Magn.* **MAG-2,** 398.
Sass, A. R., and Skurnick, I. D. (1965). *J. Appl. Phys.* **36,** 2260.
Saas, A. R., and Stewart, W. C. (1968). *J. Appl. Phys.* **39,** 1956
Sass, A. R., Stewart, W. C., and Cosentino, L. S. (1967a). *IEEE Spectrum* **4,** 91.
Sass, A. R., Stewart, W. C., and Cosentino, L. S. (1967b). *IEEE Trans. Magn.* **MAG-3,** 260.
Sass, A. R., Stewart, W. C., Dworsky, L. N., Hoffstein, V., and Schein, L. B. (1967c). *IEEE Trans. Magn.* **MAG-3,** 628.
Sass, A. R., Stewart, W. C., and Cosentino, L. S. (1969). *IEEE Trans. Magn.* **MAG-5,** 398.
Stewart, W. C., Owen, H. A., Lucas, M. S. P., and Vail, C. R. (1961). *Proc. I.E.R.E.* **49,** 1681.
Swihart, J. C. (1961). *J. Appl. Phys.* **32,** 461.
Vail, C. R., Lucas, M. S. P., Owen, H. A., and Stewart, W. C. (1960). *Solid State Electron.* **1,** 279.
Zappe, H. H. (1974). *Appl. Phys. Lett.* **25**, 424.

Note added in Proof:

5. *Continuous sheet sensing*

One of the fabrication problems facing continuous sheet memories is that at each storage site the two drive lines used to switch flux must intersect with a third "sense" line. It was discovered (see Burns *et al.*, 1963) that if a sense plane consisting of a film of normal metal was situated underneath the tin storage plane, any change in the flux trapped in the tin storage plane would give rise to an eddy current surge in the sense plane, which could be detected as a voltage transient across its corners. This technique was analyzed for sheets of square geometry by Burns and Sass (1965) and for sheets of rectangular geometry by Newhouse and Drapeau (1965), who showed that it appeared to have distinct advantages as a memory sensing technique at room temperature.

Chapter 4

Radiation Detectors

K. ROSE

Electrophysics and Electronic Engineering Division
Rensselaer Polytechnic Institute
Troy, New York

C. L. BERTIN

IBM System Development Division
Manassas, Virginia

R. M. KATZ

MITRE Corporation
Westgate Research Park
McLean, Virginia

I. Introduction

It is convenient to characterize the detection modes in superconducting films as thermal or nonthermal, depending on whether or not the film must be heated by the incident radiation to produce an output signal. The general question of detection mechanisms in superconducting films is discussed in more detail in Section II. Discussion of the particular characteristics of the thermal or bolometer mode follows in Section III, and nonthermal modes which depend on nonlinear V–I characteristics are discussed in Section IV. The performance of superconducting thin-film detectors is compared with that of other electromagnetic and nuclear radiation detectors in Sections V and VI, respectively. Section V makes a detailed comparison of superconducting detectors with other detectors representing the state of the art in bolometers and far infrared detectors. Because of its importance, the definition of a meaningful figure of merit is developed at some length in Section V. Most of these considerations apply in Section VI where the use of superconducting films as detectors of phonons and nuclear radiation is reviewed. While most of the discussion relates to discrete detectors some attention is paid in Section VII to the possible uses of superconducting films in detector arrays and imaging systems.

Although many topics closely related to radiation detection by superconducting films are not covered in this chapter since they are treated more fully in other chapters, for the sake of completeness the performance of Josephson junctions is compared with other superconducting detectors in Section V. Devices based on normal and Josephson tunneling are treated at length in Chapters 1 and 2.

II. Radiation Detection Modes in Superconducting Films

Superconducting films act as radiation detectors because the absorption of radiation changes the resistance R of the film. To affect the resistance the absorbed radiation must induce some change of the current, magnetic field, or temperature in the superconductor. Other modes of detection, such as changing the superconductor's reactance, are conceivable but have not as yet been exploited. Since the magnetic field changes produced by incident radiation are so small that they may be neglected in most circumstances, the voltage change produced across a superconductor carrying a bias current I_b may be regarded as being caused by changes induced in its temperature or current. For small changes,

$$\Delta V = I\gamma \, \Delta T + \frac{\partial V}{\partial I} \, \Delta I \tag{1}$$

where $I \approx I_b$ is the total current in the film and $\gamma = \partial R/\partial T$. Most emphasis has been placed on detection schemes using the first term in Eq. (1) which corresponds to a thermal or bolometer mode of detection. This mode will be discussed at greater length in Section III. Detection corresponding to the second term in Eq. (1) will be considered in Section IV. Actually, since the sharp dependence of the superconductor's resistance on temperature at T_c is matched by a strong dependence of resistance on current, this second term can also influence the performance of bolometers as will be discussed in Section III.

In a practical detection system the coupling of the film to the incident radiation and to the amplifier system are as significant as the detection mechanism within the film. We will consider both factors in turn.

A sufficiently thin film whose surface is perpendicular to the direction of the incident radiation with a real conductivity σ and thickness d behaves as a "half-silvered mirror" of sheet resistance $R = (\sigma d)^{-1}$. In this case, the film's absorptance (the fraction of incident power absorbed by the film) is given by

$$\alpha = 4g/(2 + g)^2 \tag{2}$$

where $g = Z_0/R$ and Z_0 is the characteristic impedance of the medium (377 Ω for free space). When $g = 1$, α reaches its maximum value of 0.5. This result is significant because it is independent of wavelength, indicating that a sufficiently thin film can be equally matched to incident radiation across the electromagnetic spectrum.

The microwave measurements of Ramey and associates (1968) suggest that this approach is valid as long as the film's thickness is comparable to its skin depth, $\delta = (\pi\mu f\sigma)^{-1/2}$. At microwave frequencies, $f = 9$ GHz, $\delta \approx 5000$ Å for a normal tin film and would drop to approximately 1500 Å when the film becomes superconducting. At shorter wavelengths the skin depth drops, but the difference between the normal and superconducting states disappears as the wavelength approaches the value for the superconducting energy gap (approximately 1 mm for tin). To shorter wavelengths a superconducting film behaves as if it were normal. The optical absorption measurements of Golovashkin and Motulevich (1965) indicate that tin should have a minimum skin depth of approximately 250 Å at 4 μ since the skin depth is proportional to $(f_{ep}/f)^{1/2}$ where f_{ep} is the electron–phonon collision frequency which rises sharply below 3 μ. These estimates of skin depth are comparable to the film thicknesses of most superconducting bolometers which are on the order of 1000 Å.

Of course, even if the film may be regarded as sufficiently thin, its substrate can significantly influence the optical coupling to the device, particularly since the index of refraction and the thickness of the substrate can present a wavelength-dependent impedance. If the wavelength of the incident radiation is short compared to the thickness of the substrate, Eq. (2) can be modified to account for this effect by replacing $(2 + g)^2$ by $(1 + n + g)^2$ where n is the substrate's index of refraction, in the denominator. In those instances where it has been tried, blackening the bolometer did not seem to significantly influence its performance.

Greater coupling to the film can be achieved by placing it in an integrating sphere. For far infrared measurements, the optimum optical coupling is probably achieved by placing the film in an integrating sphere at the end of a light pipe as described by Martin and Bloor (1961) or Zwerdling and associates (1968). In all cases it is important to shield the detector from room-temperature background radiation by the use of an appropriate optical filter.

If the bolometer is used in a simple resistance bridge, as is customary, its output voltage is reduced by a factor

$$F = R_L/(R + R_L) \tag{3}$$

where R_L is the effective load resistance or input impedance of an amplifier. Clearly, the resistance of the film must be low compared with R_L to approach unity coupling of the film to its external circuit. A low detector resistance is also desirable for high-speed measurements where the electrical time constant of the detector must be less than its response time (thermal time constant for bolometers) if the inherent speed of the detector is not to be sacrificed. Here an upper limit is set by characteristic impedances of approximately 100 Ω for practical, high-speed transmission lines. Thus, somewhat conflicting requirements on the resistance of a film detector are set by the need for maximum coupling to its output circuit and the incident radiation. In practice, the need to maximize the variation of resistance with temperature or current places an additional constraint on the film resistance.

III. The Thermal (Bolometer) Mode

The output voltage generated in a film by the incident power is a fundamental measure of a detector's response. Thus, we may characterize the performance of a bolometer by its responsivity

$$r = V_{\text{out}}/P_{\text{in}} \tag{4}$$

where V_{out} and P_{in} are generally measured in terms of rms volts and watts. Note that the responsivity does not give the minimum power which can be detected since this is set by noise in the detector or its background. In this section we will derive the responsivity from a general theory of bolometer operation and summarize the relevant parameters of various superconducting bolometers which have been developed. We will defer a more thorough discussion of the limiting noise to Section V where we make a detailed comparison with other infrared detectors.

In the absence of incident radiation, the temperature of a bolometer reaches a steady state determined by its bias current I_b. The heat transfer equation for this situation reduces to

$$K_0(T_0 - T_b) = I_b^2 R_0 \tag{5}$$

where R_0 is the resistance of the bolometer at its quiescent temperature T_0 and T_b is the temperature of the bath to which the bolometer is connected by a thermal conductance K_0. If a sinusoidally modulated pulse of power

$$P = P_0 \cos \omega t \, u(t) \tag{6}$$

is absorbed by the bolometer, its temperature changes by an amount $\Delta T = T - T_0$. The heat-transfer equation for any time-varying input power is

$$C \frac{dT}{dt} + K(T - T_b) = I_b^2 R + P \tag{7}$$

where C and K are the effective heat capacity of and the effective thermal conductance to the bolometer at temperature T. C and K have units of joules per degree Kelvin and watts per degree Kelvin, respectively. If K and I_b are not changed by this change in temperature, Eq. (5) may be used to rewrite Eq. (7) in the form

$$C \frac{d\,\Delta T}{dt} + K\,\Delta T = I_b^2 \gamma\,\Delta T + P \tag{8}$$

where γ is the temperature change of resistance defined in Eq. (1). We have used the term "effective" to emphasize the fact that the value of C usually includes more than the heat capacity of the film and K is often not the Kapitza conductance between the film and its substrate. The thermal time constant of the bolometer is determined by the largest C and lowest K in the system coupling the bolometer to the temperature bath.

If we introduce an effective thermal conductance

$$K_e = K - I_b^2\gamma \tag{9}$$

Eq. (8) has the general solution

$$\Delta T = \Delta T(0)e^{-t/\tau} + [P_0 \cos(\omega t + \phi)/K_e(1 + \omega^2\tau^2)^{1/2}] \tag{10}$$

where $\tau = C/K_e$ is the thermal time constant for the bolometer and $\tan \phi = (\omega\tau)^{-1}$. The general solution to Eq. (8) has been given since it shows that, if the bias current is so great that K_e becomes negative, the temperature will build up exponentially and the bolometer will burn out. Thermal runaway will occur above a critical bias current:

$$I_{bc} = (K/\gamma)^{1/2} \tag{11}$$

Combining the steady-state solution of Eq. (10) with the temperature dependence of resistance in Eq. (11) gives us the usual formula for the responsivity of the bolometer,

$$r = \alpha F\gamma I_b/K_e(1 + \omega^2\tau^2)^{1/2} \tag{12}$$

The factors α and F introduced in Eqs. (2) and (3) allow for the coupling of the bolometer to the incident radiation and external circuit, respectively.

The derivation given so far assumed that K was unchanged by the application of an incident signal. This is generally true for small temperature changes. If the bolometer is thermally coupled to its bath by radiation alone, straightforward manipulation of the Stefan–Boltzmann Law for blackbody radiators shows that K is proportional to T^3. Actually, a similar relation applies when considering fast temperature transients in thin superconducting films on dielectric substances or immersed in liquid helium, since thermal decay is expected to proceed by the blackbody radiation of *phonons* followed by thermal diffusion as the phonons are scattered in the medium (Jones and Pennebaker, 1963; von Gutfeld and associates, 1966). This, for example, is the origin of the Kapitza boundary resistance. In practice, of course, heat conductance by phonon radiation is lowered by partial reflection at boundaries as discussed by Little (1959).

Of more consequence is the assumption that I_b is not changed by a change in film temperature. This implicitly assumes that we are dealing with an ideal current source. If the bolometer bias circuit is composed of the bolometer in series with the load resistance R_L and a battery, changes in the bolometer resistance will change I_b. In this case, the value of K_e in Eq. (9) must be changed to

$$K_e' = K - I_b^2\gamma B \tag{9'}$$

where $B = (R_L - R)/(R_L + R)$. A more complete derivation which includes this effect is given by Kruse *et al.* (1962).

By using Eq. (6) and assuming $K \approx K_0$, they show that the equation for thermal runaway can be rewritten in the form

$$\gamma > R/(T - T_b) \tag{13}$$

Since γ decreases to zero as R approaches its normal-state value R_N, the thermal runaway is self-limiting. Nonetheless, it was generally found that $I_b \approx 0.1 I_{bc}$ gave the maximum responsivity for high-speed bolometers tightly coupled to the temperature bath (Bertin, 1968). However, for slow-speed bolometers loosely coupled to the temperature bath $I_b \approx 0.7 I_{bc}$ gave the highest responsivity (Katz, 1972).

This result is closely related to the current dependence of the resistance transition shown as the second term in Eq. (1). Bremer and Newhouse (1959), Maul (1968), and Maul *et al.* (1969) have noted that the $R(T)$ transition shifts to lower temperatures for higher bias currents. This effect is much more pronounced for high-resistance films (100 Ω/square) than it is for lower resistance films (10 Ω/square) such as Maul measured. In the case of very low resistance films (<0.1 Ω/square) such as Bremer and Newhouse studied, the effect is still weaker and is noticed primarily as a dependence of the $R(I)$ transition on temperature. The result appears as a deterioration of γ with increasing bias current which tends to fix the value of bias current for maximum responsivity if the critical current for thermal runaway is sufficiently high. Since the critical current for thermal runaway is lower for loosely coupled films, it is only slightly higher than the bias current for maximum responsivity in these films.

Examining Eq. (12) we see that α, F, γ, and I_b must be maximized and K_e must be minimized in order to maximize the responsivity of the bolometer. To retain the advantages of the detector time constant it is generally desirable to modulate the incident power at a frequency $f \lesssim (2\pi\tau)^{-1}$. Assuming a fixed K for a fixed geometry, the bias current for maximum responsivity was found to vary from 0.05 to 0.2 of I_{bc} for films whose resistance ranged from 960 Ω to 6 mΩ (Bertin, 1968). Thus, one may reasonably take $I_b \approx 0.1 I_{bc}$. For $\omega^2\tau^2 \ll 1$ and $F \approx 1$, Eq. (12) reduces to

$$r \approx 0.1\alpha(\gamma/K)^{1/2} \approx 0.1\alpha(R_N/2K\,\Delta T)^{1/2} \tag{14}$$

by using the relation

$$\gamma = \frac{dR}{dT} \approx \frac{R}{\Delta T} \approx \frac{R_N}{2\,\Delta T} \tag{15}$$

where ΔT is the transition width of the bolometer as measured from 25

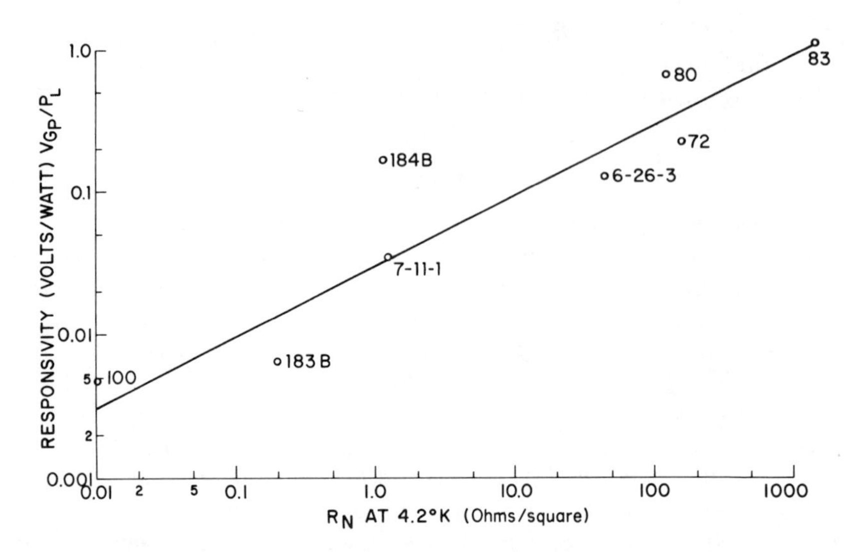

FIG. 1. Dependence of bolometer responsivity on film resistance for superconducting thin-film bolometers. [After Bertin, 1968.]

to 75% of R_N. Measurements made on superconducting tin thin-film bolometers shown in Fig. 1 tend to confirm the proportionality between r and $R_N^{1/2}$ suggested by Eq. (14). While there are many differences between films this proportionality dominates over a range of four orders of magnitude in sheet resistance. Over this range of film resistance the absorptance changes by less than an order of magnitude from 0.06 to 0.4 so that its influence on r is much weaker. The value of $\alpha = 0.06$ for the lowest resistance film is higher than would be calculated from Eq. (2) since the thickness of this film is greater than the skin depth. If the transition width ΔT were independent of R_N we would expect γ_{max} to be exactly proportional to R_N. Careful measurements found that $\gamma_{max} = 25.4R_N^{0.9}$, indicating a weak influence of R_N on ΔT.

Of the factors which may be varied in Eq. (14) to improve the responsivity, it is interesting to note that R_N can be increased by simply scribing the film to lengthen its electrical path. Ten scribes of a film would increase R_N by a factor of 100, thereby raising r by a factor of 10. Since the sheet resistance of the film measured in ohms per square is not affected by scribing, this will not affect the coupling of the film to the incident radiation. Theoretically, this improvement in responsivity is superficial; it will not change the detectivity of a film limited by its Johnson or thermal noise as will be seen in Section V.A. In practice, however, amplifier noise is much greater than device noise so that scribing can be used to optimize detector

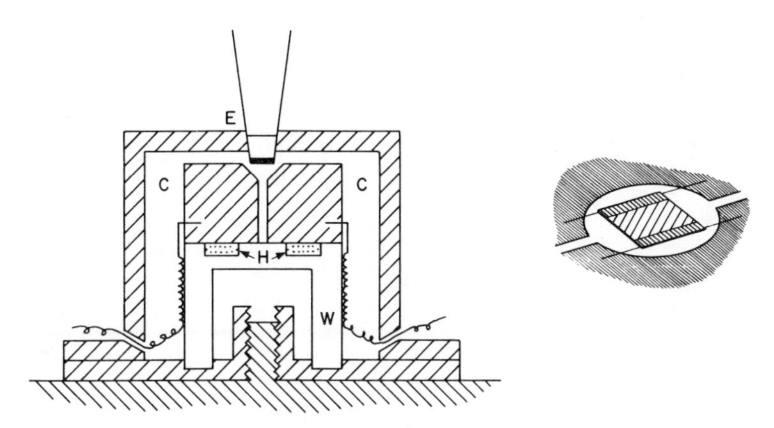

FIG. 2. Bolometer mount designed for weak coupling to the helium bath. [After Martin and Bloor, 1961.]

performance by raising the responsivity until the device noise equals amplifier noise.

The remaining factors, ΔT and K, depend intimately on a bolometer mounting which provides thermal isolation and control. The mount developed by Martin and Bloor (1961) shown in Fig. 2 is noteworthy in both respects. Here, the sensitive element is an evaporated thin film of tin deposited on E, a mica substrate 3×2 mm in area and 3 μ thick. This substrate is supported in vacuum from a divided brass cylinder C by nylon threads 10 μ in diam and about 0.5 mm long. Electrical contact to the element was made through films of lead, vacuum deposited on the nylon threads. At the operating temperature of the Sn bolometer the Pb films were well below their transition temperature, providing excellent electrical contact together with a high degree of thermal isolation. For this system K was about 3×10^{-7} W °K^{-1}. One disadvantage of such a mount for many practical applications is, of course, its frail and delicate nature. Since such problems afflict all sensitive bolometer systems this has tended to restrict the use of bolometers in general.

Temperature control of the bath to within 10^{-3}°K was provided by a simple manostat which also allowed rapid adjustment of the bolometer's operating temperature. Very fine control of the temperature was obtained by using the bolometer itself as a sensing device to control the current passing through a heater H wound on the thermal sink C. By using a nylon washer W to separate the sink from the helium bath, giving it a thermal time constant of a few seconds, it was possible to maintain the bolometer's operating temperature within 10^{-5}°K in the face of a marked variation in the incident power from a far infrared spectrometer. Judging from Fig. 5

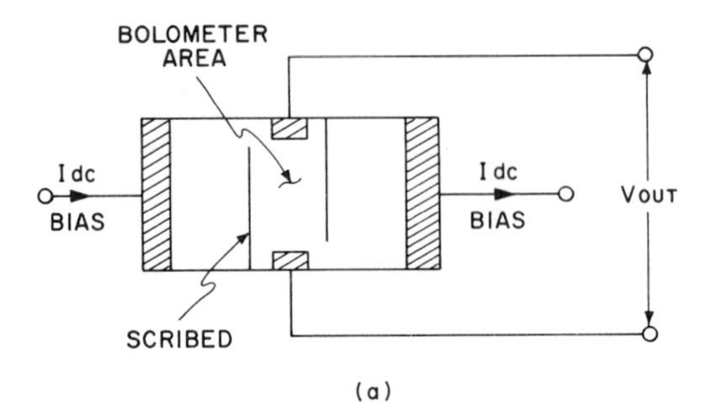

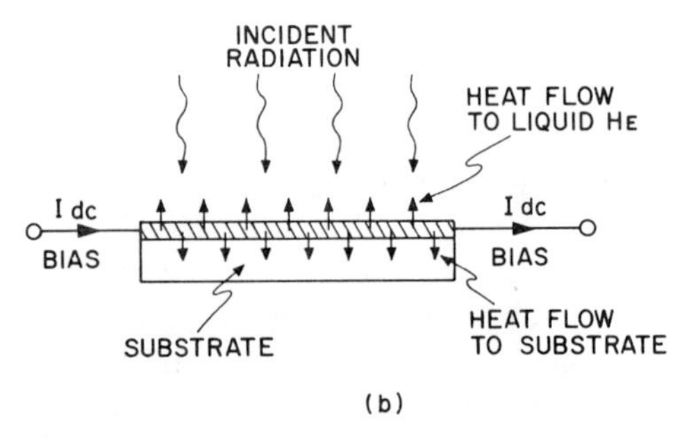

FIG 3. Bolometer mount designed for strong coupling to the helium bath. (a) Top view; (b) side view. [After Bertin and Rose, 1968.]

of the paper by Martin and Bloor (1961) $\Delta T \approx 50$ m°K for a typical film. In general, they found it more profitable to increase γ by increasing R rather than decreasing ΔT. Their optimum response was obtained with bolometers having a sheet resistance of $R_N = 30\ \Omega$/square. If one wishes to pursue the possibility of increasing r by decreasing ΔT, food for thought is given by the work of Gregory (1968) who measured superconducting transitions that were 90% complete in a temperature interval of 2×10^{-5}°K for 99.9999% pure gallium single crystals at 1.08°K by a mutual inductance method.

Since most superconducting bolometers have been developed with a maximum responsivity in mind, their thermal response time $\tau = C/K$ has generally been kept long. Nonetheless, the low heat capacities associated with thin films suggest that bolometers with very short response times could be developed. For example, consider a thin film mounted as indicated in Fig. 3 so that the heat flow into the helium bath is perpendicular to the surface of the film. Assuming that a film of area A has a thickness $d =$ 1000 Å and the heat capacity per unit volume of bulk tin $c_v \approx 0.7 \times 10^{-3}$

TABLE I

COMPARISON OF MEASURED AND CALCULATED THERMAL RELAXATION TIMES

Film	Thickness (Å)	Substrate	τ^a Measured (nsec)	τ_{bb} [b] Calculated (nsec)	τ_{am} [c] Calculated (nsec)
$In_{0.94}Sn_{0.06}$	910	z quartz	9 ± 2	7.5	8–10
$In_{0.94}Sn_{0.06}$	1000	x sapphire	16 ± 2	18	22–29
$In_{0.94}Sn_{0.06}$	1200	z sapphire	17 ± 2	18	22–29
$In_{0.94}Sn_{0.06}$	1800	z sapphire	14 ± 3		
$Pb_{0.985}Bi_{0.015}$	1400	z quartz	33 ± 3	12	12–14
$Pb_{0.985}Bi_{0.015}$	2000	x sapphire	50 ± 4	29	33–39

[a] The measured values of τ are normalized to a 1000-Å film thickness.
[b] Theoretical estimates based on blackbody.
[c] Theoretical estimates based on acoustic mismatch.

J °K^{-1} cm^{-3}, and that the Kapitza boundary resistance per unit area corresponds to $K_A \approx 0.5$ W °K^{-1} cm^{-2} gives

$$\tau = C/K = c_v d/K_A \approx 14 \quad \text{nsec} \tag{16}$$

Thermal response times to ruby and GaAs laser pulses ranging from 9 to 50 nsec have been observed in superconducting indium alloy films vacuum deposited on crystal substrates by von Gutfeld and associates (1966). Their results are listed in Table I. In this case, the time constants correspond to the emission of phonons from the film into a dielectric substrate rather than liquid helium. In this connection, it is interesting to note that the thermal response times obtained by Martin and Bloor (1961) ranged from 10 to 100 msec.

High-speed operation with thermal time constants less than 20 nsec has been achieved for superconducting bolometers made from thin tin films of 1-cm² area vacuum deposited on room-temperature glass substrates and mounted as shown in Fig. 3. The responses of a superconducting bolometer and a crystal detector to microwave pulses from an X-band klystron with rise and fall times Δt of approximately 45 nsec were compared and found to be identical, except that because of its low resistance the bolometer was found to be 5 dB more sensitive when both detectors were driving a 50-Ω load. Thus, $\tau \approx \Delta t/2.2 \approx 20$ nsec (Millman and Taub, 1965), but this is only an upper limit on the thermal time constant for these bolometers, since the measured fall time is limited by the speed with which the klystron can be switched on and off. More important, the temperature dependence of the responsivity was found to be the same within a factor of 2 for input radiation at wavelengths of 3 cm and 1 μ for low-resistance films as shown

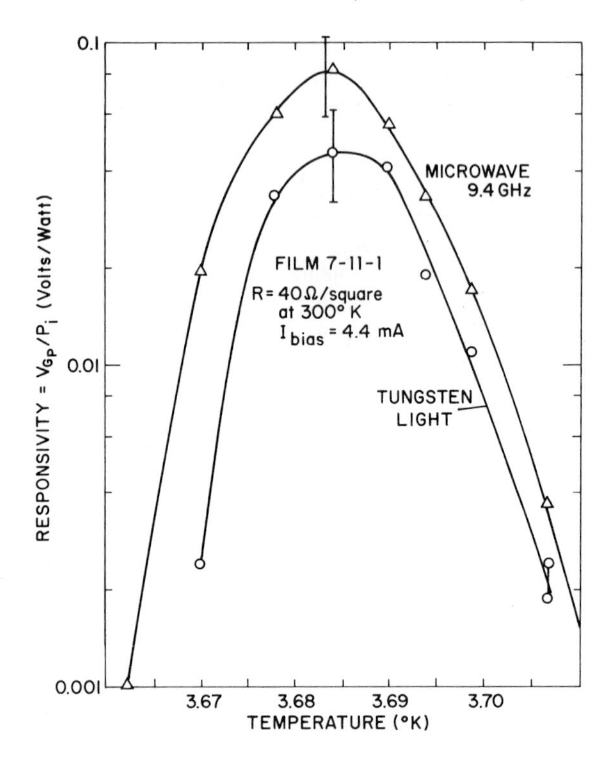

FIG. 4. The dependence of responsivity on infrared and microwave radiation for a superconducting thin-film bolometer. [After Bertin, 1968.]

in Fig. 4. Considering the sources of error involved in the calibration of power, particularly at optical wavelengths, this is quite reasonable agreement. Higher resistance films exhibited an anomalous enhanced responsivity at lower temperatures as will be discussed further in Section IV.

Historically, the initial work on superconducting bolometers seems to have been carried out by Andrews and co-workers in the 1940s on the basis of an earlier proposal (Andrews, 1938) which was made independently by Goetz (1939). The first work was carried out with lead and tin films in a switching mode in which the films were cooled well below their transition temperature and biased by a current slightly less than critical so that incident radiation would snap the film into the normal state (Brucksch *et al.*, 1941; Andrews *et al.*, 1941). Lead films, 10–500 μ thick, in their transition region were able to detect 2×10^{-11} W with $\tau = 1$ sec. Later work was carried out on the element tantalum, using a blackened aluminum foil as a heat sensitive surface (Andrews *et al.*, 1942) and NbN alloys (Andrews *et al.*, 1946), following the discovery by Horn that these alloys had critical temperatures from 14 to 17°K. This was a significant development since these bolometers could be operated above liquid-helium temperatures by using liquid hydrogen as a coolant. Fuson (1948) has reviewed in detail the properties of a large number of these devices which consisted of NbN flakes 2.5×0.5 mm^2 in area and 6 to 25 μ thick fastened to a $\frac{1}{8}$-in.-diam copper rod by bakelite lacquer. Their time constant depended on the thickness of the lacquer coating and was roughly 0.7–4.0 msec for thin coatings on the order of 25 μ and 4.0–20.0 msec for thick coatings on the order of 120 μ. Further development of high-temperature superconducting bolometers would probably be desirable from the viewpoint of developing practical infrared detection systems, but an upper limit near 20°K is set by the critical temperatures of present superconductors.

Subsequent studies have usually been made with systems using thin tin films (Martin and Bloor, 1961; Williams, 1961; Judge, 1968; Maul, 1968; Bertin, 1968). These detector systems have generally been variants of the coupling and mounting schemes discussed previously. The variants of most importance are whether the film is tightly or loosely coupled to the temperature bath and whether or not the film is scribed. Martin and Bloor's system was novel in that the element was placed in an ac bridge driven by an oscillator at 800 Hz and amplitude modulated at the chopping frequency of the incident radiation, 10 Hz. Gerig (1965) devised a detection scheme using a lead ring coupled by mutual inductance to the tank circuit of a marginal oscillator. This system was able to detect an equivalent heat input of 5.5×10^{-12} W with an integration time constant of 1 sec.

Franzen (1963) proposed a nonisothermal bolometer whose resistance change was produced by the growth of a normal region in the center of the device. While this device would appear to have some attractive features its properties have not, as yet, been verified experimentally. Various forms of this device seem to have been constructed (Konovodchenko *et al.*, 1968; Cavallini *et al.*, 1969); however, it appears that their sensitivity is lower than that of an ordinary bolometer (Pankratov, 1970).

IV. The Nonthermal (Current) Mode

In the course of bolometer studies on thin films with high surface resistance, an unexpected enhanced mode of detection was observed at microwave frequencies (Bertin and Rose, 1968). Figure 5 compares the temperature dependence of the microwave responsivity for three superconducting films whose normal-state resistance R_N ranges from 1.3 to 1500 Ω/square. (The temperature dependence of responsivity of the lowest resistance film was

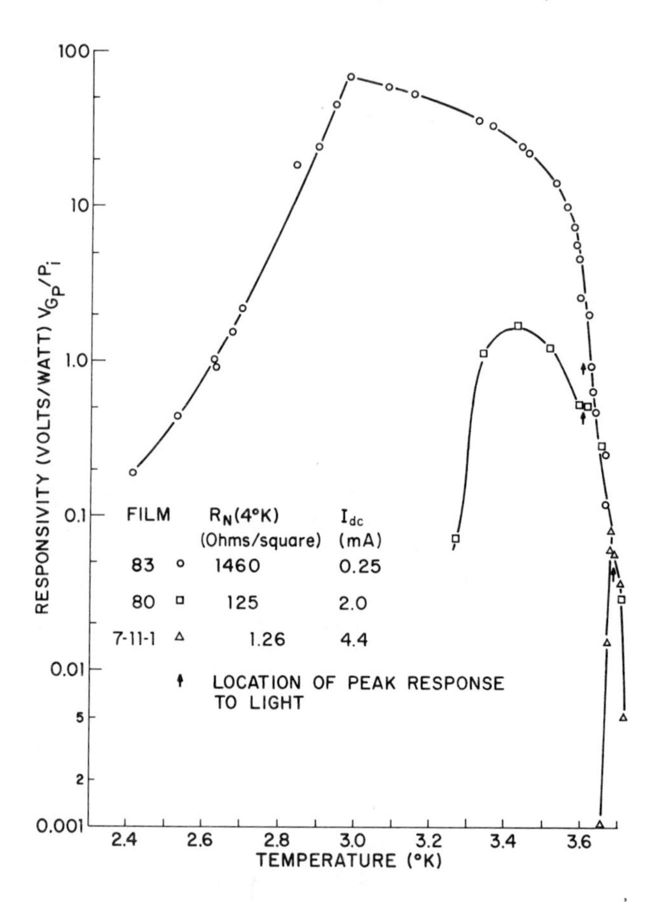

FIG. 5. The dependence of responsivity on temperature for three films covering a wide range of sheet resistance. Note the increasing enhanced mode effect for higher film resistances. [After Bertin, 1968.]

shown on an expanded temperature scale in Fig. 4.) The temperature at which the peak optical response occurs is indicated by upward arrows for each film. In each case the observed optical responsivity follows the temperature variation of $\gamma = dR/dT$ quite closely and has the sharply peaked shape shown for the lowest resistance film. Thus, for the highest resistance film, the maximum value of γ was $10^4\ \Omega\ °K^{-1}$ and occurs at 3.58°K which coincides with the peak in the responsivity for light. At 3.0°K, the value of γ and the responsivity to light have decreased by more than two orders of magnitude while the responsivity to microwave has increased by two orders of magnitude. This clearly suggests a nonthermal mode of operation.

Examination of Eq. (1) suggests that the mechanism which is responsible

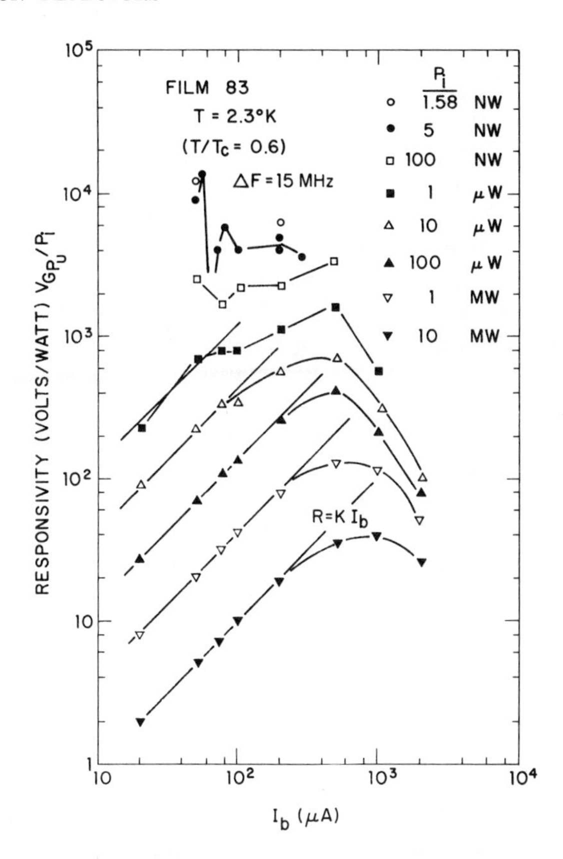

FIG. 6. The dependence of enhanced mode responsivity on bias current for a high-resistance film at several incident power levels. [After Bertin, 1968.]

is the nonlinear dependence of voltage on current of the superconducting film. If this were the case, the dependence of the responsivity on various parameters should be predictable from the V–I characteristic of the film. For example, if the V–I characteristic is antisymmetric, V can be expanded in odd powers of the total current $I = I_b + i$ where $i = I_p \sin \omega t$ is the alternating current induced in the film by the microwave signal. For this type of characteristic one can show that the responsivity should be proportional to the bias current I_b for $I_b \ll I_p$ (Bertin and Rose, 1971b). This is verified in Fig. 6 where the responsivity is plotted as a function of bias current for various input power levels at a fixed temperature. As one would expect, the proportionality breaks down at low input power levels.

Similarly, one can show that the responsivity should be independent of the input power level provided $I_p \ll I_b$, and this has also been verified.

One often observes that the responsivity falls off in proportion to the inverse square root of the incident power at high power levels, and this can be derived from a piecewise-linear approximation to the actual characteristic. Qualitatively, the principal features of the dependence of the responsivity on bias current and input correlate quite well with the films, V–I characteristics.

Quantitative correlation, based on a calculation of the pulsed direct voltage output expected when a sinusoidally modulated current is applied to the V–I characteristics as measured in the presence of a bias current, has proved more difficult. This is partly because of the inherent difficulty of the calculation for a resistive characteristic, but also because both the real and imaginary parts of the superconducting film's complex conductivity may play a role in the interaction of the film with a microwave signal (Rose and Sherrill, 1966). Curves generated by digital computer calculations carried out on the basis of the dc V–I characteristic agreed with the experimental responsivity curves within a factor of 2, at lower bias currents and incident power levels, for a high-resistance film. For a low-resistance film, where the imaginary part of the complex conductivity would be expected to play a greater role, the agreement was only within a factor of 10 but could be brought into agreement within a factor of 2 by including the more complex nonlinearity produced by the imaginary part of the complex conductivity.

Returning to Figs. 5 and 6 we can note several characteristics of the current mode. In the first place, the magnitude of the effect increases as the surface resistance of the film increases. (This mode was not observed in films with sheet resistance less than 40 Ω/square at 300°K.) Secondly, it has a very high responsivity, approaching 1.3×10^4 V W^{-1}, which enables it to detect a pulsed signal of 1.6 nW with a bandwidth of 15 MHz. Thirdly, it is relatively independent of temperature in its region of maximum sensitivity. As might be expected, the bias current for maximum enhanced mode response is not the same as that for maximum bolometer response for a given film.

How far the wavelength response of the current mode extends into the millimeter and infrared regions of the spectrum has not yet been established. Although essentially the same response was measured at 3 cm and 8.6 mm, an upper limit is surely set by the superconducting energy gap which would correspond to a cutoff wavelength of about 1 mm for tin (Rose, 1969). If this (rather than parasitic shunt capacitances within the film) sets the lower wavelength limit for this type of detector, it should be possible to build current mode detectors sensitive out to about 200 μ. The fact that the film is able to respond directly to the signal current in the current mode implies that this mode is inherently fast. This was verified by observing heterodyning at 9.35 GHz over the entire V–I characteristic. This suggests that for maximum sensitivity superconducting films operating in the current mode could be used as converters in a superheterodyne receiver.

So far, the current mode has been explained in terms of the nonlinear V–I characteristics of superconducting films having a high sheet resistance. The physical origin of these characteristics has not been explained, although a number of possibilities exist. A priori, possible mechanisms are the nonuniform thickness of the film (resulting in a range of critical currents for different regions of the film), mixed-state resistance, and Josephson tunneling between grains in the film. Intermediate-state effects are unlikely; microwave studies of similar films in the presence of perpendicular magnetic fields give definite evidence for Type-II behavior (Soderman, 1969). Effects due to nonuniform film thickness would not be surprising, since electron microscopy shows that high-resistance films of this type, vacuum deposited on room-temperature substrates, exhibit substantial lumpiness. As might be expected, their fabrication requires careful control of the deposition process.

It is likely that the current-mode detection mechanism is a complex combination of influences of the mixed state, Josephson currents, and film lumpiness or weak links. For example, Anderson and Dayem (1964) have shown that mixed-state vortices driven across weak-link bridges through which a current is flowing can give rise to Josephson steps in the V–I characteristic. Now, the Anderson–Dayem effect disappears in the presence of a 0.1-G magnetic field, while current-mode V–I characteristics were unaffected by normal magnetic fields up to 70 G (Bertin, 1968). On the other hand, Saxena *et al.* (1972) have studied junction arrays prepared by compressing aluminum powder in a cylindrical tube. They have observed both a direct voltage, proportional to an applied rf field below T_c that resembles the current mode, and structure in the V–I characteristics that shifts with frequency that resembles Josephson behavior. The direct voltage is washed out by a magnetic field of about 100 G. They conclude by suggesting that this mechanism may account for the enhanced (current mode) sensitivity observed by Bertin and Rose (1971b). More detailed study of the V–I characteristics of high-resistance films is needed to establish the relative importance of these influences on current-mode detection.

Recently, Ayer (1974) has established the existence of Josephson effects in the current mode. A linear dependence of critical current on temperature, steps in the V–I characteristic which scale with frequency, and the expected variation of step height with power level have all been observed. It appears that, at least in some regions, granular tin films behave like a correlated array of Josephson junctions (weak links) in series.

These results are undoubtly related to other observations of rectification and other nonlinear effects in superconducting films. Lazarev and associates (1941) observed a rectified voltage when a direct and an alternating current were applied simultaneously to a superconducting wire. Various workers have observed nonlinear effects in NbN (Andrews and Clark, 1946; Andrews *et al.*, 1947; Lebacqz, 1949), and Bodmer (1950) reported measuring a mode of detection in addition to the bolometer mode, which seemed to be rectification due to a nonlinear resistance in the film. Since the work of Lazarev various investigations have been made of nonlinear effects induced in superconductors by currents and magnetic fields. For example, Nethercot and von Gutfeld (1963) generated the second harmonic of 9.2-GHz radiation in tin films and foils biased by a parallel magnetic field, and Sherrill and Rose (1964) generated the third harmonic of 9.4-GHz radiation in unbiased tin films similar to those in which the current mode was observed.

V. Comparative Performance as Electromagnetic Radiation Detectors

A. General Considerations

So far, the only figure of merit we have used to describe radiation detectors is their responsivity r. Such a description is incomplete since the sensitivity of a particular detector is determined by the minimum power it can detect and this, in turn, is limited by noise generated within the detector. For this reason it is convenient to introduce the noise equivalent power (NEP):

$$\mathrm{NEP} = v_N/r(\Delta f)^{1/2} \tag{17}$$

where v_N is the rms noise voltage measured in a bandwidth Δf. The NEP is a measure of the minimum radiation intensity needed to produce an output with a signal-to-noise ratio of unity. Δf is taken as 1 Hz and the units of NEP are watts per hertz$^{1/2}$.

Since the NEP of all known detectors is proportional to the square root of their active area A (Jones, 1949; Jones, 1953), this factor needs to be removed when comparing detectors having different areas; otherwise the NEP of any given detector can be reduced an arbitrary amount simply by reducing its area. For this reason Jones (1949) introduced a new figure of merit, the detectivity.

$$D^* = A^{1/2}/\mathrm{NEP} \tag{18}$$

If detectors are compared on the basis of D^*, note that the most sensitive detector will have the largest value. Because it is often desirable to compare the minimum equivalent input power which was detected experimentally with that expected from a particular noise source it is convenient to work with a minimum detectable power

$$P^* = (D^*)^{-1} = \mathrm{NEP}/A^{1/2} \tag{19}$$

where P^* has the dimensions watts centimeters^{-1} hertz$^{-1/2}$.

To proceed further one needs to determine the noise sources which apply to a particular detector. There are many possible sources of noise for a superconducting bolometer, and we will consider them in turn below. More detailed discussions of these noise sources are given in Jones (1953) and Kruse *et al.* (1962).

1. *Background radiation or photon noise:* P_R^*

In the absence of all other noise sources, bolometers and other thermal detectors are ultimately limited by fluctuations in the power emitted by their surroundings. These surroundings may be represented by a blackbody at a temperature T_s. If the detector is at a quiescent temperature T_0, there is an additional source of noise due to fluctuations in the power which it would emit as a blackbody. Integrating these rates of energy emission over all wavelengths in the blackbody spectrum, the mean-square noise power of the fluctuations which would be measured in a bandwidth Δf for a thermal detector of area A is

$$(\overline{\Delta P_R{}^2}) = 8k\sigma(T_s{}^5 + T_0{}^5)A\ \Delta f \epsilon^{-1} \tag{20}$$

where k and σ are the Boltzmann and Stefan–Boltzmann constants and ϵ is the radiant emittance of the detector. As defined in Eq. (2) $\epsilon = \alpha$. $\overline{\Delta P^2}$ has dimensions of watts^2. The corresponding NEP is given by

$$\mathrm{NEP} = (\overline{\Delta P^2}/\Delta f)^{1/2} \tag{21}$$

Note that the NEP is proportional to $A^{1/2}$ as predicted.

Photoconductors and other quantum detectors, which are background limited by fluctuations in the rate of photon emission from the background rather than the rate of energy emission, generally have lower values of P_R^*. This is primarily because the background radiation is filtered by the long wavelength cutoff inherent in a photoconductor. If a detector is background limited its P_R^* can always be lowered by altering the blackbody background radiation with a cooled optical filter.

For a thermal detector,

$$P_R^* = 0.25 \times 10^{-16}(T_s{}^5 + T_0{}^5)^{1/2}\epsilon^{-1/2} \tag{22}$$

so that for $\epsilon = 1$ and $T_0 = 4°\mathrm{K}$, $P_R^* = 3.6 \times 10^{-11}$ W cm^{-1} Hz$^{-1/2}$ for $T_s = 290°\mathrm{K}$ and $P_R^* = 1.1 \times 10^{-15}$ W cm^{-1} Hz$^{-1/2}$ for $T_s = 4°\mathrm{K}$. This indicates the importance of shielding a superconducting bolometer from room-temperature radiation if one wishes to achieve a respectable sensitivity. (Note that this source of detector *noise* is caused by *fluctuations* in the background radiation. The background radiation will warm the bolometer, adding to the *signal*.) Unfortunately, the high values of detectivity implied by a cryogenic background have not been achieved in practice because of the limitations imposed by other forms of noise.

2. *Johnson noise:* $P_J{}^*$

While photon noise results from fluctuations in the radiation field, Johnson noise *is* the radiation field in the detector at its operating temperature. Indeed, the Johnson noise spectrum of a resistor can be shown to be the one-dimensional form of the blackbody radiation spectrum. The rms noise voltage across a resistor R at temperature T_0 measured in a bandwidth Δf is given by

$$v_\mathrm{N} = (4kTR\ \Delta f)^{1/2} = 1.28 \times 10^{-10}(R\ \Delta f)^{1/2}(T_0/295°\mathrm{K})^{1/2} \qquad (23)$$

This corresponds to

$$P_J{}^* = (v_\mathrm{N}/r)(A\ \Delta f)^{-1/2} \approx 1.28 \times 10^{-9}(K\ \Delta T/A)^{1/2}(T_0/295°\mathrm{K})^{1/2} \qquad (24)$$

where the expression for r from Eq. (14) has been used.

Since the thermal time constant of the film is $\tau = C/K$ where $C = c_\mathrm{v}\ dA$ Eq. (24) can be rewritten in the form

$$P_J{}^* \approx 1.28 \times 10^{-9}(c_\mathrm{v}d\ \Delta T/\tau)^{1/2}(\alpha F)^{-1}(T_0/295°\mathrm{K})^{1/2} \qquad (25)$$

where the factors α and F have been included to allow for imperfect coupling.

A number of important conclusions can be drawn from Eqs. (24) and (25). In the first place, we see that $P_J{}^*$ is independent of the film resistance. This means that for a bolometer limited by Johnson noise no ultimate advantage is gained by scribing the film to raise R in order to raise the responsivity, since the signal-to-noise ratio at the outset of the detector will remain the same: From Eq. (25) we see that $P_J{}^*$ is independent of A for a given time constant τ, as expected. We also see that in order to lower the Johnson noise limit we need either to lower ΔT or K or, equivalently, to raise τ. For $T_0 = 4°\mathrm{K}$, $\Delta T = 50$ m°K, and $\alpha = F = 1$, Eq. (24) gives $P_J{}^* \approx 7.5 \times 10^{-14}$ W cm^{-1} Hz$^{-1/2}$ for $K = 3 \times 10^{-7}$ W/°K. This is near the value estimated by Martin and Bloor (1961) after allowance is made for their differences in A, Δf, and α. On the other hand, for $K \approx 0.5$ W °K^{-1} and $A = 1$ cm^2 (Bertin, 1968), $P_J{}^* \approx 2.4 \times 10^{-11}$ W cm^{-1} Hz$^{-1/2}$, and the ratio of $P_J{}^*$ agrees with the estimated ratio of $\tau^{-1/2}$ for these two cases within an order of magnitude.

3. *Temperature noise:* $P_T{}^*$

Einstein (1904) derived a formula for the mean-square temperature fluctuations of a body in thermal contact with another body having a much larger heat capacity. From this one can derive a formula for the minimum detectable power $P_T{}^*$ for a bolometer at temperature T_0:

$$P_T{}^* = (4kT_0{}^2K/A)^{1/2} = 2.2 \times 10^{-9}(K/A)^{1/2}(T_0/295) \tag{26}$$

For $T_0 = 4°\mathrm{K}$, $P_T{}^* \approx 6.7 \times 10^{-14}$ W cm^{-1} Hz$^{-1/2}$ for $K = 3 \times 10^{-7}$ W °K^{-1} and $A = 0.06$ cm^2 (Martin and Bloor, 1961) and $P_T{}^* \approx 2.1 \times 10^{-11}$ W cm^{-1} Hz$^{-1/2}$ for $K = 0.5$ W °K^{-1} and $A = 1$ cm^2 (Bertin, 1968).

4. *Pickup and amplifier noise*

While these are not, strictly speaking, inherent limitations on the performance of a detector, careful consideration must be given to both factors in an experimental situation. The problem of pickup is severe with both bolometric and current-mode detectors since they are equally sensitive over a wide range of frequencies. Thus, the large electrical transients accompanying a pulsed light source such as a laser can impair even relatively low-sensitivity measurements. Pickup can lead to erroneous conclusions regarding the noise limitations of a superconducting detector. For example, Andrews and associates (1946) reported a very large current-dependent noise in their NbN bolometers, most of which was later found to be demodulation of radio signals from a local station (Andrews and Clark, 1946).

The design of low-noise preamplifier circuits is of great importance for all low-level detectors and has received considerable investigation (Woodward and Silvermetz, 1965). This is of particular importance since superconducting detectors have been limited by amplifier noise rather than their inherent noise in many instances. When calculating the noise in a system one generally specifies a particular amplifier or coupling circuit by its noise figure as a function of source resistance and signal frequency. The noise figure or noise factor F is customarily defined† as

$$F = N_o/(Gk290\ \Delta f) \tag{27}$$

where N_o is the available output noise power measured in the system bandwidth (more exactly, the system's noise bandwidth) Δf. G is the power gain of the system and 290°K is the reference temperature of the input noise. The noise figure is often quoted in decibels and, since we are dealing

† See Standards on Electron Tubes: Definition of Terms, 1962 (1963).

with power ratios,

$$F_{\mathrm{dB}} = 10 \log_{10} F \tag{28}$$

F as defined in Eq. (27) can be interpreted as the ratio of the total noise power at the output terminals to that part of the total output noise power which would be produced by Johnson noise in a 290°K resistor at the input terminals. Thus, Eq. (27) can be rewritten in the form

$$\begin{aligned} N_{\mathrm{o}} &= G(\text{input noise power}) + (\text{excess noise power}) \\ &= Gk290\ \Delta f + Gk290(F - 1)\ \Delta f \end{aligned} \tag{29}$$

In many cases, the noise at the input terminals is not due to Johnson noise in a 290°K resistor. In this case, an equivalent noise temperature T_i is introduced to represent the source, and Eq. (29) may be rewritten in the form

$$N_{\mathrm{o}} = GkT_{\mathrm{i}}\ \Delta f + Gk290(F - 1)\ \Delta f \tag{30}$$

or

$$N_0/GkT_{\mathrm{i}}\ \Delta f = 1 + (F - 1)(290/T_{\mathrm{i}}) \tag{31}$$

From the form of Eq. (29) we see that we may also define an equivalent noise temperature T_{e} for our amplifier system by (Mumford and Scheibe, 1968)

$$T_{\mathrm{e}} = 290(F - 1) \tag{32}$$

To calculate the noise figure or equivalent noise temperature for a particular amplifier system its power gain and output noise power need to be determined (Mumford and Scheibe, 1968). Martin and Bloor (1961) calculate a noise figure for their amplifier system which includes an input transformer; however, their noise figure is based on the ratio given in Eq. (31) rather than that given in Eq. (27).

5. *Other noise sources:* $P_{\mathrm{B}}{}^*$

In this category we group all noise sources which do not fall into one of the above categories. Possible sources include noise due to the contacts or fluctuations (due to helium bubbling) in the bath temperature. Both tend to fall off with increasing detection frequency. Strandberg and Kierstead (1966) measured the low-frequency noise spectrum of a superconducting bolometer from 20 Hz to 20 kHz as shown in Fig. 7. They observed that the spectral density of the output noise varied as f^{-a} with $a \approx 2$. Moreover, the magnitude of the noise density depended on whether or not the sample was immersed in liquid helium and whether or not the temperature of the helium bath was below the λ point so that the helium was in a superfluid

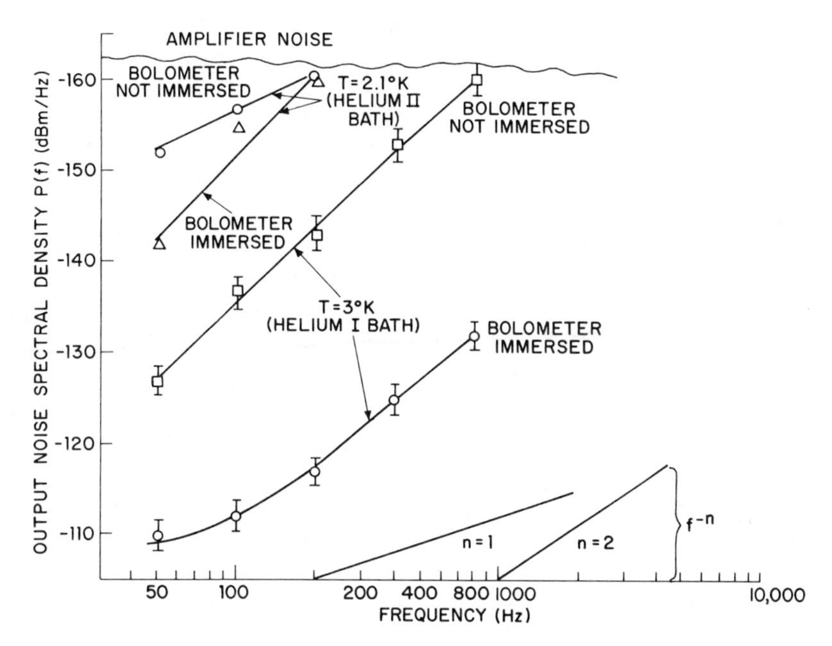

FIG. 7. Output noise spectrum for a tin bolometer. [After Strandberg and Kierstead, 1966.]

state. Since the noise density had a very low characteristic frequency (approaching zero) and decreasing when the sample was removed from the helium and when the helium was superfluid (so that bubbling of the helium was suppressed), they concluded that the observed noise occurs from the detection by the bolometer of the acoustic noise in bubbling liquid helium. [Other workers (Burton, 1968) have found it helpful to attach a loudspeaker to the output of the bolometer when attempting to minimize noise, since the sound of bubbling helium was easily identified.]

This output noise power N_o can be converted to an equivalent minimum detectable power $P_B{}^*$ by means of the responsivity. Since

$$P_{out} = V_{out}{}^2/R \tag{33}$$

and

$$P_{in} = V_{out}/r \tag{34}$$

then using Eqs. (17) and (18),

$$P_B{}^* = (RP_{out}/A\ \Delta f)^{1/2}r^{-1} \tag{35}$$

Now $(P_{out}/\Delta f)$ is the output spectral density of the noise power which ranges from 10^{-14} to 10^{-19} W Hz$^{-1/2}$ since -110 dBm $= 10^{-11} \times 10^{-3}$ W.

We can now calculate P_B^* by assuming typical values for their superconducting bolometers which were tin films approximately 1000 Å thick and 3 cm^2 in area deposited on sapphire substrates with a resistance $R \approx R_N/2 = 35\ \Omega$ and a responsivity $r \approx 23\ V\ W^{-1}$ at a bias current of 5 mA (Maul and Strandberg, 1969; Maul, 1968). This gives values of P_B^* ranging from 10^{-8} to $5 \times 10^{-11}\ W\ cm^{-1}\ Hz^{-1/2}$. These are reasonable values of P_B^* since $K \approx 2.8\ W\ {}^\circ K^{-1}$ for these films.

The question of a noise inherent to superconducting bolometers associated with, for example, domain motion in the intermediate state has plagued the superconducting bolometer even since its conception. In particular, fears were raised by the current-dependent noise observed by Andrews and associates (1946) which were not completely allayed by the subsequent discovery that it was largely pickup (Andrews and Clark, 1946). For example, Lalevic (1960) developed a theory which correlated the noise to be expected in a superconducting bolometer with the value of the surface energy or, equivalently, the number of domains in a superconductor. He concluded that a small interface energy would lead to a large number of domains which would tend to smooth out variations in the curve of resistance versus temperature. On this basis, he was able to explain Andrews' observation in a hard superconductor such as NbN (Andrews *et al.*, 1946). Such estimates for hard superconductors should be viewed with some caution since they were made before people were generally aware of the Ginzburg–Landau theory and its prediction of Type-II superconductivity (Abrikosov, 1957). Nonetheless, fluctuations in the resistive transition have certainly been observed; Baird (1959) made a detailed investigation of the resistive transition in tantalum and found definite resistance levels in the intermediate region. Using a longer specimen added to the resistance states available, thereby producing smaller resistance fluctuations.

On the other hand, Martin and Bloor (1961) were able to reach 30 times the level of Johnson and temperature noise which they estimated for their superconducting bolometer before they were limited by amplifier noise. From this they concluded that any additional source of noise associated with the superconducting transition could not be greater than about 10 times the Johnson noise. Following the initial work of Strandberg and Kierstead (1966), Maul has made a study of the excess noise level observed in the output of tin thin-film superconducting bolometers from 100 Hz to 4 MHz (Maul, 1968; Maul and Strandberg, 1969). In addition to the low-frequency component of the noise spectrum, he found a higher frequency component with the same frequency spectrum as the thermal noise but with 10 times the amplitude. This higher frequency component could be reduced to that predicted on the basis of temperature fluctuations by the application of magnetic fields of a few gauss. They feel that neither of these components was due to contact noise since equivalent results were obtained from four-point measurements. Since similar noise spectra were obtained with tin films deposited on liquid-nitrogen-cooled and room-temperature substrates, the excess noise does not appear to be affected by grain size in the film. As might be expected, some films displayed a more pathological behavior. While the source of this excess noise is not yet explained, it does appear that the presence of excess noise is not a significant problem for the development of superconducting bolometers.

Pankratov *et al.* (1970) have studied the noise of a superconducting tin bolometer in both isothermal and nonisothermal modes of operation. The devices were a $10 \times 1\ \text{mm}^2$ rectangle with a resistance of $1\ \Omega$ at the isothermal operating point, deposited on mica and attached at the ends or by nylon threads to a copper base, corresponding to a thermal conductance of about $1\ \mu\text{W}\ {}^\circ\text{K}^{-1}$. They established that temperature fluctuations of liquid helium in the cryostat were a source of excess noise below 10 Hz but this effect was eliminated by placing a sheet of thermal insulation (Teflon) between the bolometer base and the helium reservoir. In the isothermal mode, calculations of the noise due to the statistical nature of the heat exchange of the bolometer with its surroundings by heat conduction agreed reasonably well with the experimental results. Radiation and Johnson noise could be ignored because they were small. At 10 Hz the equivalent noise power was $2 \times 10^{-13}\ \text{W Hz}^{-1/2}$. Excess noise, nonlinearly dependent on the current, was observed in the nonisothermal mode so that its threshold sensitivity was much worse.

Recently, Katz and Rose (1972) have made a comparative study of the noise limitations in the thermal and current modes in the same thin-film device. Thin tin films were studied with normal-state sheet resistances on the order of $100\ \Omega$/square which were scribed to produce total normal-state resistance of $10\ \text{k}\Omega$ and mounted in vacuum. The resulting bolometers had response times and sensitivities similar to those of Martin and Bloor (1961). The dominant noise in the thermal mode appears to be Johnson noise above the thermal cutoff frequency when the film impedance and chopping frequency are optimized for state of the art amplifiers. In the same films the dominant noise in the current mode is $1/f$ or current noise below 100 kHz. This is not necessarily a significant limitation since the speed of response of the current mode is so much greater than that of the thermal mode that one can chop the incident radiation at a higher frequency where device performance is limited by amplifier noise.

In conclusion, we note that since all of these noise sources are independent, the total noise power for a situation in which many sources are present should be calculated by taking the square root of the sum of the squares of the noise powers for each source.

The importance of comparing different radiation detectors on the basis of an appropriate figure of merit cannot be overemphasized. Ideally, a figure of merit should be chosen whose value represents the upper limit of performance which may be expected from a given class of detector. That is, all of the parameters which can be varied for a given class of detector should be factored out of the figure of merit so that its value does not depend on such variables but reflects only the inherent limitations of the detector. We have already seen that responsivity is an inadequate figure of merit since it can be increased arbitrarily in the presence of thermal noise without decreasing the minimum detectable power level. Similarly, NEP is an inadequate figure of merit because it can be reduced arbitrarily by focusing the same amount of radiation onto a smaller area detector. Even D^* is not a completely adequate figure of merit since it too can be changed by changing the time constant τ for a bolometer.

To allow the comparison of detectors having widely different time constants, Jones (Jones, 1953) noted that all known radiation detectors could be placed into two main classes, depending upon the dependence of NEP on τ. We can show that the NEP for superconducting bolometers limited by thermal (Johnson or temperature) noise in the presence of sinusoidally modulated radiation may be written in the form

$$(\mathrm{NEP})^{-1} = C_1\tau^{1/2}/2A^{1/2}(1 + \omega^2\tau^2)^{1/2} \tag{36}$$

where C_1 is a constant. This is the defining formula for a class-IIa detector according to Jones' classification scheme. Class-II detectors include thermocouples, bolometers, photographic negatives, photoconductive cells which are limited by current noise, and the human eye. [A class-II detector is classed as being class IIa if its frequency response corresponds to a single time constant as shown in Eq. (36). If the factor $\tau^{1/2}$ is missing in the numerator of Eq. (36) the detector would be a class-Ia detector. Examples of class-I detectors are detectors limited by photon noise (e.g., photoconductors, photoemissive detectors, and the Golay cell).] It should be noted that our viewpoint in the above classification differs slightly from Jones (1953) since he considered measurements made in a bandwidth Δf limited by the time constant of the detector while we will always normalize our results to a reference bandwidth $\Delta f = 1$ Hz. It is worth noting that one can have a higher D^* at a high frequency f by using a slower bolometer for which $\tau > 1/f$ rather than a faster bolometer for which $\tau \lesssim 1/f$.

That (NEP) is proportional to $(A/\tau)^{1/2}$ for superconducting bolometers limited by Johnson or temperature noise at frequencies low enough so that $\omega\tau \ll 1$ follows directly from Eqs. (24)–(26) and the definition of P^* in Eq. (19). We see that in both cases NEP is proportional to $K^{1/2} = (C/\tau)^{1/2}$ and hence is proportional to $(A/\tau)^{1/2}$ since C is proportional to A. It is interesting to note, however, that τ may or may not depend on A depending upon how the bolometer is mounted. If it is tied by a filament to the bath as indicated in Fig. 2, K is independent of A and, consequently, τ is directly proportional to A. On the other hand, if the bolometer is mounted as indicated in Fig. 3, K is proportional to A and τ is independent of A. Thus, fast superconducting bolometers should have a time constant which is independent of their geometry, which can be a factor of importance in the design of high-speed detector systems.

Before turning to a detailed comparison of superconducting bolometers with other detectors, a few notes regarding waveform calibration may be in order. As mentioned at the beginning of Section III, the responsivity is strictly defined as the ratio of rms output voltage to rms signal voltage for a radiation signal varied sinusoidally in time (Jones, 1953). For maximum sensitivity, measurements of the detector output are often made with a narrow-band amplifier (such as a lock-in amplifier) which will recover only the fundamental component of a given signal. Unfortunately, the signal waveforms from photodetectors are generally square waves (or at least trapezoidal waves) because it is easier to switch the radiant energy source off and on at some rate than it is to provide linear amplitude modulation. Thus, if the detector generates a square wave of peak-to-peak amplitude V_{Gpu}, this corresponds to a fundamental sine wave of amplitude $V_{Gp} = (2/\pi)V_{Gpu}$ which has an rms value $V_r = (\sqrt{2}/\pi)V_{Gpu}$. As long as the responsivity is independent of the incident power level, the only meaningful comparison is on the basis of a square-wave output no matter whether one describes the input and output in terms of their amplitude, rms value, or other average. So long as the incident power is square-wave modulated at a frequency which is low compared with τ^{-1}, the responsivity will reach a steady-state value given by Eq. (12) for $\omega\tau = 0$. In this case, if the output is measured in terms of the amplitude of the fundamental sine wave V_{Gp}, the correct value of the responsivity would be given by

$$r = V_{Gpu}/P_{\mathrm{i}} = (\pi/2)(V_{Gp}/P_{\mathrm{i}}) \tag{37}$$

where P_{i} is measured as the peak-to-peak value of the incident power. That is, P_{i} is the average value of incident power which would be measured if the source were on continuously. Thus, the values of r given in Figs. 1, 4, and 5 should be multiplied by $\pi/2$ to get the true rms responsivity. Since the values in Fig. 6 were measured with a wide bandwidth amplifier so that the shape of the signal was preserved, the values of r are correct as given. If sinusoidally modulated input power is used one must apply the concept of rms input with some care, since the incident power can never be negative while the average of the sinusoidally varying output voltage might be taken to be zero. In this case, the rms value of the incident power $P_{\mathrm{i}} =$

$P_0(\cos \omega t + 1)$ is not $P_0/\sqrt{2}$ but $\sqrt{\frac{3}{2}}P_0$. Fortunately, as long as signals of identical shape are compared so that $V_G = V_{Gp}[\cos(\omega t + \phi) + 1]$ in this case, the value of the responsivity will be the same provided $(\omega\tau)^2 \ll 1$. Unfortunately, the literature is not always consistent in this respect which leads to some difficulty when comparing devices.

B. Bolometers

The static and dynamic characteristics of several bolometers are compared in Table II. This list includes semiconducting bolometers as well as superconducting bolometers and is arranged in approximate chronological order according to the publication date of the article or thesis containing the work. As inspection of Table II shows, a large range of characteristics is covered. Brief comments about each reference follow.

1. *Smith et al. (1957)*

The parameters of the NbN bolometer quoted here are based on a review of the work of Andrews and associates. This is early work which represents something of a compromise between speed and sensitivity. Note that while its time constant is comparable to the semiconductor bolometers its responsivity is much lower. (In this connection, it is important to realize that the exponential rise in resistivity in semiconductors caused by the freeze-out of carriers into shallow donor or acceptor levels can lead to values for $\gamma = \partial R/\partial T$ which exceed those for superconductors.) The great virtue of the NbN detector is its operating temperature which is much higher than that of any of the other detectors in Table II. Low-frequency fluctuations and pickup problems severely limited performance.

2. *Boyle and Rogers (1959)*

This type of bolometer is made from a thin flake of graphite. Along with the Golay cell it was often used in far infrared spectroscopy prior to the development of the doped-germanium bolometers (Tinkham, 1964). The numbers quoted here are taken from Low (1961) who was interested in comparing his bolometer with the carbon bolometer.

3. *Martin and Bloor (1961)*

This is probably the most sensitive superconducting bolometer developed to date and has been used as an illustrative example throughout the chapter. The value of r listed in Table II is $2\sqrt{2}$ times the value quoted in Martin and Bloor (1961) in order to bring their formula for r into agreement with Eq. (12).

TABLE II

STATIC AND DYNAMIC CHARACTERISTICS OF BOLOMETERS

Reference	Material	Area (cm^2)	T_0 (°K)	I_b (A)	R (Ω)	γ (Ω °K^{-1})	K (W °K^{-1})	r (V W^{-1})	τ (s)
1. Smith *et al.* (1957)	NbN	0.01	14.4	3×10^{-2}	0.2	25	5×10^{-3}	13.5	5×10^{-4}
2. Boyle and Rodgers (1959)	C	0.2	2.1		1.2×10^{6}		3.6×10^{-5}	2×10^{3}	10^{-2}
3. Martin and Bloor (1961)	Sn	0.06	3.7		10–30	190	3×10^{-7}	850	10^{-2}
4. Low (1961)	Ge:Ga	0.15	2.15	6.5×10^{-5}	1200	-1.8×10^{5}	1.8×10^{-4}	4×10^{3}	4×10^{-4}
5. Judge (1968)	Sn scribed	0.85	3.7	4×10^{-6}	74.5	4.4×10^{3}	1.9×10^{-7}	2×10^{4}	25
6. Maul (1968)	Sn scribed	3.1	3.91	5×10^{-3}	56	5.2×10^{3}	2.76	1.4	25×10^{-9}
7. Bertin (1968)	Sn								
(a) Film 80		1	3.63	2×10^{-3}	66	1.3×10^{3}	1.7	1	$\approx 10^{-8}$
(b) Film 83		1	3.59	2.5×10^{-4}	1200	1.9×10^{4}	1.7	1.7	$\approx 10^{-8}$
8. Zwerdling *et al.* (1968)	Ge:Ga, Sb	0.05	1.65	1.5×10^{-6}	2.6×10^{6}	-5.8×10^{6}	1.6×10^{-4}	4×10^{4}	2.6×10^{-4}

4. *Low* (*1961*)

This bolometer has been developed by Low into a practical device which has already made significant contributions to infrared astronomy (Low, 1969). Apparently, the time constant quoted here must be viewed with some caution since Zwerdling *et al.* (1968) reports that generally available bolometers of this type seem to operate efficiently only at very low chopping frequencies ≤ 30 Hz and, in at least one case (Merriam *et al.*, 1967) the responsivity dropped off to 4% of its zero-frequency value at 100 Hz. By careful attention to the details of mounting the device and filtering the incident radiation, Low has approached the thermal noise for his device.

5. *Judge* (*1968*)

Judge has made a careful investigation of the system considerations involved in bolometer design. To maximize the responsivity, K was reduced to a very low value by suspending the bolometer on a nylon net. As a result, the thermal time constant was quite long, so that the bolometer was particularly sensitive to acoustic noise arising from helium bubbling. It was found that temperature fluctuations in the bath of 1 m°K could cause voltage fluctuations in the detector on the order of 10 mV if a simple tin detector on a glass substrate was placed directly on the primary heat sink. Judge found that a secondary heat sink could be introduced between the primary heat sink and the substrate with enough heat capacity to substantially damp the thermal fluctuations. The responsivity quoted in Table II was measured directly rather than simply calculated from the static characteristics of the bolometer as is often done. Considering the basic similarity of this bolometer to the Martin and Bloor bolometer, it is interesting to note the substantial differences in r and τ that were observed.

6. *Maul* (*1968*)

This bolometer was designed for high speed and has quite a high thermal conductance indicating close coupling to the helium bath. Since it is scribed, it is possible to obtain high values of R and γ even with relatively low values of sheet resistance. The responsivity of 1.4 V W^{-1} is obtained by correcting the value 23 V W^{-1} to reflect a coupling coefficient $\alpha \approx 0.06$ (see end of Section III) for these films of approximately 1 Ω/square at 4.2°K.

7. *Bertin* (*1968*)

These bolometers were likewise designed for high speed but, since they are not scribed, achieve a high value of γ by means of a high sheet resistance. Enough data were obtained on each film studied to allow comparison

TABLE III

COMPARISON OF MEASURED AND CALCULATED RESPONSIVITIES

Film	R_N (4.2°K) (Ω/square)	α Calculated	r Calculated (V W^{-1})	r Measured (V W^{-1})
83	1915	0.11	0.3	1.7
72	250	0.38	0.53	0.35
80	131	0.4	0.6	1.0
7-11-1	5.6	0.058	0.044	0.066

of the measured and calculated responsivities. The results are shown in Table III for a wide range of film resistance. r (calculated) is taken from Eq. (12) in the limits $\omega^2\tau^2 \ll 1$ and $F \approx 1$. α was calculated from R using Eq. (2) and the other values were measured. r (measured) was taken to be $(\pi/2)(V_{Gp}/P_i)$ and was taken from measurements at microwave frequencies. With K chosen to be 1.7 W °K^{-1} the measured and calculated values for r agree within a factor of 2 (except for film 83) indicating among other things the validity of our expression for the absorptance α in lower resistance films.

8. *Zwerdling et al. (1968)*

This reference gives an extensive detailed description of the design of a semiconducting bolometer system which has achieved a significant improvement over the Low bolometer by careful attention to mounting, filtering, and the choice of material. Most of this "improvement," however, seems to be the result of using a detector with a smaller area as will be seen when figures of merit are compared in Table IV. Responsivity and NEP measurements were made in the far infrared using a blackbody radiation source.

Comparing superconducting bolometers 3, 5, 6, and 7 described in Table II we notice that these designs include all four combinations of scribed or unscribed Sn films and tight or loose coupling to the helium bath. InSb electron bolometers (Kinch, 1968), in which the temperature of the electrons is decoupled from the temperature of the lattice so that the incident radiation only affects the electron temperature, have not been included in Table II because their response is strongly wavelength dependent. Their response is compared with that of current mode detectors in Section V.C.

From the results listed in Table II we can calculate various figures of merit for these bolometers. As we saw in Section V.A the most appropriate figure of merit would be

$$M_2 = 2.12 \times 10^{-10}(Af)^{1/2}(\mathrm{NEP})^{-1} = 8.46 \times 10^{-11}A^{1/2}\tau^{-1/2}(\mathrm{NEP})^{-1} \quad (38)$$

TABLE IV

FIGURES OF MERIT FOR BOLOMETERS

References	Limiting noise	NEP Measured W $Hz^{-1/2}$	NEP Calculated W $Hz^{-1/2}$	D^* Measured cm $Hz^{1/2}$ W^{-1}	P^* Measured W $Hz^{-1/2}$ cm^{-1}	P^* Calculated W $Hz^{-1/2}$ cm^{-1}	M_2 Measured	M_2 Calculated
1. Smith *et al.* (1957)	Fluctuations, pickup		3×10^{-11}	3.3×10^{9}		3×10^{-10}		12
2. Boyle and Rogers (1959)	Granularity	10^{-11}	10^{-13}	4.5×10^{10}	2.2×10^{-11}	2.2×10^{-13}	38	3800
3 Martin and Bloor (1961)	Amplifier	$\approx 10^{-12}$	$\approx 10^{-13}$	2.5×10^{11}	4.0×10^{-12}	4.0×10^{-13}	200	2000
4. Low (1961)	Amplifier	2.4×10^{-12}	4.3×10^{-13}	2.0×10^{11}	5×10^{-12}	8.5×10^{-13}	850	4500
5. Judge (1968)	Amplifier	2.6×10^{-12}	2.6×10^{-13}	3.8×10^{11}	2.6×10^{-12}	2.6×10^{-13}	6	60
8. Maul (1968)	Thermal		8.0×10^{-10}			4.5×10^{-10}		1200
7. Bertin (1968)								
(a) Film 80	Amplifier	1.4×10^{-8}	1.5×10^{-10}	7.1×10^{7}	1.4×10^{-8}	1.5×10^{-10}	60	5500
(b) Film 83	Amplifier	8.1×10^{-9}	3.1×10^{-10}	1.2×10^{8}	8.1×10^{-9}	3.1×10^{-10}	100	2600
8. Zwerdling *et al.* (1968)	Amplifier	4.6×10^{-12}		4.9×10^{10}	2.1×10^{-11}		260	

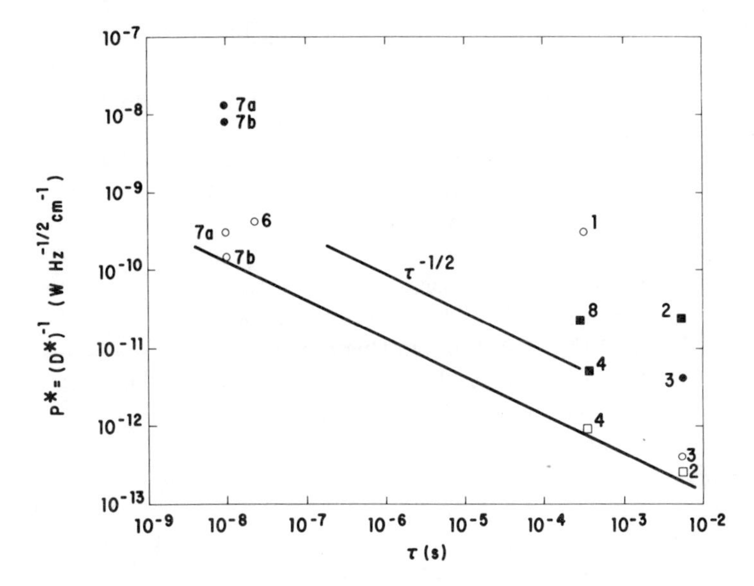

Fig. 8. Comparative performance of superconducting (○,●) and semiconducting (□,■) bolometers on the basis of calculated (○,□) and measured (●,■) values. The numbers refer to Table IV where more detailed characteristics are given.

where $f = (2\pi\tau)^{-1}$ as introduced by Jones (1953). This is, in fact, the figure of merit which Low (1961) feels is most appropriate for purposes of comparison. Nevertheless, since other figures of merit are more common in the literature we will calculate NEP, D^*, and P^*, as well. The results are given in Table IV.

Table IV lists both measured and calculated values for NEP and uses them to derive measured and calculated values of P^*. Only the measured value of D^* is given, but both measured and calculated M_2 values appear. In each case the limiting-noise mechanism is listed; thermal noise includes both Johnson and temperature noise. The calculated figures of merit indicate performance levels if the only limiting noise were the element's Johnson and temperature noise.

Examination of Table IV provides an instructive comparison of figures of merit. For instance, while Judge's bolometer has one of the highest values of D^* (lowest NEP or P^*), it achieves this by means of a very long time constant, so that its inherent figure of merit M_2 is very low. On the other hand, Bertin's value of calculated M_2 indicates that it could achieve higher values of D^* (lower values of NEP and P^*) than any other bolometer if its time constant was increased.

The measured and calculated values of P^* are plotted as a function of τ on a log–log scale in Fig. 8. The most obvious feature of Fig. 8 is its wide range of P^* and τ. However, by superimposing lines corresponding to P^* proportional to $\tau^{-1/2}$ on Fig. 8 we can, in effect, compare the bolometers on the basis of M_2, with detectors lying below a given line having greater merit than those lying above the line. It is apparent from Fig. 8, and also from the calculated M_2 values in Table IV, that the best semiconductor and superconductor devices are all quite comparable in performance under device thermal-noise-limited performance. Also shown is the extent to which amplifier noise is a limitation on the performance of cryogenic detectors, while the gap in devices which operate with time constants τ between nanoseconds and milliseconds reflects, in part, the difficulty in obtaining thermal time constants in this range.

Katz (1974) has recently developed medium-speed bolometers with 10 μsec response times by scribing films tightly coupled to the helium bath in order to raise their electrical time constant. Scribing raised the bolometer impedance, responsivity, and noise voltage, correspondingly. However, since the sensitivity of the high-speed bolometer was limited by amplifier noise, lengthening its response time improved its effective sensitivity. It was not possible to reduce the time constant of slow-speed bolometers by mounting them so that they would be coupled more tightly to the thermal reservoir. In all attempts the thermal conductance was limited by an anomalous boundary resistance resembling that observed by Judge (1968).

C. Nonbolometric Detectors

In this section we compare the sensitivity of the nonthermal current mode of detection discussed in Section IV with other wavelength detectors. The InSb and doped-Ge devices to which we compare our results all have values of τ between 1 μsec and 100 nsec but vary greatly in their sensitivity as a function of the wavelength of the incident radiation. Thus, a plot of D^* as a function of λ, as given in Fig. 9, is appropriate for comparing their performance.

The detectors with which we compare our results are of two types: shallow impurity photoconductors and hot electron photoconductors. In impurity photoconductors the incident radiation raises carriers from impurity levels to the conduction or valence bands. The response of such detectors is limited to wavelengths less than a cutoff wavelength set by the energy difference between the band and the impurity level. Wavelengths as long as 120 μ can be detected by using shallow impurity levels in Ge.

The values shown for Ge:Au and Ge:Zn are representative of commercially available detectors and are taken from a review by Levinstein (1965). The values shown for Ge:In are taken from a review by Smith (1965) who compares them with the InSb Putley detector.

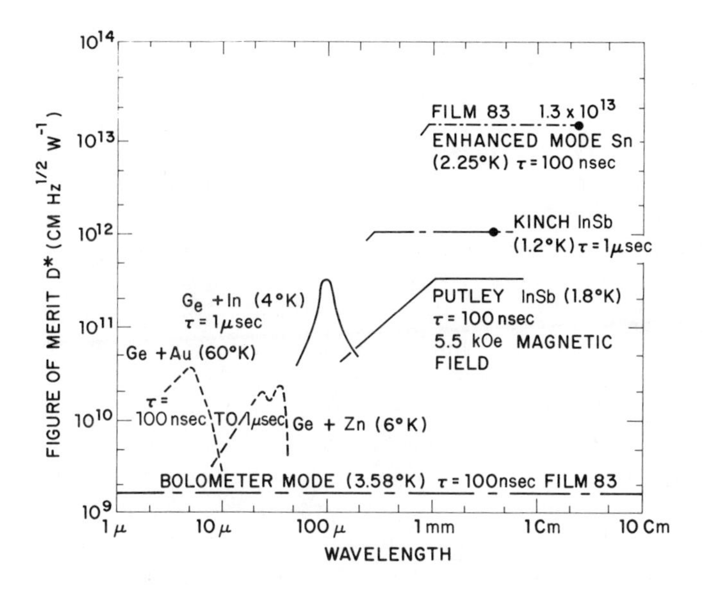

FIG. 9. Comparative performance of wavelength-dependent detectors.

To develop photoconductors that were responsive at longer wavelengths, shallower impurity levels were needed. This would require, according to the hydrogenic model, bands with a low effective mass such as the conduction band of InSb. Unfortunately, it was found that in such semiconductors the impurity levels tend to merge with the band, leading to metallic behavior at all temperatures (Putley, 1965). Fortunately, however, the electrons in these merged bands can have their temperature significantly altered by the absorption of radiation (hot electron photoconductivity). This effect has been studied by Rollin (1961) and developed into practical detectors by Putley (1960, 1961), Kinch and Rollin (1963), and Kinch (1968) among others. The subject is reviewed at length in a paper by Putley (1965). A short wavelength limit is set on this mechanism by decreasing free carrier absorption for $\omega\tau > 1$, where τ in this instance is the mean free collision time of electrons in InSb. In practice, the response falls off for wavelengths shorter than 1 mm but can be raised somewhat by the application of a 7- to 10-kG magnetic field which splits the conduction band into Landau levels.

The detector performance in the current mode is limited by amplifier noise as was Putley's detector. Putley's detector, in the presence of a 5.5-kG magnetic field has $r = 200$ V W^{-1}, a response time $\tau = 100$ nsec, $A = 0.1$ cm^2, and $D^* = 6.5 \times 10^{10}$ cm Hz$^{1/2}$ W^{-1}. Thus, the limiting-noise voltage introduced by Putley's amplifier in a 1-Hz bandwidth must be

$$V_{\mathrm{N}} = r\,\mathrm{NEP} = A^{1/2} r D^{*-1} \approx 1\ \mathrm{nV} \tag{39}$$

or $V_{\mathrm{N}} \approx 2\ \mu\mathrm{V}$ in the 4-MHz bandwidth corresponding to a rise time of $\tau = 100$ nsec (Dreher, 1967). Using the peak responsivity given in Fig. 6 of $1.3 \times 10^{4}\ \mathrm{V\ W^{-1}}$ for film 83 and comparing the current-mode detector with the Putley detector on the basis of the same amplifier noise in a 1-Hz bandwidth we find $D^* \approx 1.3 \times 10^{13}\ \mathrm{cm\ Hz^{1/2}\ W^{-1}}$. Since the results for the current mode were obtained at microwave frequencies, an indication has been given of the lower wavelength cutoff which would be expected for the enhanced mode if it were limited by the wavelength corresponding to the superconducting energy gap in tin.

Although the properties of Josephson junctions as far infrared radiation detectors is outside the scope of this chapter some remarks in passing by way of comparison certainly seem warranted. Superconducting point contact Josephson junctions are extremely sensitive, achieving NEP $= 5 \times 10^{-13}\ \mathrm{W\ Hz^{-1/2}}$ as measured at 4 mm with $\tau \leq 10$ nsec at wavelengths corresponding to photon energies in the neighborhood of the energy gap for a particular superconductor (Grimes *et al.*, 1968). Film 83 in the enhanced mode has achieved somewhat lower values of NEP $\approx 7.7 \times 10^{-14}\ \mathrm{W\ Hz^{-1/2}}$ at 3 cm on the basis of the above calculations of D^*. However, as noted at some length earlier, comparison on the basis of NEP is deceptive since it gives an advantage to detectors having a smaller active area. It is difficult to compare film 83 with an area of 1 cm^2 with the point contact Josephson junctions mentioned above since the active area in the latter case is indeterminate. An any event, it appears that superconducting films operating in the current mode may offer significant promise as millimeter and submillimeter wave detectors.

VI. Other Detectors

A. Phonon Detectors

Since bolometers respond to incident energy the form in which the energy is delivered matters little. As a result, one would expect the same considerations which apply to superconducting bolometers as detectors of electromagnetic radiation or photons to apply to superconducting bolometers as detectors of vibrational radiation or phonons. One difference of importance is that the range of phonon energies is less than about 50 meV which would correspond to photon wavelengths lengths longer than 25 μ.

Thus, one is dealing with the detection of quanta whose energies are comparable to those of the far infrared where detectors are generally unavailable. von Gutfeld and Nethercot (1964) and Andrews and Strandberg (1966) have used superconducting thin-film bolometers to detect heat pulses at low temperatures in quartz and sapphire. We have already reported the short response time of the bolometers used by von Gutfeld and Nethercot in Table I (von Gutfeld *et al.*, 1966). Andrews (1966) compared the response of a superconducting bolometer to coherent 9.0-GHz longitudinal phonons in an x-cut quartz rod with the response determined by the conventional reentrant cavity technique. He found that the microwave reentrant cavity technique introduced a factor of 0.4 dB/cm in addition to the residual attenuation of 0.6 dB/cm measured by the bolometer. This suggests that superconducting bolometers would be superior detectors for *absolute* acoustic attenuation measurements at microwave frequencies because they are phase insensitive and thus would not be affected by the experimentally unknown phase averaging of the microwave field.

B. Nuclear Radiation Detectors

Wood and White (1969) have pointed out that as a result of statistical limitations (Dearnaley and Northrop, 1966) maximum energy resolution of a gas or semiconductor nuclear particle spectrometer is achieved when the mean energy loss by the particle per liberated charge carrier in the spectrometer is minimized. In semiconductor spectrometers, for example, this mean energy loss is a few times the semiconductor's energy gap. Since the energy gap for the creation of an excitation in a superconductor is on the order of one-thousandth the semiconductor energy gap, it would appear that superconductors should be useful in nuclear spectroscopy. An additional advantage for γ-ray spectroscopy, in particular of superconducting detectors made of lead, for example, over semiconducting detectors made of silicon, is their higher mass density and atomic number.

Early experiments on nuclear radiation by Andrews and associates (1949) indicated that NbN strips having dimensions $3.5 \times 0.4 \times 0.006$ mm biased in the middle of their transition region at 15.5°K could detect individual α particles on the basis of comparison with an ionization counter. Later, Spiel and associates (1965) studied superconducting–normal-state transitions induced by 5.3 MeV α particles from a ^{210}Po source in thin In films 34 μ wide. The radius r_c of the cylindrical volume driven into the normal state could be estimated from the lower critical currents of the bombarded film by assuming that the width of the film had been reduced by $2r_c$. They found that r_c rose from 3 to 14 μ as the bath temperature was raised to T_c.

Close to T_c self-terminating voltage pulses with rise times of 15 nsec and fall times of 70 nsec were observed at low current densities. For higher current densities, the boundary of the normal region initiated by an α particle spread to the ends of the film as a result of Joule heating.

Sherman (1962) considered the sizes of the normal cores which would be produced by nuclear radiation in a proposal for the use of superconductors as nuclear radiation detectors. He found $r_c \approx 0.8$ to $4\ \mu$ for 60-MeV fission fragments, depending on the initial temperature produced in the film, and $r_c \approx 0.15\ \mu$ for 5-MeV α particles as well as rise and fall times of 10^{-2} and 10^3 nsec for the resistance pulses which would be produced. As a result of further investigations, Spiel and Crittenden (Crittenden, 1969) now believe that the anomalously large critical radii they observed were caused by the detection of phonons transmitted by the ballistic flight from the irradiation path deep within the crystal quartz substrates they used. By coating their substrates with electron-polymerized "varnish" before depositing the film they obtained critical radii in agreement with expectations. Regardless of the mechanism, larger critical radii are desirable since this allows the use of wider films as particle detectors.

In any event, the switching action produced by driving a current-carrying superconducting strip normal, while useful as a particle detector or energy threshold discriminator, would not produce a response proportional to particle energy loss. In this regard further development of superconducting bolometers might be desirable. An alternate approach developed by Wood and White (1969) is to use pulses induced in the quasiparticle current of a superconducting tunnel junction. Excess current pulses up to 23 μA in amplitude were observed in tin tunnel junctions of area 7×10^{-4} cm^2 bombarded by 5-MeV α particles. They estimated that the maximum energy loss per tunneling quasiparticle was approximately 0.145 eV which implied a statistically limited energy resolution a factor $(0.145/3.6)^{1/2} = 0.20$ smaller than that obtained for a silicon spectrometer.

Cavallini *et al.* (1969) have operated a superconducting bolometer as a molecular beam detector with a sensitivity, for argon, of 7×10^6 molecules sec^{-1}. This corresponds to a NEP of 3×10^{-13} W Hz$^{-1/2}$.

VII. Detector Arrays and Imaging Systems

Since arrays of thin-film bolometers or other superconducting detectors have not yet been constructed, this section is necessarily speculative and brief. Nonetheless, one of the most potentially attractive features of superconducting thin-film devices is their ability to be fabricated in large arrays by the use of the standard techniques of vacuum metallization and photolithography already developed for semiconductor microcircuits. Thus, once a desirable detector element has been found, it should be possible to fabricate vector and matrix arrays of these elements in a relatively straightforward fashion. One-dimensional (vector) arrays of elements could be used at the output of a spectrometer to facilitate high-speed spectroscopy or as lines scanned by a rotating mirror in an imaging system. The development of two-dimensional (matrix) arrays would allow direct imaging. On the basis of present-day metallization techniques, element and circuit sizes as small as those achievable with semiconductor devices certainly seem feasible. Indeed, the ability to vacuum deposit the elements of high sensitivity should make the cost per element quite low, which is a factor of considerable importance in the design of large arrays.

Practical problems to be overcome in any array are those of electrical and, particularly for bolometers, thermal cross talk between elements. Likewise, consideration must be given to the relative merits of amplifying the signal before or after switching between the elements of the array. The choice would depend in the main on the complexity of each circuit and the cross talk generated by the switching element. Many of these problems are ones which have been faced already in the design of superconducting computer memories as discussed in Chapter 3. In some ways the problems are simpler with respect to imaging since relatively modest matrix arrays containing 10^4 elements or less would be acceptable for most purposes. On the other hand, a high packing density corresponding to element areas of 25 mils^2 would also be desirable.

More direct imaging schemes, perhaps involving the use of magneto-optical techniques to monitor changing fluxon patterns (De Sorbo and Newhouse, 1962) might be envisioned but have not yet been developed. One analysis of such a scheme (Smith, 1969) suggests that such a scheme might involve a conversion loss on the order of 50 dB.

VIII. Conclusions

In conclusion, it appears that superconductors offer definite advantages over other types of detectors in several areas. As bolometers superconducting films appear to be superior to other types in terms of speed or sensitivity, while the current mode offers promise of superior far infrared detection. The speed and sensitivity of the superconducting bolometer would also appear to offer advantages for the detection of phonons and nuclear radiation as well, while the simplicity of the superconducting film makes it an attractive element in arrays. As is often the case, there already exist semiconductor devices which are competitive or superior in many of the most important applications. Thus, while the superconducting bolometer could probably be developed into a superior detector for far infrared spectroscopy, this is not presently an area of major technological interest. Likewise, while the problems of cryogenic cooling are often faced equally by competitive semiconductor detectors, it does restrict the use of superconducting detectors in many situations. In any event, before the potential promise of superconducting detectors can be realized, a substantial amount of development is needed.

References

Abrikosov, A. A. (1957). *Sov. Phys. JETP* **5**, 1174.

Anderson, P. W., and Dayem, A. H. (1964) *Phys. Rev. Lett.* **13**, 195.

Andrews, D. H. (1938). *Amer. Phil. Soc. Year Book*, p. 132.

Andrews, D. H. (1942). *Rev. Sci. Instr.* **13**, 281.

Andrews, D. H., and Clark, C. W. (1946). *Nature* **158**, 945.

Andrews, D. H., and Clark, C. W. (1947). *Phys. Rev.* **72**, 161 (A).

Andrews, D. H., Brucksch, W. F., Jr., Ziegler, W. T., and Blanchard, E. R. (1941). *Phys. Rev.* **59**, 1045 (L).

Andrews, D. H., Milton, R. M., and DeSorbo, W. (1946). *J. Opt. Soc. Amer.* **36**, 353 (A).

Andrews, D. H., Fowler, R. D., and Williams, M. C. (1949). *Phys. Rev.* **76**, 154 (L).

Andrews, J. M., Jr. (1966). Quarterly Progress Report, Vol. 82, p. 12. MIT, Research Laboratory of Electronics, Cambridge, Massachusetts.

Andrews, J. M., Jr., and Strandberg, M. W. P. (1966). *Proc. IEEE* **54**, 523.

Ayer, W. J., Jr. (1974). "Electromagnetic Radiation Detection by Granular Thin Film Superconductors." Doctoral thesis, Rensselaer Polytechnic Institute, Troy, New York.

Baird, D. (1959). *Can. J. Phys.* **37**, 120.

Bertin, C. L. (1968). Doctoral thesis. Rensselaer Polytechnic Institute, Troy, New York.

Bertin, C. L., and Rose, K. (1968). *J. Appl. Phys.* **39**, 2561.

Bertin, C. L., and Rose, K. (1971a). *J. Appl. Phys.* **42**, 163.

Bertin, C. L., and Rose, K. (1971b). *J. Appl. Phys.* **42**, 631.

Bodmer, M. (1950). Doctoral thesis. Johns Hopkins University, Baltimore, Maryland.

Boyle, W. S., and Rodgers, K. F. (1959). *J. Opt. Soc. Amer.* **49**, 66.

Bremer, J. W., and Newhouse, V. L. (1959). *Phys. Rev.* **116**, 309.

Brucksch, W. F., Jr., Ziegler, W. T., Blanchard, E. R., and Andrews, D. H. (1941). *Phys. Rev.* **59**, 688 (A).

Burton, C. H. (1968). Private communication.

Cavallini, M., Gallinaro, G., and Scoles, G. (1969). *Z. Naturforsch.* **24a**, 1850.

Crittenden, E. C., Jr. (1969). Private communication.

Dearnaley, G., and Northrop, D. C. (1966). "Semiconducting Counters for Nuclear Radiations." E. and F. N. Spon Ltd., London.

DeSorbo, W., and Newhouse, V. L. (1962). *J. Appl. Phys.* **33**, 1004.

Dreher, T. (1967). "Fast Pulse Techniques." E-H Research Laboratories, Oakland, California.
Einstein, A. (1904). *Ann. Phys. (Leipzig)* **14,** 354.
Franzen, W. (1963). *Phys. Rev.* **53,** 596.
Fuson, N. (1948). *J. Opt. Soc. Amer.* **38,** 845.
Gerig, J. S. (1964). Doctoral thesis. Stanford University, Stanford, California.
Goetz, A. (1939). *Phys. Rev.* **55,** 1270 (L).
Golovashkin, A. I., and Motulevich, G. P. (1965). *Sov. Phys. JETP* **20,** 44.
Gregory, W. D. (1968). *Phys. Rev.* **165,** 556.
Grimes, C. C., Richards, P. L., and Shapiro, S. (1968). *J. Appl. Phys.* **39,** 3905.
Jones, R. C. (1949). *J. Opt. Soc. Amer.* **39,** 327.
Jones, R. C. (1953). *In* "Advances in Electronics" (L. Marton, ed.), Vol. II, pp. 2–96. Academic Press, New York.
Jones, R. E., and Pennebaker, W. B. (1963). *Cryogenics* **3,** 215.
Judge, D. L. (1968). Doctoral thesis. University of Southern California, Los Angeles, California.
Katz, R. M. (1974). Doctoral thesis. Rensselaer Polytechnic Institute, Troy, New York.
Katz, R. M., and Rose, K. (1972). *Proc. IEEE* **60,** 00.
Kinch, M. A. (1968). *Appl. Phys. Lett.* **12,** 78.
Kinch, M. A., and Rollin, B. V. (1963). *Brit. J. Appl. Phys.* **14,** 672.
Konovodchenko, V. A., and Domitriyev, F. N. (1968). *Tr. Fiz. Tekh. Inst. Akad. Nauk Ukr. SSR* **III,** 95.
Kruse, P. W., McGlauchlin, L. D., and McQuistan, R. B. (1962). "Elements of Infrared Technology." Wiley, New York.
Lebacqz, J. V., Clark, C. W., Williams, M. C., and Andrews, D. H. (1949). *Proc. IRE* **37,** 1147.
Lalevic, B. (1960). *J. Appl. Phys.* **31,** 1234.
Lazarev, B. G. *et al.* (1941). *Zh. Eksp. Teor. Fiz.* **11,** 573.
Levenstein, H. (1965). *Appl. Opt.* **4,** 639.
Little, W. A. (1959). *Can. J. Phys.* **37,** 334.
Low, F. J. (1961). *J. Opt. Soc. Amer.* **51,** 1300.
Low, F. J. (1969). 1969 *Int. Conv. Dig.*, p. 44.
Martin, D. H., and Bloor, D. (1961). *Cryogenics* **1,** 159.
Maul, M. K. (1968). Doctoral thesis. Massachusetts Institute of Technology, Cambridge, Massachusetts.
Maul, M. K., and Strandberg, M. W. P., and Kyhl, R. L. (1969). *Phys. Rev.* **182,** 522.
Merriam, J. D., Eisenman, W. L., and Naugle, A. B. (1967). *Appl. Opt.* **6,** 576.
Millman, J., and Taub, H. (1965). "Pulse, Digital, and Switching Waveforms." McGraw-Hill, New York.
Mumford, W. W., and Schiebe, E. H. (1968). "Noise Performance Factors in Communication Systems." Horizon House—Microwave Inc., Dedham, Massachusetts.
Nethercot, A. H., Jr., and von Gutfeld, R. J. (1963). *Phys. Rev.* **131,** 576.
Pankratov, N. A., Zaytsev, G. A., and Krebtov, I. A. (1970). *Radio Eng. Electron. Phys.* **15,** 1652.
Putley, E. H. (1960). *Proc. Phys. Soc. (London)* **76,** 802.
Putley, E. H. (1961). *J. Phys. Chem. Solids* **22,** 241.
Putley, E. H. (1965). *Appl. Opt.* **4,** 649.
Ramey, R. L., Kitchen, W. J., Lloyd, J. M., and Landes, H. S. (1968). *J. Appl. Phys.* **39,** 3883.
Rollin, B. V. (1961). *Proc. Phys. Soc.* **77,** 1102.

Rose, K. (1969). *Cryogenics* **9**, 227.
Rose, K., and Sherrill, M. D. (1966). *Phys. Rev.* **145**, 179.
Saxena, A., Crow, J. E., and Strongin, M. (1972). Int. Conf. Low Temp. Phys., 13th, Paper HaSl.
Sherman, N. K. (1962). *Phys. Rev. Lett.* **8**, 438.
Sherrill, M. D., and Rose, K. (1964). *Rev. Mod. Phys.* **36**, 312.
Smith, C. W. (1969). Master's Project Report, Rensselaer Polytechnic Institute, Troy, New York.
Smith, R. A. (1965). *Appl. Opt.* **4**, 631.
Smith, R. A., Jones, F. E., and Chasmar, R. P. (1957). "The Detection and Measurement of Infrared Radiation." Oxford, London and New York.
Soderman, D. A. (1969). Doctoral thesis. Rensselaer Polytechnic Institute, Troy, New York.
Spiel, D. E., Boom, R. W., and Crittenden, E. C., Jr. (1965). *Appl. Phys. Lett.* **7**, 292.
Strandberg, M. W. P., and Kierstead, J. D. (1966). Quart. Progr. Rep. **82**, 11. MIT, Research Laboratory of Electronics, Cambridge, Massachusetts.
Tinkham, M. (1964). *Science* **145**, 240.
von Gutfeld, R. J., and Nethercot, A. H., Jr. (1964). *Phys. Rev. Lett.* **12**, 641.
von Gutfeld, R. J., Nethercot, A. H., Jr., and Armstrong, J. A. (1966). *Phys. Rev.* **142**, 436.
Williams, R. A. (1961). Doctoral thesis. Ohio State University, Columbus, Ohio.
Wood, G. H., and White, B. L. (1969). To be published.
Woodward, A. E., and Silvermetz, D. (1965). *In* "Handbook of Military Infrared Technology" (W. L. Wolfe, ed.), p. 583. Office of Naval Research, Department of the Navy, Washington, D. C.
Zwerdling, S., Smith, R. A., and Theriault, J. P. (1968). *Infrared Phys.* **8**, 271.

Chapter 5

Refrigerators and Cryostats for Superconducting Devices

WALTER H. HOGAN

Cryogenic Technology, Inc.
Waltham, Massachusetts

I. Introduction

Refrigeration for the preservation of food or for personal comfort is well established, and the end user rarely concerns himself with the how or why; it is sufficient to specify what is wanted. Low-temperature refrigeration has, until recently, been mostly the interest of those in the gas industry concerned with the liquefaction and separation of gases, and of a few physicists in the laboratory.

Today, many physicists and engineers must become familiar with low-temperature technology and refrigeration in a way that was never required of the food processor or the builder. Those concerned with electrooptics, microwave technology, astronomy, microscopy, high-energy physics, and power generation and distribution have a greater need to understand this interfacing technology. Those concerned with superconducting devices, and associated equipment, which typically operate in the 20–1.5°K temperature region, have an even greater need for understanding of fundamentals.

During the past decade, basic research has uncovered numerous opportunities for commercial and industrial applications of superconducting devices. Many, disclosed in preceding chapters, are in the development stage. Others, such as superconducting solenoids associated with masers in satellite communications, are in commercial use.

This chapter reviews the basic and practical thermodynamics of refrigeration, particularly as it applies to the low-temperature range of 20–1.5°K. It will deal specifically with the properties of helium, which is the only refrigerant for this temperature range, and discuss many of the design considerations relating to liquid-helium cryostats.

The chapter is intended to be tutorial to physicists and engineers whose primary interest is with superconducting devices rather than with cryogenics, but who wish to understand, use, and evaluate low-temperature-refrigeration techniques.

II. Thermodynamics

A. General

Thermodynamics is a collection of basic concepts and four simple laws, which are used to study any system in which energy changes occur. A thermodynamic system is a space enclosed by a boundary which contains the material to be studied. The boundary may be real or imaginary, and may change shape and volume with time, but it must always be defined. Thermodynamics is concerned with the energy content of the mass within the boundary, with the energy which crosses the boundary in the form of work or heat, and with the energy content of mass which crosses the boundary.

Although many of the founders of thermodynamics and many successful builders of heat engines believed that heat was a substance called *caloric,* the proposition that heat is not a substance is no longer considered a hypothesis but an expression of certainly proven fact. The evolvement of thermodynamics, however, has given us many terms involving heat which stem from the concept of caloric, e.g., latent heat, sensible heat, heat capacity, specific heat, heat content.

Today, it is preferred to think of heat only as energy in transit across a boundary. By analogy, the water in a cloud turns into rain and becomes a puddle; only in transit is it called rain. Similarly work is energy in transit across a boundary.

The transfer of energy from one body to another can occur across a boundary in the form of heat or work. The changes in energy content are best described in terms of pressure, volume, temperature, internal energy, enthalpy, and entropy, all of which are interrelated. They are microscopically statistical in nature or are macroscopic quantities with no meaning at the microscopic level.

1. *Pressure of a gas*

From the *Kinetic Theory of Gases* (Loeb, 1938), a gas is composed of gas molecules considered as discrete particles having mass and velocity; they move in straight lines until collision. In the ideal, they are too far apart for either attractive forces or repulsive forces to play any part, and hence gases always tend to expand. Pressure results from a series of impacts of the molecules, and the consequent exchange of momentum, with any object experiencing the pressure.

Consider a fixed volume confining a gas. The molecular motion is random and with a Maxwellian velocity distribution. However, we can start with the same assumption with which Joule (1851) started in his calculation—he assumed that a gas pressed against the faces of a rectilinear volume as if all the molecules moved with a mean molecular velocity, and that one-third of the molecules moved parallel to each of the three directions of the edges, so that each face was impinged by only one-third of the molecules.

If there are N molecules per unit volume, and the sides of the rectilinear volume are a, b, and c, there will be $N{:}abc$ molecules in the volume. One-third of these, $\frac{1}{3}N{:}abc$, will be moving in the direction c against faces ab. A molecule moving in the direction c will travel a distance $2c$ between the two ab faces between each impact on one face ab; with a velocity v, the number of impacts will be $v/2c$ in a unit time.

The total number of impacts on one face ab in unit time by all particles is

$$\tfrac{1}{3}Nabc(v/2c) = \tfrac{1}{6}Nvab$$

If we consider each impact with the face ab to be ideally elastic (although it is not necessary to do so), the velocity of separation is equal to the velocity of approach v, and the magnitude of the impulse from the molecule of mass m is $2mv$.

The force on the face ab is equal to the number of impacts times the impulse or momentum change of each impact:

$$F = \tfrac{1}{6}Nvab2mv = \tfrac{1}{3}Nmv^2ab$$

The pressure on the face ab, the gas pressure, is

$$P = F/ab = \tfrac{1}{3}Nmv^2 \qquad (1)$$

Since $Nm = \rho$, the density

$$P = \tfrac{1}{3}\rho v^2$$

2. *Boyle's law*

Considering again the case of a volume V with N molecules per unit volume, the total number of molecules is determined and constant:

$$NV = C, \quad \text{a constant,} \qquad \text{and} \qquad N = C/V$$

On substitution into (1), we obtain

$$P = Cmv^2/3V \qquad \text{and} \qquad P \propto 1/V, \qquad \tfrac{1}{3}(Cmv^2) \quad \text{remaining constant}$$

This is an expression of Boyle's law which states that the pressure of a perfect gas varies inversely as the volume, with temperature remaining constant (which we will show is a function of v^2).

3. *Charles' law*

The last equation also made clear that if the volume were held constant, the pressure was then proportional to v^2, the mean square of the molecular velocity.

An empirical law, discovered nearly simultaneously by Gay-Lussac (1802) and Dalton (1802), expresses the dependence of the pressure of a gas on its temperature, and is contained in the amended form of Boyle's law:

$$P = C\rho(1 + \alpha\theta) \qquad (2)$$

where P and ρ denote the pressure and density as before and θ is the temperature Celsius. C is a constant and α the thermal coefficient of expansion.

Taking the temperature θ to be zero and using $P = \tfrac{1}{3}\rho v^2$, we obtain the value of C:

$$C = \tfrac{1}{3}v_1^2$$

where v_1^2 is the square of the mean molecular speed at $\theta = 0°C$.

On substitution back we have

$$v^2 = v_1^2(1 + \alpha\theta) \tag{3}$$

Thus, we find that the square of the molecular speed of gas molecules, and therefore the kinetic energy of its molecular motion increases proportionately with the temperature. Or, more particularly, the kinetic energy of the molecular motion is the mechanical measure of temperature.

The relationship (3) allows for a position of an absolute zero of temperature for gaseous bodies to be determined. Translational molecular motion has disappeared when $(1 + \alpha\theta) = 0$.

The Celsius temperature scale is an arbitrary scale using the freezing point of water as 0, and the boiling point as 100. The relationship between this arbitrary scale and absolute zero had to be determined experimentally.

Experiment shows that the thermal coefficient of expansion of a perfect gas is constant and has the value

$$\alpha = 1/273.15 \tag{4}$$

and therefore for absolute zero,

$$\theta = -273.15°\text{C}$$

We can now rewrite (2) as

$$P = C\rho(273.15 + \theta), \qquad (273.15 + \theta) = T°\text{K (absolute)} \tag{5}$$

$$P = C_1\rho T \tag{6}$$

and say that the pressure of a perfect gas is directly proportioned to the absolute temperature, density remaining constant. This is normally referred to as Charles' law.

4. *Pressure, temperature, and energy*

Since the density ρ measures the mass of gas contained in a unit volume, the magnitude K, the kinetic energy,

$$K = \tfrac{1}{2}\rho v^2$$

is nothing else than the amount of kinetic energy possessed by the molecules in unit volume.

Using the formula for pressure $P = \frac{1}{3}\rho v^2$ we have the simple relation between the two:

$$K = \tfrac{3}{2}P \tag{7}$$

The pressure and the total kinetic energy of a gas in a unit volume stand to each other in an invariable relation which is independent of the temperature.

The kinetic energy of a molecule is $\frac{1}{2}mv^2$ and using (3)–(5) we have

$$\tfrac{1}{2}mv^2 = [\tfrac{1}{2}mv^2(1/273.15)]T = C_2T$$

The absolute temperature and the mean kinetic energy of a gas molecule stand to each other in an invariable relation which is independent of the pressure.

5. *Specific heats*

To increase the temperature of a unit mass of gas, held at constant volume, by 1 deg requires an input of energy denoted by c_V, the specific heat at constant volume. The molecular kinetic energy has been increased by a fixed amount—so has the pressure. For any temperature increase, heat added Q_V is

$$Q_V = c_V(T - T_0)$$

To increase the temperature of a unit mass of gas, held at constant pressure, by 1 deg requires an input of heat denoted by c_P, the specific heat at constant pressure. The molecular kinetic energy has been increased by the same fixed amount, but the volume has changed to allow the pressure to remain constant. For any temperature increase, the heat added, Q_P is

$$Q_P = c_P(T - T_0)$$

The difference between these magnitudes

$$(c_P - c_V)(T - T_0)$$

is a measure of the work done by a unit mass of gas when its temperature is increased from T_0 to T and is allowed to expand from the pressure P it acquired at constant volume to its original pressure P_0 isothermally, or

$$(P - P_0) = \rho(c_P - c_V)(T - T_0) \tag{8}$$

The thermal energy put into a unit volume of gas, held at constant density to increase its temperature is

$$Q = \rho c_V(T - T_0)$$

From (7) the change in kinetic energy in the unit volume is

$$K - K_0 = \tfrac{3}{2}(P - P_0)$$

which from (8), may be written

$$K - K_0 = \tfrac{3}{2}\rho(c_P - c_V)(T - T_0)$$

So, the ratio of the change in kinetic energy per unit volume to the heat added per unit volume is

$$(K - K_0)/Q = \tfrac{3}{2}(c_P - c_V)/c_V$$

In a monatomic gas, such as helium, we may expect that all the heat, or thermal energy, added is converted into translational kinetic energy (rather than part rotational and vibrational) and so

$$K - K_0 = Q \qquad \text{and} \qquad (c_P - c_V)/c_V = \tfrac{2}{3}$$

For this to be so, the ratio of the specific heats must be

$$c_P/c_V = \tfrac{5}{3} = 1.66$$

which is empirically verified.

We can now see that C_1 in Eq. (6) is the same as $(c_P - c_V)$ in (8). If we call $(c_P - c_V) = R$, the gas constant, then (6) becomes

$$P = \rho RT$$

Since ρ is the density in moles per unit volume and n is the number of moles,

$$\rho = n/V \qquad \text{and} \qquad PV = nRT$$

6. *Internal energy, enthalpy, and entropy*

We have seen that, by a consideration of gas kinetics, we can relate the molecular kinetic energy, pressure, specific volume, temperature and the three specific heats, c_V, c_P, and R with heat and work.

There are other thermodynamic properties that are useful in thinking about and making calculations of thermodynamics. The addition of heat to a gas in a confined volume results in an increase in the temperature of the gas; that is, an increase of its specific internal energy u. The increases of temperature or internal energy with the addition of heat is qualitatively related by the specific heat at constant volume:

$$dQ_V = c_V\, dT = du$$

In fact, we can now make the definition of specific heat at constant volume in terms of the rate of change of two properties:

$$c_V = \left.\frac{\partial u}{\partial T}\right|_V$$

The addition of heat to a gas in an expandable volume, held at constant pressure, results in an increase in the temperature of the gas, and therefore in its internal energy, and also in the production of some work,

$$dQ_P = c_V\,dT + R\,dT = c_P\,dT$$

The property of the gas to change its internal energy and to do work with a change in temperature is sometimes referred to as its total heat capacity, or more properly, its enthalpy, h:

$$dQ_P = c_P\,dT = dh = du + R\,dT = du + d(PV)$$

Now, we can define the specific heat at constant pressure, in terms of the rate of change of two other gas properties

$$c_P = \left.\frac{\partial h}{\partial T}\right|_P$$

We can plot diagrams using pressure as the ordinate and volume as the abscissa to make a useful determination of work as the area of the diagram. It would be useful to have another diagram which would show the heat added or subtracted from a system, directly as the area of the diagram. For this, the coordinates must be such that their product will give heat units. If the ordinates are in absolute temperature, the abscissas must be in heat units per degree, that is Q/T, so that $T \times Q/T = Q$, the amount of heat added or subtracted from the system. The property Q/T is called entropy S:

$$dS = dQ/T \qquad \text{or} \qquad dQ = T\,dS$$

The values given to pressure P, volume V, and temperature T, are usually absolute; although the absolute zero of P and T can only be approached but not reached.

The values of the specific heats, c_V, c_P, and R, are constant for any given gas in the ideal state; for helium, this means down to about 50°K. As the gas approaches being a vapor or a liquid or is at high pressures, the values of the ratio of c_P to c_V and the value of n change, but the relationship $PV = nRT$ for an ideal gas can be expressed as

$$PV/nRT = Z = 1$$

For a nonideal gas the value of Z changes from unity, and the value of Z, the compressibility, is found in empirically derived tables. Similarly, the values of c_P and c_V change as a function of temperature and pressure. (See Figs. 6–8.)

In practice, analyses of thermodynamic systems are concerned with energy balances and as such are concerned essentially with changes in the thermodynamic-state point properties. The reference points for values of specific internal energy u, enthalpy h, and entropy S may therefore be chosen arbitrarily, rather than with absolute properties.

Heat Q, which refers to energy crossing a boundary from one temperature level to a lower one either by conduction or radiation (there is no other mode of transfer) and which increases or decreases enthalpy, and work W, which refers to energy crossing a boundary with no consequential change in entropy, are always measured in incremental terms $\pm dQ$, $\pm dW$. The appropriate equation connecting them is

$$\text{gain of energy} = \text{heat transferred inward} - \text{external work done by the system}$$

or

$$du = dQ - dW$$

Having related the gas properties P, V, T, c_V, c_P, R, u, h, and S, we can now look at what happens to a gas when it expands.

7. *Adiabatic expansion*

From the derivation (Section II.A.1) we know that the pressure of a gas is expressed by

$$P = \tfrac{1}{3}Nmv^2$$

where v is the mean molecular velocity relative to the surface experiencing the pressure.

Consider gas in an adiabatic cylinder, that is, perfectly insulated against heat loss or gain, closed by a moveable piston, of area A and moving with a velocity w away from the gas molecules.

The absolute molecular velocity before impact is then $(v + w)$ and after impact is $(v - w)$. The change in kinetic energy of a molecule will be

$$\tfrac{1}{2}m(v - w)^2 - \tfrac{1}{2}m(v + w)^2 = -2mvw$$

That is to say, the kinetic energy after impact with the moving wall is less than before impact. We know that the total number of impacts in unit time is $\frac{1}{6}NvA$, so that the total loss in kinetic energy in the gas, which is equal to the work done, $-dW$, on the piston in unit time is (the minus sign is a convention that is used when the work is done *by* the gas *on* the piston, rather than *by* the piston *on* the gas)

$$-dW = \tfrac{1}{6}NvA - 2mvw = -\tfrac{1}{3}Nmv^2wA$$

$$-dW = -\tfrac{1}{3}Nmv^2\,dV$$

$$-dW = -P\,dV$$

If we consider the cylinder to hold n moles of gas, we can now see what happens to the properties P, V, T, and W as the gas expands adiabatically.

Since there is no heat input

$$du = 0 - dW, \qquad nc_V\, dT = -P\, dV \tag{9}$$

Also

$$PV = nRT \qquad \text{and} \qquad dT = (P\, dV + V\, dP)/nR$$

Substitution of the value of dT into (9) gives

$$P\, dV + nc_V \frac{P\, dV + V\, dP}{nR} = 0, \qquad RP\, dV + c_V P\, dV + c_V V\, dP = 0$$

$$R \frac{dV}{V} + c_V \frac{dV}{V} + c_V \frac{dP}{P} = 0, \qquad (R + c_V) \frac{dV}{V} + c_V \frac{dP}{P} = 0$$

$$c_P \frac{dV}{V} + c_V \frac{dP}{P} = 0$$

while integration gives

$$c_P \ln V + c_V \ln P = \text{const} = C$$

where

$$c_P/c_V = k, \qquad k \ln V + \ln P = C, \qquad \ln PV^k = C$$

Therefore

$$PV^k = C, \qquad P = C/V^k$$

$$W = \int_{V_1}^{V_2} P\, dV = C \int_{V_1}^{V_2} \frac{dV}{V^k}$$

$$W = C[V^{-k+1}/(-k+1)]_{V_1}^{V_2} = C(V_2^{1-k} - V_1^{1-k})/(1-k)$$

Since $PV^k = C = P_1V_1^k = P_2V_2^k$,

$$W = (P_1V_1 - P_2V_2)/(k-1)$$

Since $PV = nRT$,

$$W = [nR(T_1 - T_2)]/(k-1)$$

Using two simultaneous equations

$$P_1V_1/T_1 = P_2V_2/T_2 \qquad \text{and} \qquad P_1V_1^k = P_2V_2^k$$

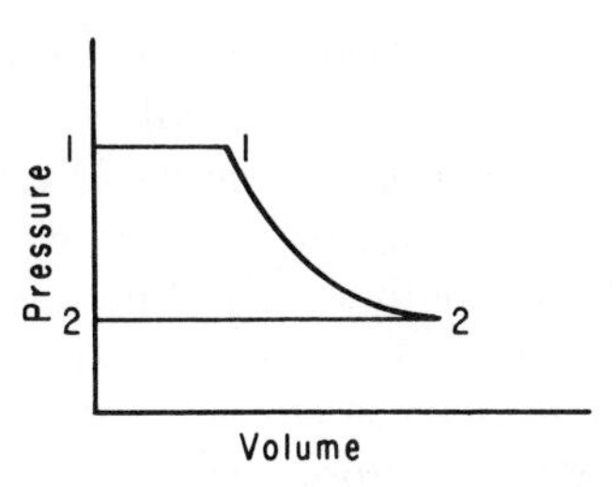

FIG. 1. Pressure volume or engine indicator diagram.

we can derive the following relationships:

$$T_2 = T_1(V_1/V_2)^{k-1} = T_1(P_2/P_1)^{(k-1)/k}$$

$$P_2 = P_1(V_1/V_2)^k = P_1(T_2/T_1)^{k/(k-1)}$$

$$V_2 = V_1(P_1/P_2)^{1/k} = V_1(T_1/T_2)^{1/(k-1)}$$

We have looked at the work of expansion and related that to the consequent changes of pressure, volume, and temperature. The work is equal to the area under the curve 1–2 in Fig. 1.

In a piston-and-cylinder–type expansion engine, the work per cycle is equal to the area under the lines 1–1 and 1–2, minus the area under the line 2–2:

$$\begin{aligned} W &= P_1V_1 + [(P_1V_1 - P_2V_2)/(k-1)] - P_2V_2 \\ &= [k/(k-1)](P_1V_1 - P_2V_2) \\ &= [k/(k-1)]P_1V_1[1 - (P_2/P_1)^{(k-1)/k}] \\ &= [k/(k-1)]nR(T_1 - T_2) \end{aligned}$$

B. LAWS OF THERMODYNAMICS†

The two principles on which classical thermodynamics is based are the first and second laws of thermodynamics. In order, however, to develop a complete exposition, three more postulates are required. These are the zeroth law, the third law, and the state principle. A number of equivalent statements may be given for any law.

1. *Zeroth law*

The zeroth law can be expressed in terms of temperature equality as follows: If two bodies are at the same temperature as a third body, then

† See Van Wylen and Sonntag, 1965.

they are at the same temperature as each other. The zeroth law enables us to classify the status of all systems into a set of classes such that any two states belonging to one class are equal in temperature, whereas any two states belonging to different classes are unequal in temperature. If heat can flow spontaneously from one body to another with no effects external to the systems, we will say the first body is at a higher temperature. We may now define a temperature scale in terms of the magnitude of a selected property of a proscribed system, such as the height of the meniscus in the capillary of a mercury thermometer, or the electrical resistance of a length of platinum wire.

2. *First law*

The first law of thermodynamics relates to the energy content of the mass within the thermodynamic boundary, with the energy fluxes which cross the boundary in the form of mechanical work or heat, and with the energy content of mass which enters or leaves the system across the boundary.

For a nonflow system, where no mass crosses the boundary, the first law may be expressed in terms of energy as follows: The total flow of energy into (or out of) a system equals the increase (or decrease) in the internal energy of the system:

$$Q = u_{\mathrm{f}} - u_{\mathrm{i}} - (-W) \qquad \text{or} \qquad Q = u_{\mathrm{f}} - u_{\mathrm{i}} + W$$

where u_{f} is the final and u_{i} the initial internal energy, Q the heat entering the system, and W the work leaving the system. This combines a statement of energy conservation with the equivalence of heat and mechanical work.

For a flow situation where mass crosses the boundary, the energy content of the mass must be included, as pictorially described in Fig. 2:

$$\begin{aligned} Q + \sum m_{\mathrm{e}}[h_{\mathrm{e}} + (v_{\mathrm{e}}^2/2g) + Z_0] = W &+ \sum m_l[h_l + (v_l^2/2g) + Z_l] \\ &+ m_{\mathrm{f}}[u_{\mathrm{f}} + (v_{\mathrm{f}}^2/2g) + Z_{\mathrm{f}}] \\ &- m_{\mathrm{i}}[u_{\mathrm{i}} + (v_{\mathrm{i}}^2/2g) + Z_{\mathrm{i}}] \qquad (10) \end{aligned}$$

where subscripts e and l refer to the mass properties entering and leaving the system; subscripts f and i refer to the final and initial mass properties within the system; m refers to the mass, h the specific enthalpy, u the specific internal energy, v the velocity, g the gravitational constant, and Z the specific potential energy. Q and W refer to the heat entering and the work leaving, respectively. The convention is adopted that Q is positive when it enters a system and negative when it leaves, and the opposite for work, being positive when it leaves or is done by the system.

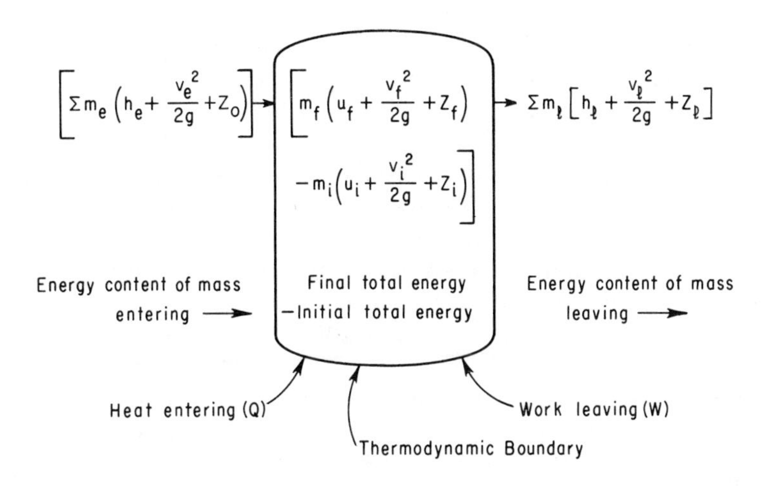

FIG. 2. Pictorial description of first law of thermodynamics.

In thermodynamics systems using helium as the working fluid, there is typically no change in the potential-energy term, so (10) reduces to

$$Q + m_e[h_e + (v_e^2/2g)] = W + m_l[h_l + (v_l^2/2g)] + m_f(u_f) - m_i(u_i)$$

and frequently the velocity term for the mass is negligible, further simplifying the equation.

Primary contribution to the internal-energy term describing the contents of the system may come from the material in the piping, heat exchangers, and other apparatus within the system boundary, rather than from the working fluid itself, particularly during transient conditions such as the cool down of a refrigerator system.

3. *Second law*

The first law of thermodynamics relates to the energy balance around a system and to the conservation of energy. It holds equally well for processes occurring in opposite directions, i.e., converting work into heat and heat into work. The second law of thermodynamics relates to the permissible direction for a process.

The Clausius expression of the second law is as follows: Heat cannot of itself pass from a colder to a hotter body. This is an expression of the fact that automatic flow of heat (by conduction and radiation) is always from the hotter to the colder body. A body cannot be made hotter by making a colder one colder, except by the performance of work.

It is apparent from the first law that work is equivalent to heat, and in a simple case of work done against friction, the transformation of work into heat is 100% efficient and may continue indefinitely.

To study the opposite case of the transformation of heat into work we must devise a process which will enable the transformation to continue indefinitely. The isothermal expansion of a gas in a cylinder will eventually stop when the pressure in the cylinder is the same as that outside the cylinder. Driving a fan from a jet issuing from a compressed gas bottle will stop when the pressure is lost. What is needed then is a series of procesess in which the system is repeatedly brought back to its initial state, i.e., a cyclic process.

Consider a cyclic process in which an amount of heat Q_1 is absorbed at one temperature and another amount of heat Q_2 is rejected at a lower temperature and a net amount of work W is done by the system.

The thermal efficiency of this transformation is

$$\text{thermal efficiency} = \text{work output/heat input}$$

$$\eta = W/Q$$

Applying the first law to one complete cycle, remembering there is no change in internal energy since the system was returned to its initial state:

$$Q_1 - Q_2 = W$$

and therefore

$$\eta = (Q_1 - Q_2)/Q_1 = 1 - (Q_2/Q_1)$$

The efficiency will only be 100% if Q_2 is zero, which would require a sink at absolute zero. This would be a violation of the third law of thermodynamics.

Another expression of the second law is as follows: It is impossible to construct an engine that, operating in a cycle, will produce no effect other than the extraction of heat from a reservoir and the performance of an equivalent amount of work. Therefore, the transformation of heat into work cannot be effected with 100% efficiency.

4. *Third law*

The temperature of a substance can be lowered by a series of adiabatic, or isentropic and isothermal changes.

Figure 3 illustrates a series of changes in a gas between two pressure levels. First an adiabatic expansion causing a drop in temperature, followed

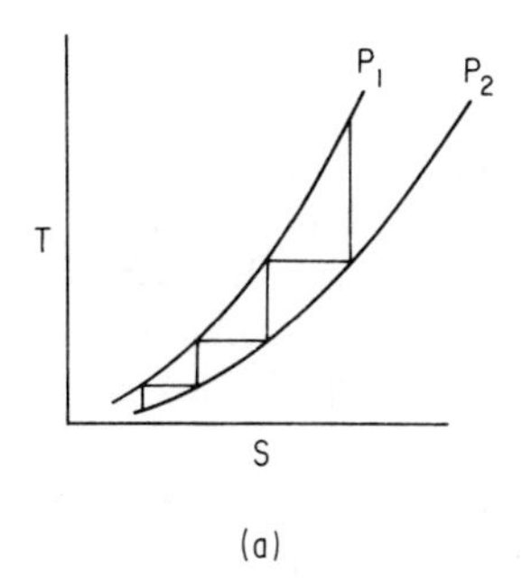

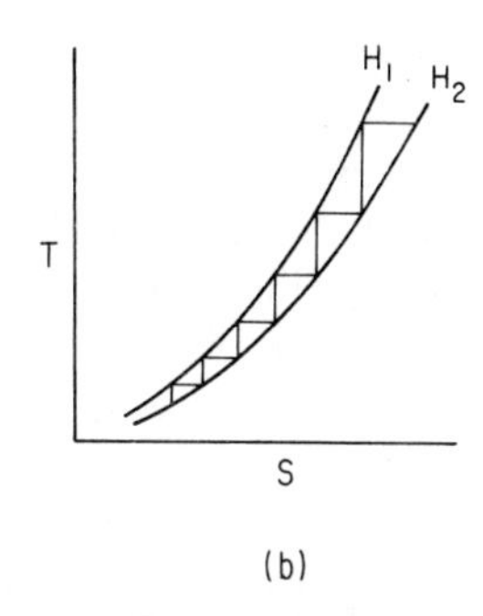

FIG. 3. Temperature–entropy diagrams showing adiabatic and isothermal step changes (a) of pressure and (b) of magnetic field.

by an isothermal recompression. The lowest temperatures so far attained have been by adiabatic demagnetization of salts. As illustrated in Fig. 3b, the adiabatic demagnetization from H_1 to H_2 causes a temperature drop. Isothermal magnetization followed again by adiabatic demagnetization causes a further temperature drop.

In both these processes, the temperature drop tends to be roughly proportional to the initial temperature. If the first step halved the temperature, and the next step halved again, it would clearly take an infinite number of steps to reach absolute zero. Furthermore, as a matter of experience, each succeeding step is more difficult than the preceding one, consequently it would become increasingly more difficult, reaching impossibility, to approach the infinity of steps. Hence the third law: It is impossible, by any procedure, no matter how idealized, to reduce any system to the absolute zero of temperature in a finite number of operations.

5. *The state principle*

Thermodynamic systems vary in complexity. The degree of complexity is a function of the number of independent properties that must be fixed in order to fix the thermodynamic state of the system. The state principle postulates the following concerning the independent properties of a system: The stable equilibrium thermodynamic state of a system contained by a proscribed boundary and subject to prescribed fields at its boundary is fully determined in its energy.

The four laws of thermodynamics and the state principle are interdependent statements which taken together constitute classical thermodynamics. All general relations of classical thermodynamics can be derived from them.

C. THE CARNOT CYCLE

In Fig. 4 is indicated a piston and cylinder confining a quantity of gas. The gas can be put through a series of process steps to perform a thermodynamic cycle. The steps of this cycle are shown in the accompanying pressure–volume (P–V) diagram, which describes the work process, and in the temperature–entropy (T–S) diagram which describes the heat process.

Starting at position *a*, the piston is moved slowly to position *b*, while the cylinder is in contact with a heat source at temperature T_1. This step provides a reversible isothermal expansion of the gas. During this step a quantity of heat Q_1 was absorbed by the system, equal to the area, from T_1 to T_0, under the line *ab* on the *T–S* diagram. Also a quantity of work was done by the gas on the piston, equal to area under the line *ab* on the *P–V* diagram.

The cylinder is now isolated from the heat source and the piston is allowed to move slowly to position *c*, thereby providing a reversible adiabatic expansion of the gas. During this step no heat was extracted and the temperature of the system dropped from T_1 to T_2. A quantity of work was done by the gas on the piston, equal to area under the line *bc* on the *P–V* diagram.

The cylinder is now connected to a heat sink at temperature T_2, and the piston moved slowly from position *c* to position *d*, providing reversible isothermal compression of the gas. During this step a quantity of heat Q_2, equal to the area under the line *cd* on the *T–S* diagram, was rejected by the system. Also a quantity of work was done by the piston on the gas, equal to the area under the line *cd* on the *P–V* diagram.

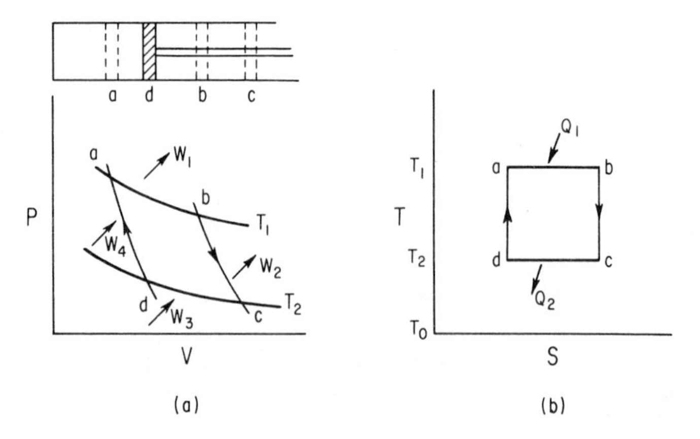

FIG. 4. Pressure–volume and temperature–entropy diagrams describing the carnot cycle.

In the final step, the cylinder is removed from the heat sink, and the piston is moved from position d to a, providing reversible adiabatic compression of the gas. During this step a quantity of work was done by the piston on the gas, equal to the area under the line da on the P–V diagram, and the temperature of the system was increased from T_2 to T_1.

The system is now back to its original condition, and during the cycle, a quantity of heat Q_1 was absorbed at temperature T_1 and a quantity of heat Q_2 was rejected at temperature T_2. The difference $Q_1 - Q_2$, represented by the enclosed area in the T–S diagram, was converted into work, the magnitude of which is represented by the enclosed area in the P–V diagram.

The process just described is the simplest example of the Carnot cycle, and an engine operating in this reversible way would be a Carnot engine.

Since the cycle described is reversible, the steps could be performed in the reverse direction, starting at position c and moving to position b to provide an adiabatic compression, rejecting heat at T_1, then an adiabatic expansion while moving to position d, followed by an isothermal expansion, absorbing heat from the lower temperature T_2 while moving to position c.

The first process is a Carnot heat engine; the second process describes a Carnot refrigerator, in which a quantity of work, equal to the area of the P–V diagram was used to pump a quantity of heat Q_2 from T_2 to be rejected, together with the heat equivalence of the work input, to form Q_1 at the temperature T_1.

If such an effective engine producing work were connected to an equally effective engine pumping heat, since for both processes $Q_2 = Q_1 - W$, such a combination could work forever between the heat source and the heat sink, then we could picture a perpetual motion machine of the first kind. Such a combination has never been found.

If we could define for the refrigerator a process more effective than that described, and were to combine this with a heat engine as described, then we could have a combination capable of continuously, and indefinitely, producing energy. This would be a perpetual motion machine of the second kind. Such a combination has never been found.

Since such combinations have not been found, we are led to assume that the reversible process described for both the Carnot heat engine and the Carnot refrigerator describes the most ideal process, against which all cycles can be compared to establish an efficiency with respect to ideality.

We can analyze the system described in terms of its temperature T, pressure P, volume V, and specific heat of the gas c_V. Remembering $PV = nRT$, and the first-law requirement

$$dQ = du + dW = c_V\,dT + nRT\,\frac{dV}{V}$$

Along the isothermal from a to b in Fig. 4, the work done is equal to the heat entry, which is

$$nRT_1 \int_a^b \frac{dV}{V} = nRT_1 \ln \frac{V_b}{V_a}$$

Similarly along the isothermal from c to d, the work done is equal to the heat rejection, which is

$$-nRT_2 \int_c^d \frac{dV}{V} = nRT_2 \ln \frac{V_d}{V_c}$$

Along the adiabatics, the heat entry is zero, and the work done from b to c is $c_V(T_1 - T_2)$ and from d to a is $-c_V(T_1 - T_2)$.

So the efficiency of the cycle is

$$\eta = \frac{\text{work done in cycle}}{\text{heat taken in at } T_1} = nR\left(\frac{T_1 \ln V_b/V_a - T_2 \ln V_d/V_c}{T_1 \ln V_b/V_a}\right) = \frac{Q_1 - Q_2}{Q_1}$$

But $V_b/V_c = V_a/V_d$, since they are an adiabatic between isothermals. Therefore we find

$$V_c/V_d = V_b/V_d \qquad \text{and} \qquad \eta = (T_1 - T_2)/T_1$$

Also

$$\eta = 1 - T_2/T_1 = 1 - Q_2/Q_1$$

Therefore

$$T_2/T_1 = Q_2/Q_1 \qquad \text{and} \qquad T_1/T_2 = Q_1/Q_2$$

Thus, the ratio of any two temperatures on the absolute, or Kelvin, scale is the same as the ratio of the heats absorbed and rejected by a Carnot heat engine operating between reservoirs at these temperatures. Also, for the efficiency to be 100%, the heat rejected Q_2 must be zero, and therefore the sink temperature T_2 must be zero.

Just as the amount of work that can be transformed from an amount of heat is dependent on the temperatures of the source and sink, so too the amount of heat that can be pumped up by a given work input is dependent on the temperatures of the source, at lower temperature, and the sink, at higher temperature. The effectiveness of the refrigerator is measured by its coefficient of performance, which is

$$\text{COP} = \text{heat taken in at } T_e/\text{work done in cycle} = T_2/(T_1 - T_2)$$

III. Properties of Helium

Helium exists in two stable isotopic forms, ^{3}He and ^{4}He. The relatively rare and expensive isotope ^{3}He is only used where temperatures below 1°K are required. The phase diagram for ^{4}He is shown in Fig. 5, in which the

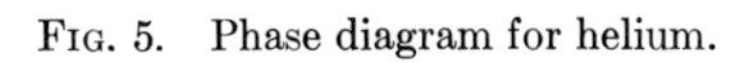

FIG. 5. Phase diagram for helium.

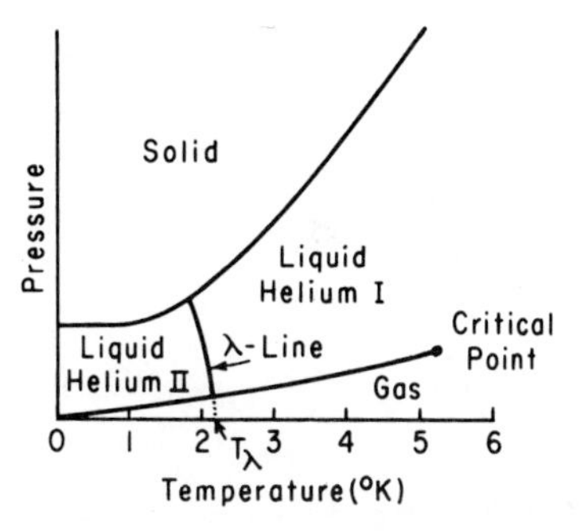

pressure scale is exaggerated for clarity. The critical point is at 5.19°K and 2.26 atm pressure. Liquid ^{4}He exists in two phases, He I and He II, separated by a phase boundary line called the λ line. This transition occurs, at saturated vapor condition, at 2.172°K, which is called the λ point. In the He II phase, the interesting phenomenon of superfluidity and heat transfer by second sound occur, however, since most superconducting devices require refrigeration only down to 3°K, neither the properties of ^{3}He nor He II will be considered here. But, since most superconducting devices do require refrigeration below 20°K, helium is the only practical refrigerant, and in most cases, the only practical working fluid for the whole refrigeration cycle.

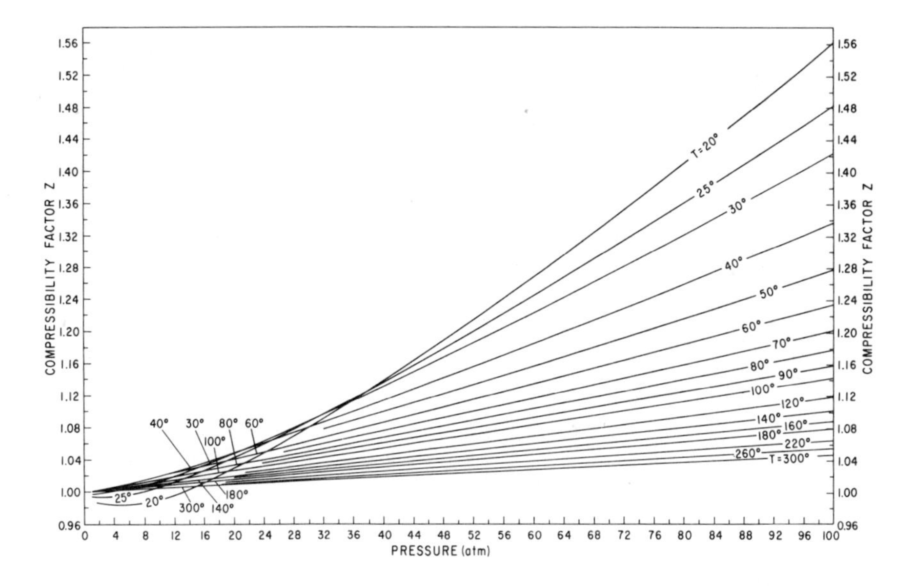

FIG. 6. Compressibility factor for helium.

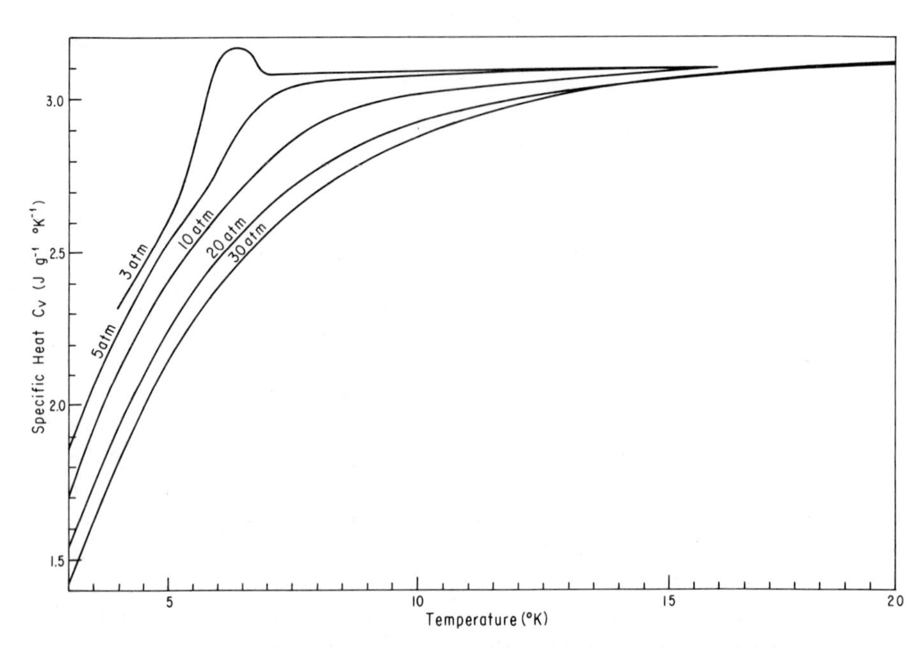

FIG. 7. Variation of specific heat of helium at constant volume. [After Nat. Bur. Stand., Institute for Materials Research, Cryogenics Division, Boulder, Colorado.]

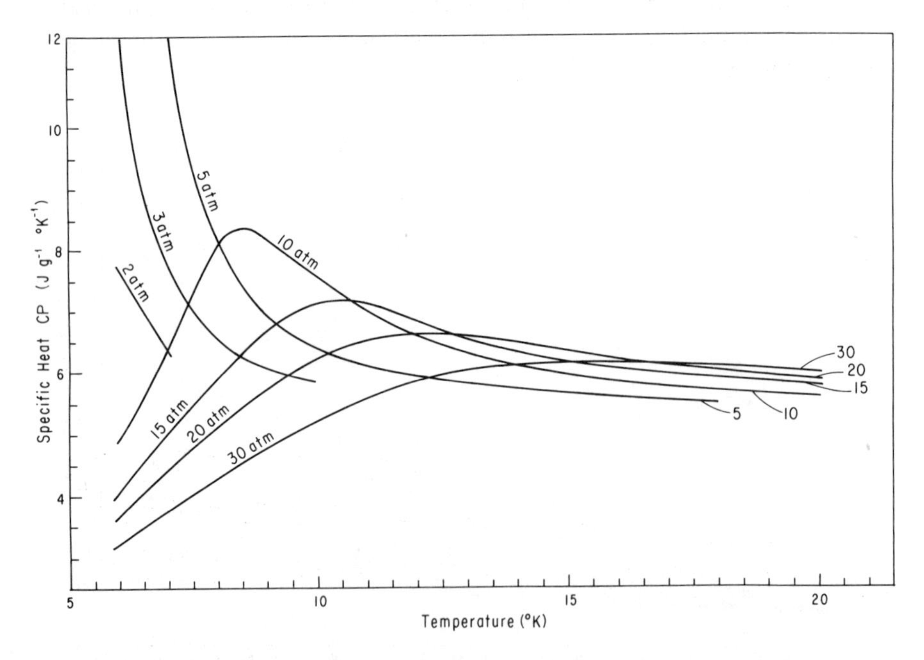

FIG. 8. Variation of specific heat of helium at constant pressure. [After Nat. Bur. Stand., Intitute for Materials Research, Cryogenics Division, Boulder, Colorado.]

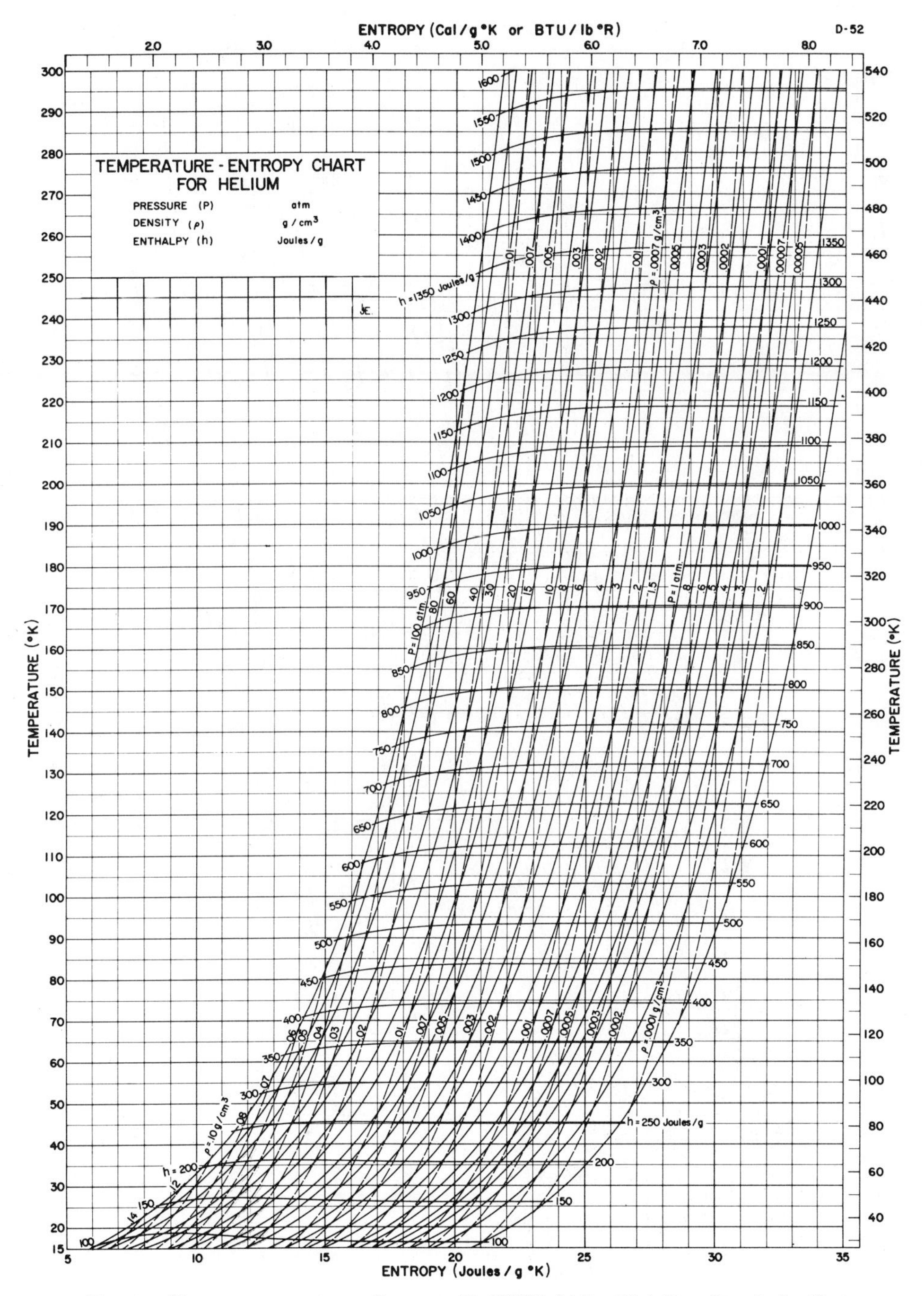

Fig. 9. Temperature–entropy diagram, 12–300°K. [After Nat. Bur. Stand., Institute for Materials Research, Cryogenics Division, Boulder, Colorado.]

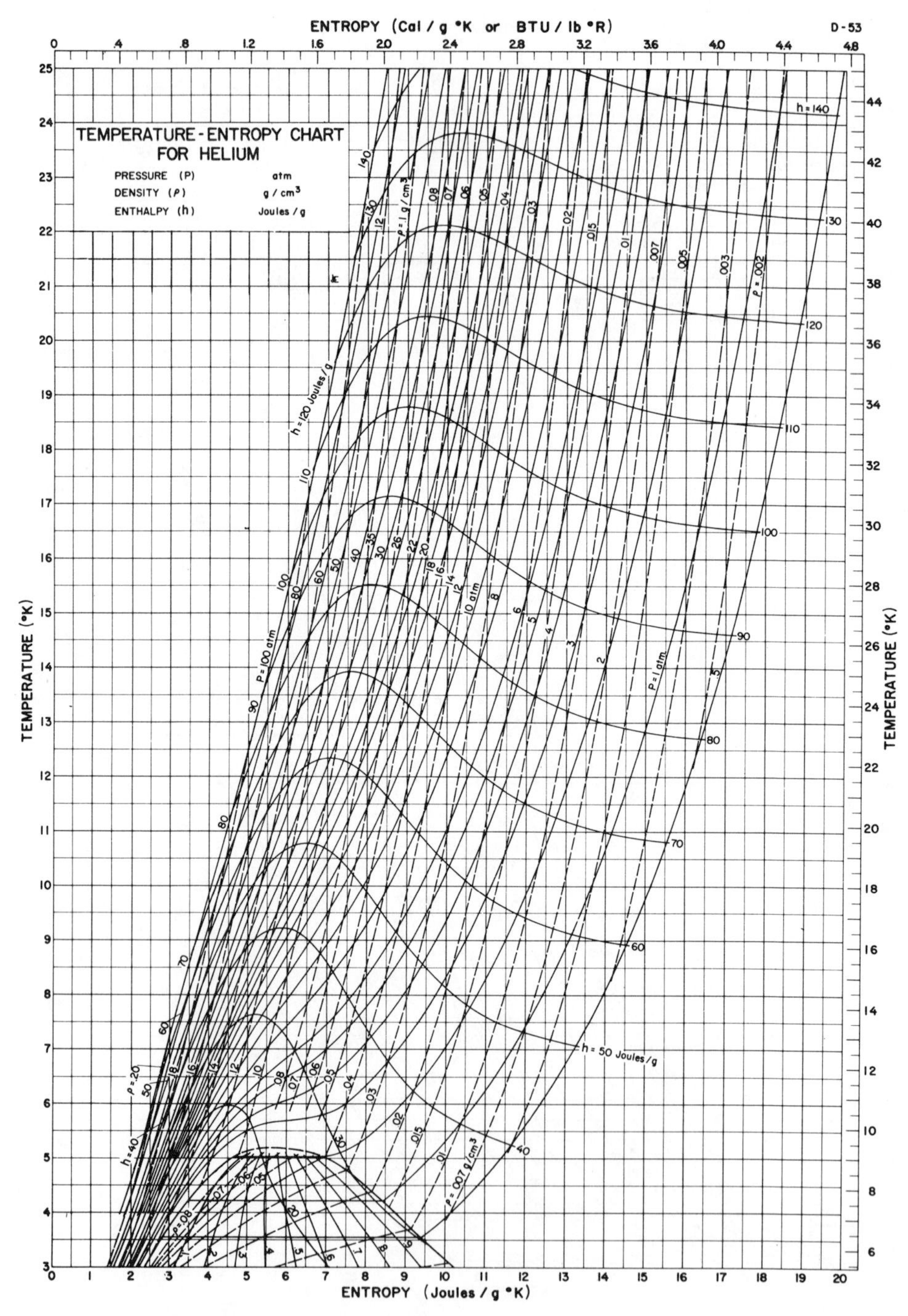

Fig. 10. Temperature–entropy diagram, 3–25°K. [After Nat. Bur. Stand., Institute for Materials Research, Cryogenics Division, Boulder, Colorado.]

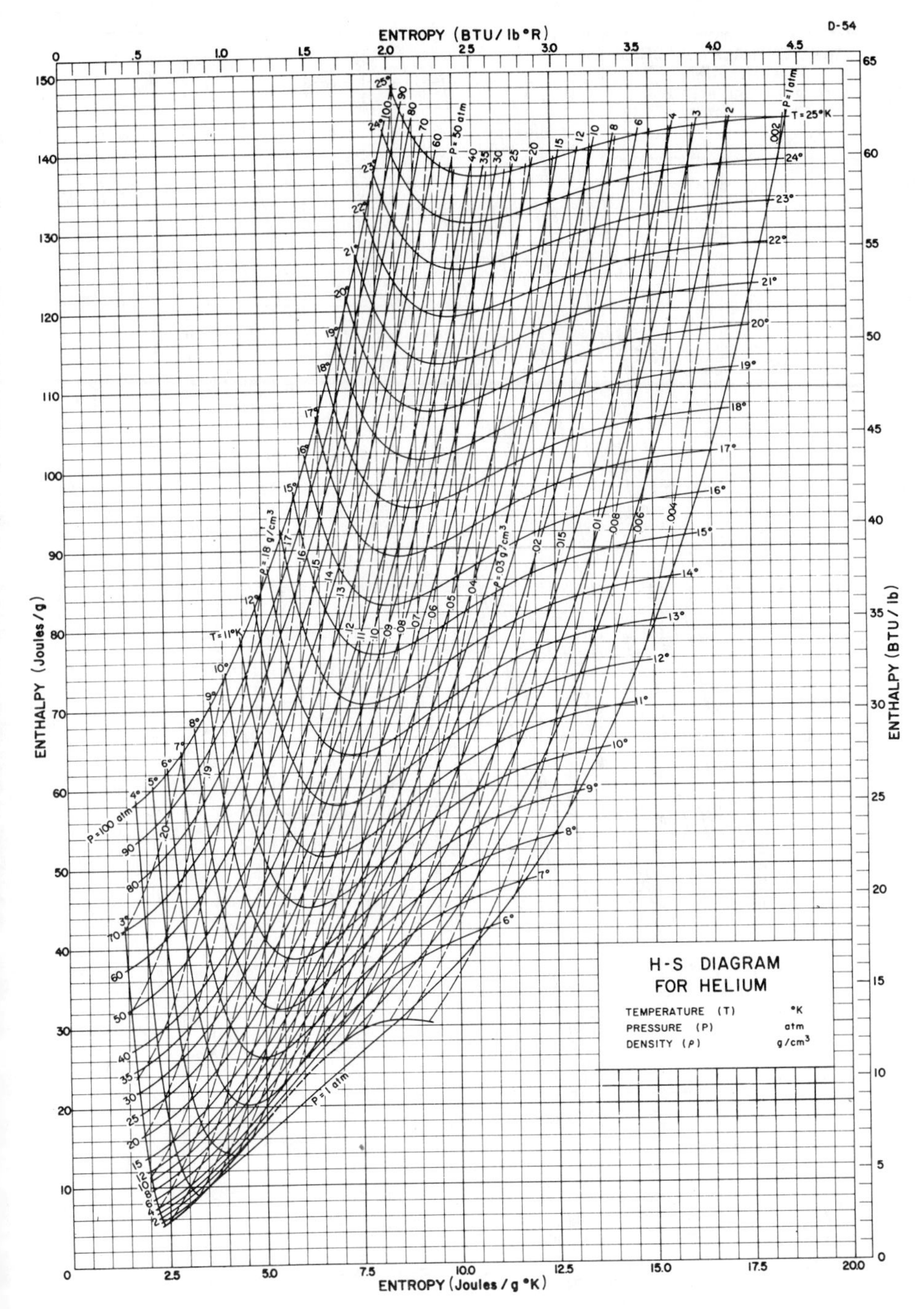

Fig. 11. Enthalpy–entropy diagram, 3–25°K. [After Nat. Bur. Stand., Institute for Materials Research, Cryogenics Division, Boulder, Colorado.]

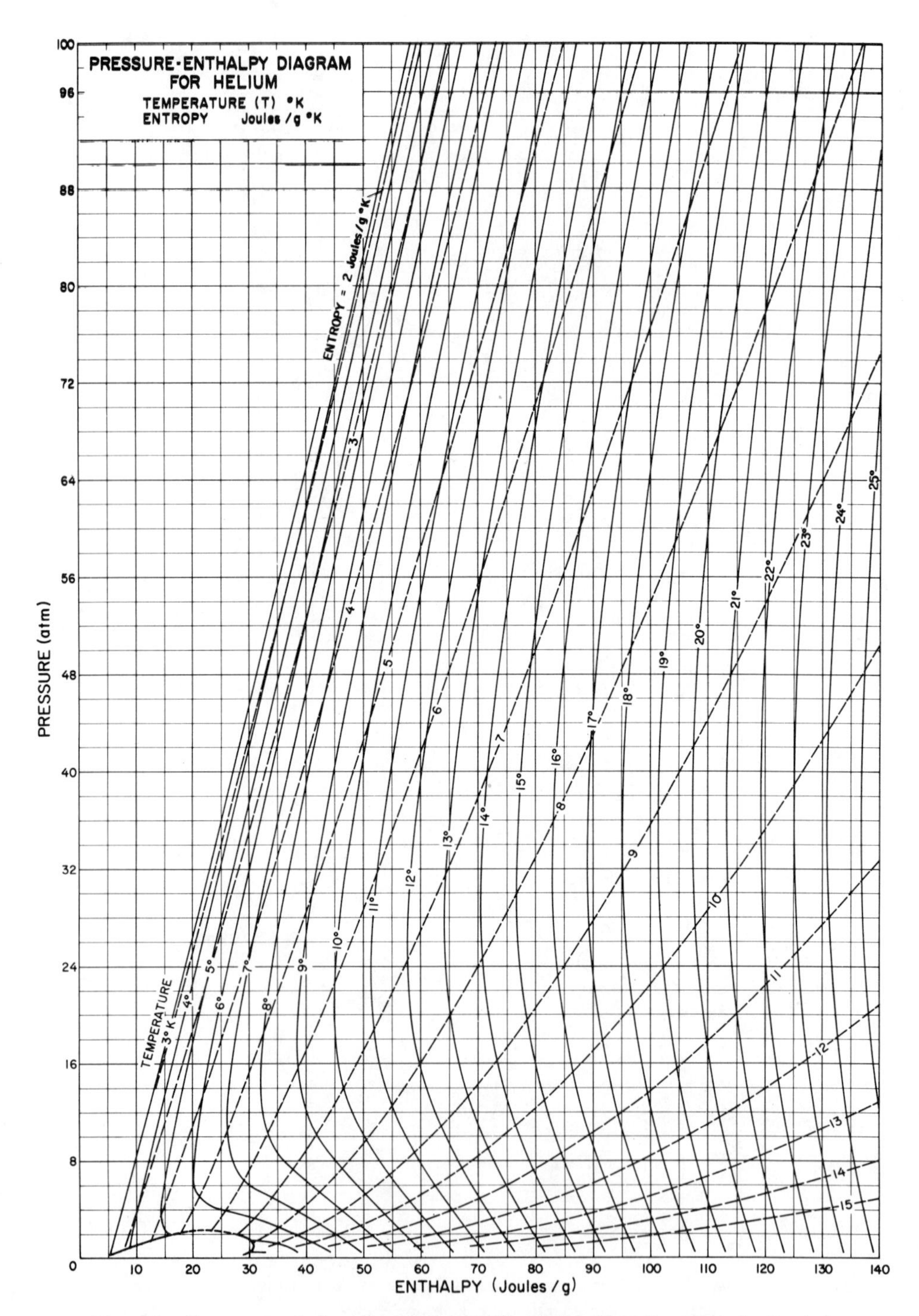

FIG. 12. Pressure–enthalpy diagram, 3–25°K. [After Nat. Bur. Stand., Institute for Materials Research, Cryogenics Division, Boulder, Colorado.]

A. Gaseous Helium

At normal temperatures and pressures, helium obeys, closely enough for engineering calculations, the ideal gas law $PV = nRT$, with units as follows:

pressure P in atmospheres or newtons per square meter,
volume V in liters or cubic meters,
the gas constant R as 0.0205 liter·atm/g·°K or 2.08 J/g·°K

the mass n in grams and the temperature T in degrees Kelvin.

When dealing with high pressures and temperatures below that of liquid nitrogen, one must consider deviations from the ideal gas law. A typical form of the equation of state is as follows:

$$PV = nRT[1 + B(n/v) + C(n/v)^2] = nRTZ$$

where B and C are virial coefficients and Z is called the compressibility factor. Figure 6 shows the compressibility factor for helium gas from 1 to 100 atm and from 20 to 300°K (Wade, Wadd Tech. Rep. No. 60–56).

Refrigeration cycles rarely employ pressures above 30 atm, and it is only at temperatures below about 25°K that deviations from the ideal have an important effect. Figures 7–14 are self-explanatory and are taken from

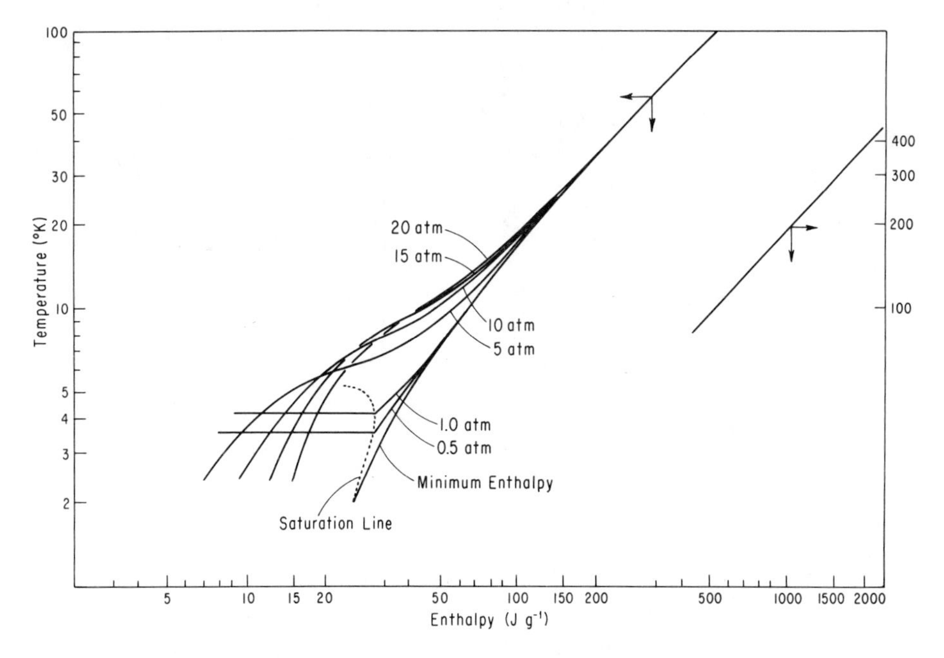

Fig. 13. Temperature–enthalpy diagram for helium. [After Nat. Bur. Stand., Institute for Materials Research, Cryogenics Division, Boulder, Colorado.]

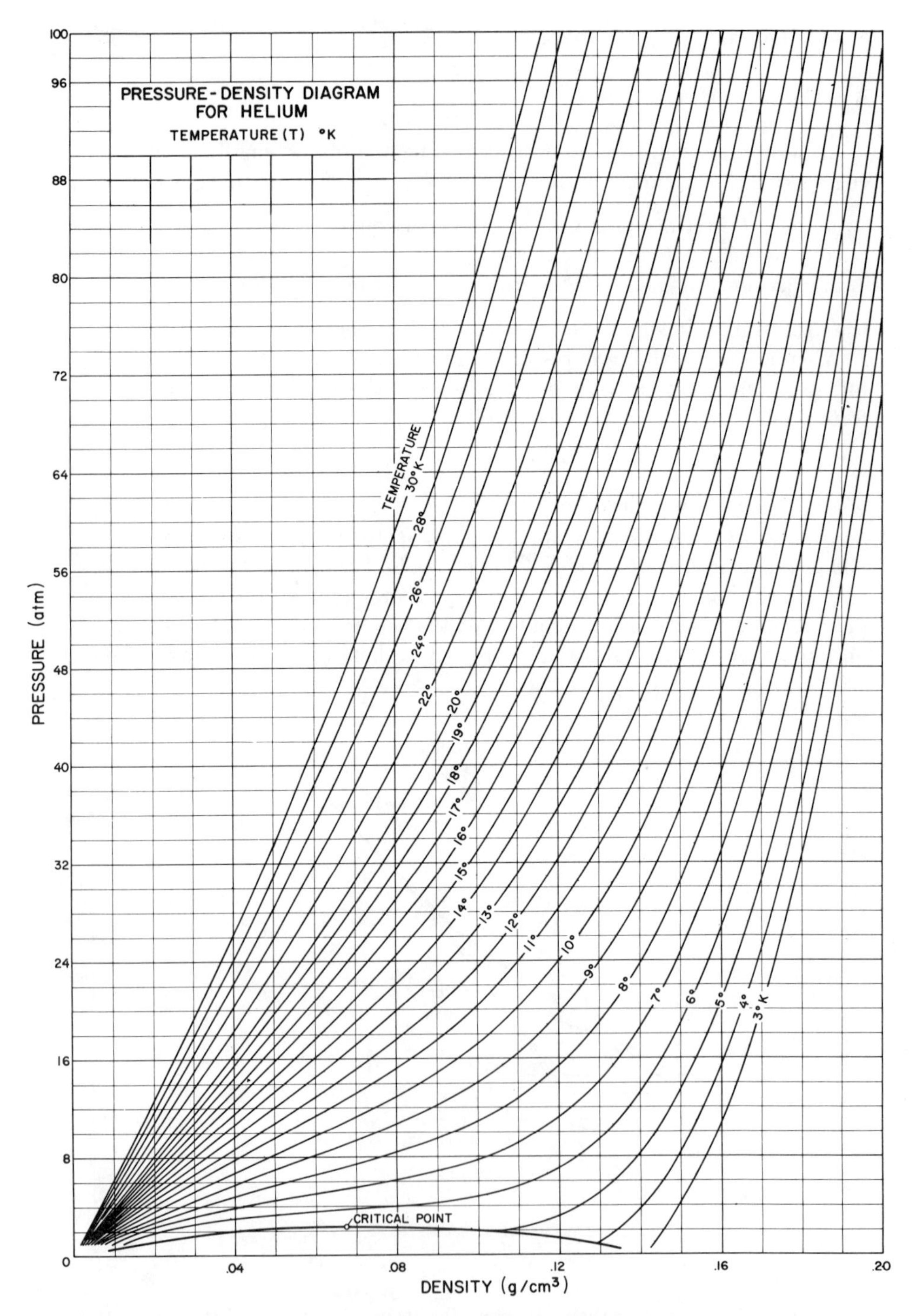

FIG. 14. Pressure–density diagram, 3–30°K. [After Nat. Bur. Stand., Institute for Materials Research, Cryogenics Division, Boulder, Colorado.]

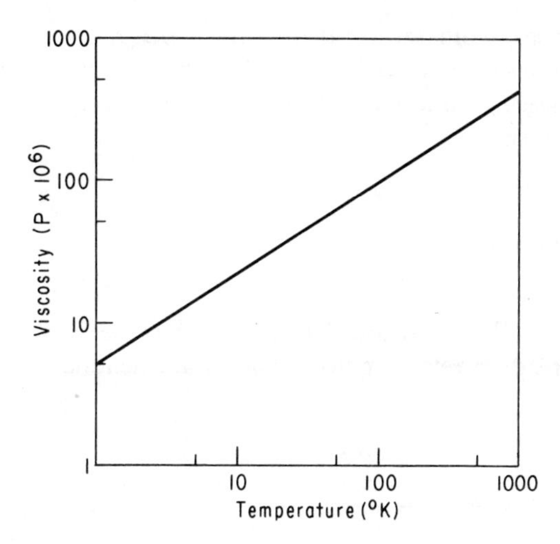

FIG. 15. Viscosity of gaseous helium at pressures near atmosphere. [After Keesom, 1942].

the National Bureau of Standards publications (Wade, Wadd Tech. Rep. No. 60–56). They show what happens to gas temperature during isentropic and isenthalpic expansion, and they are necessary for the design of thermodynamic cycles and the design of heat exchangers.

Figure 15 shows the viscosity temperature, and Fig. 16 (Scott, 1959) shows the thermal conductivity–temperature data for helium gas. Both viscosity and thermal conductivity are weak functions of pressure at

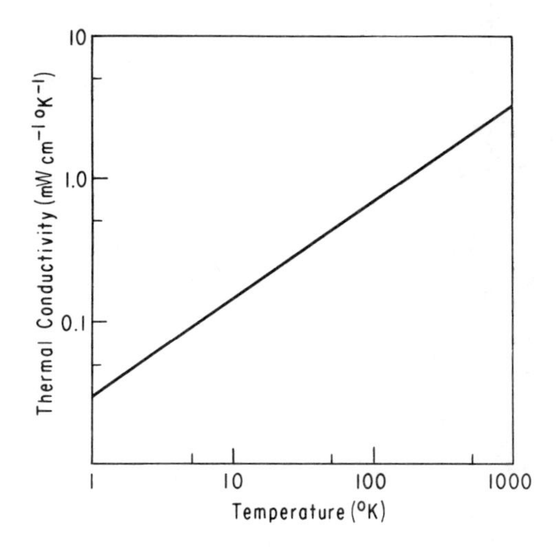

FIG. 16. Thermal conductivity of gaseous helium. [After Keesom, 1942.]

temperatures near room temperature. The data given are from Keesom. Near the two-phase liquid–vapor region, both viscosity and thermal conductivity are strong functions of pressure; however, no good data are available at this time.

B. Liquid Helium

Figures 17–22 (Wade, Wadd. Tech. Rep. No. 60–56, Scott, 1959) give all the pertinent data relevant to normal liquid helium.

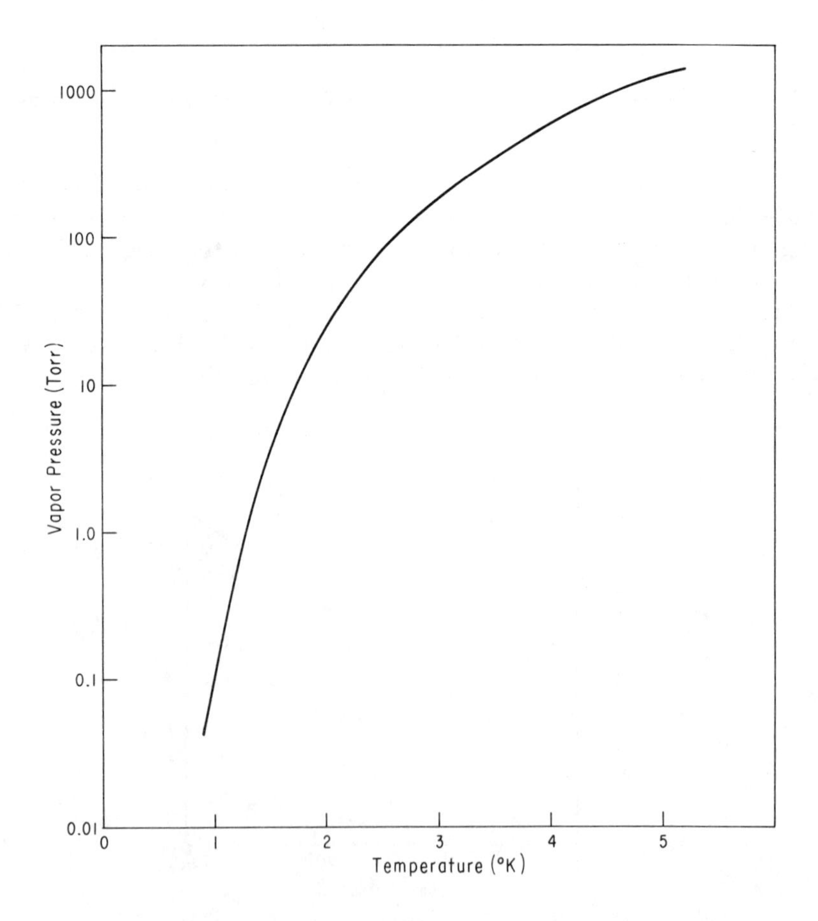

Fig. 17. Equilibrium vapor pressure of helium. [After Scott, 1959.]

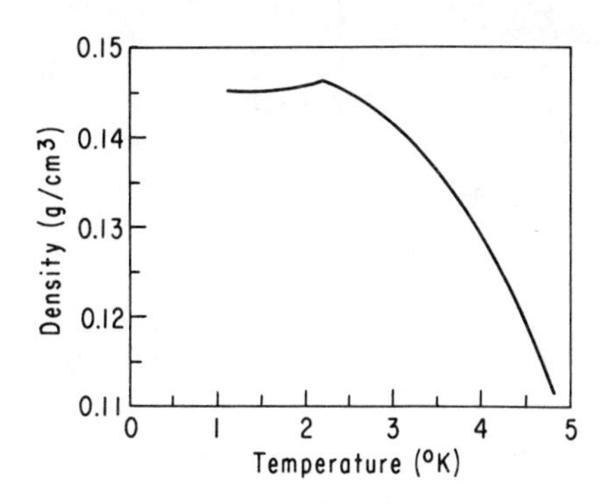

FIG 18. (Above) Density of saturated liquid helium. [After Keesom, 1942.]

FIG. 19. (At right) Specific heat of liquid helium. [After Keesom, 1942.]

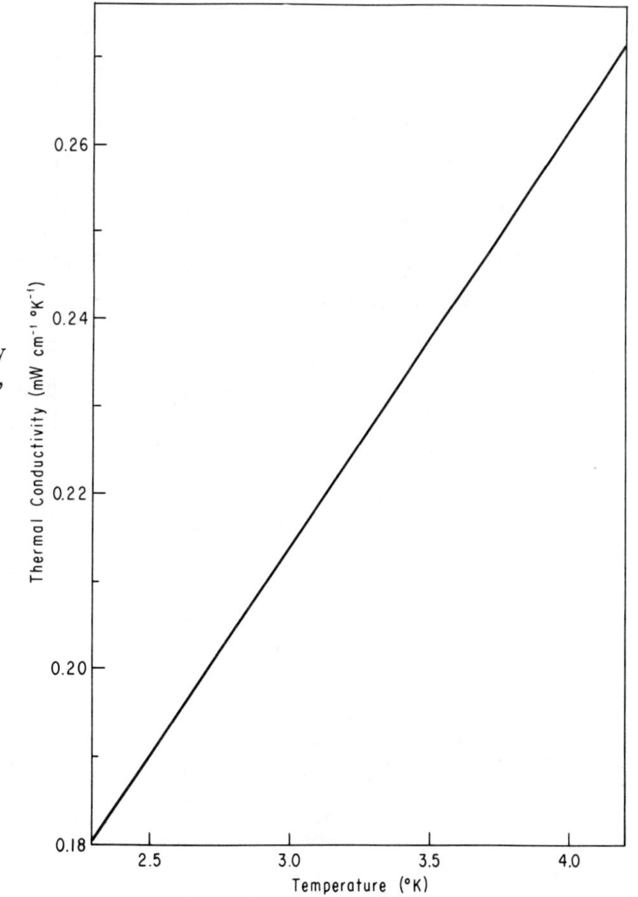

FIG. 20. Thermal conductivity of liquid helium. [After Keesom, 1942.]

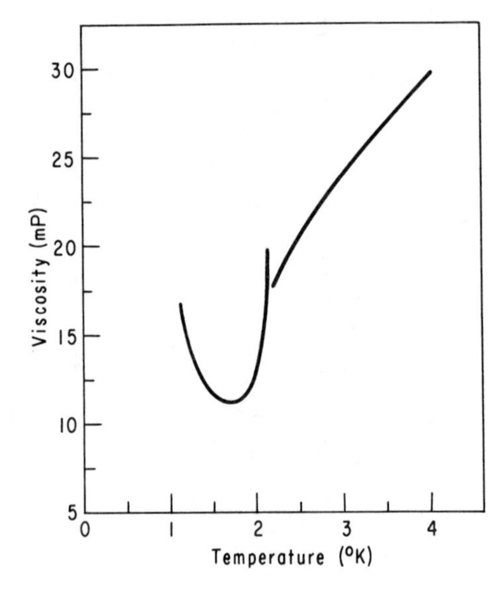

Fig. 21. Viscosity of saturated liquid helium. [After Keesom, 1942.]

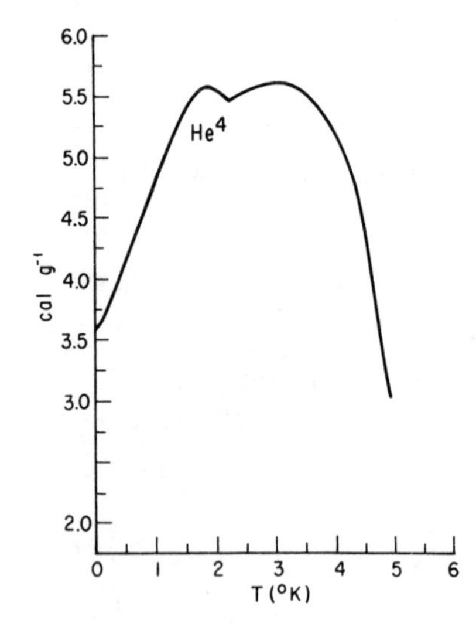

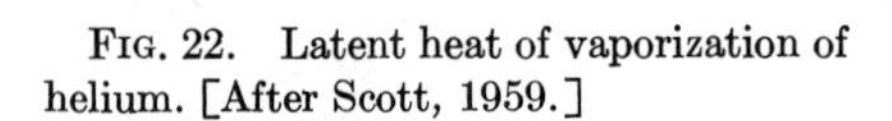

Fig. 22. Latent heat of vaporization of helium. [After Scott, 1959.]

IV. Refrigeration Cycles

As a practical matter, there are only two methods that can be used for obtaining cooling much below 20°K, as is required for cooling superconducting devices. These are:

(a) open-cycle Dewar systems which use the latent heat of evaporation, and perhaps some of the sensible heat, of liquid helium, and;

(b) closed-cycle systems, which use closed-loop gas-expansion cycles, in which the ultimate refrigerant is helium.

In the laboratory, open-cycle systems are used because of the obvious advantages of simplicity, flexibility, and the initially low cost provided by open Dewars of liquid helium. Usually parts at liquid-helium temperature are thermally insulated using liquid nitrogen, but many systems today use the sensible heat of helium boiloff gas to intercept heat leaks.

For operation in the field or where long-time continuous operation is required, the logistic and handling problems of liquid helium, as well as the cost of the cryogens, restrict the use of open-cycle systems. Today, closed-cycle refrigeration systems have been, or can be, built for practically every application of superconductivity.

A. Ideal Cycles

All cryogenic refrigeration systems depend on gas expansion, where the gas does work on itself or on some work extraction device. They also depend on the use of efficient heat exchangers to separate the region at which compression of the gas is effected, usually ambient temperature, from the lower temperature region at which the gas is expanded.

1. *Isentropic expansion*

The expansion of gas in such a way as to extract work also causes a drop in its temperature, and the amount of refrigeration made available is equal to the work produced. Three mechanical processes are used for extracting work from the gas. The first and most common utilizes a reciprocating piston as the expansion device, in the manner of a steam engine. The work may also be extracted by means of rotary or turbine devices, a technique that is best suited to large systems. The third approach is for the gas to do work on itself, by forcing gas through a vent as the pressure is reduced.

Figure 23a is a typical temperature–entropy diagram on which an ideal isentropic expansion is indicated. Expansion isentropically and adiabatically, from the initial pressure P_i and initial temperature T_i to the final pressure P_f results in a temperature drop to T_2. Refrigeration is made available by warming the gas, at constant pressure, up to its initial temperature. The amount of refrigeration, per unit mass of gas, is equal to the specific enthalpy change $h_1 - h_2$. It is also proportionately equal to the

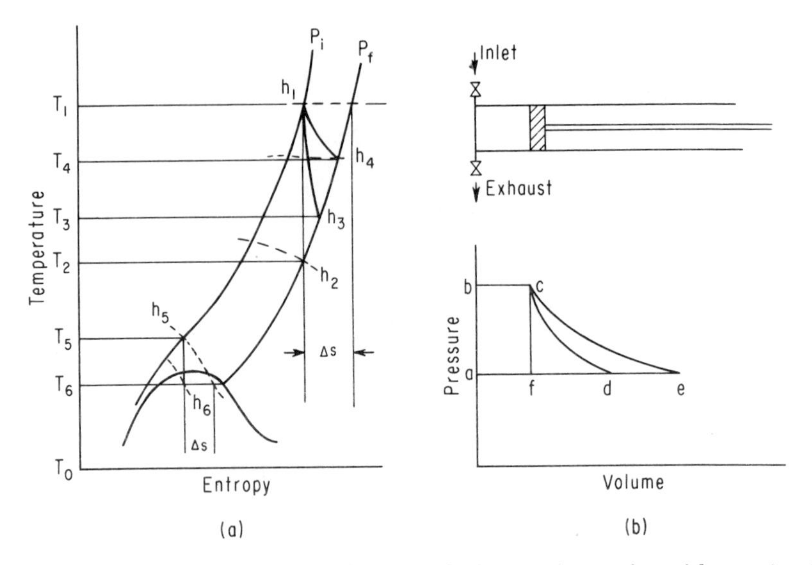

FIG. 23. Schematic piston and cylinder kind of expansion engine with associated (a) temperature–entropy and (b) pressure–volume diagrams showing various modes of gas expansion.

area under the P_f line from h_2 to h_1, down to $T = 0$. The work produced per cycle in this isentropic–adiabatic expansion is shown schematically as the enclosed area *abcd* on the pressure–volume diagram (Fig. 23b).

The maximum amount of work per unit of gas, and therefore the maximum refrigeration, is produced if the expansion is isothermal. This is, in effect, an infinite number of infinitely small isentropic–adiabatic expansions each followed by a constant pressure heat input to restore the infinitely small temperature drop at each stage. The refrigeration produced is proportionately equal to the area under the constant temperature T_1 line, which is $T_1 \Delta S$, on the T–S diagram, Fig. 23a. The work produced per cycle is proportionately equal to the enclosed area *abce* on the P–V diagram, Fig. 23b. In practice the achievement of isothermal expansion is very difficult, and in fact deviations from the adiabatic condition are more usually due to thermal and pressure drop losses rather than the application of useful refrigeration, and result in a temperature drop from T_1 to T_3 rather than T_2.

For conventional refrigeration, wherein the refrigerant load is all at uniform temperature, the lower temperature associated with the adiabatic expansion is wasted. However, for a system in which the cooling load is a continuously varying function, the refrigeration can be utilized effectively down to the exhaust temperature, e.g., the cooling of a Joule–Thomson loop in a refrigerator or liquefier.

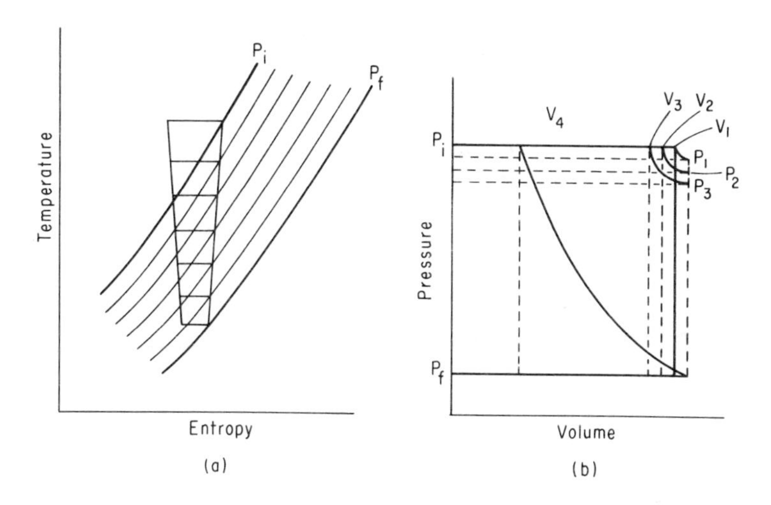

FIG. 24. Schematic representation of (a) temperature–entropy and (b) pressure–volume diagrams of a gas expansion from a fixed volume (Simon expansion).

As mentioned above, the gas can also expand doing work against itself by forcing gas through a vent as the pressure is reduced. This is shown schematically as line h_1 to h_4 on the T–S diagram, and the work per cycle shown as area *abcf* on the P–V diagram. This process is also isentropic in part and adiabatic. This can best be illustrated by reference to Fig. 24.

Expansion from a constant volume cannot be shown properly on a T–S diagram, since that kind of diagram describes what happens to a unit mass of gas. However, we can make a representation as shown in Fig. 24a, in which the width of the isentrope is representative of the mass undergoing expansion. The first gas undergoes only an infinitesimal expansion. Some fraction undergoes a full expansion. This process is better described by use of a P–V diagram as in Fig. 24b, if we imagine the volume to be divided by a large number of imaginary planes or pistons. All of the gas behind plane V_1 is expanded adiabatically from P_i to P_1. All of the gas behind plane V_2 is expanded adiabatically from P_i to P_2. All of the gas behind plane V_3 is expanded adiabatically from P_i to P_3, and so on. Finally, some fraction of the gas, say that behind plane V_4, is fully expanded adiabatically from P_i to P_f. We could imagine plane V_4 to be a piston pushing on the gas between planes V_1 and V_4, as the gas behind it expands.

This type of expansion is often called a Simon expansion, after the man who first liquefied helium by this method. It is also the kind of expansion employed in many practical low-temperature expanders. Any shape of P–V diagram can be viewed as we have looked at the rectangular diagram of Fig. 24b.

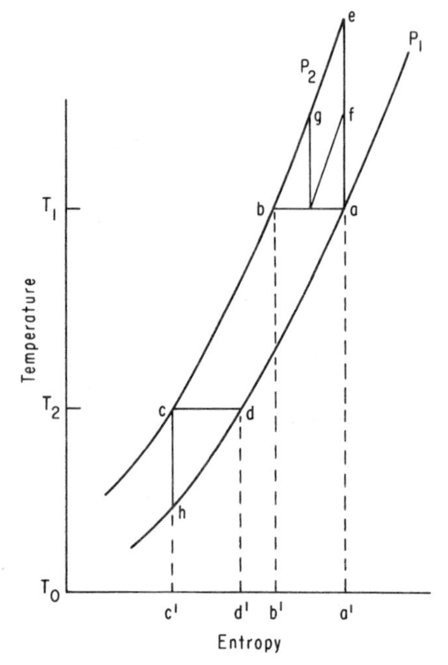

FIG. 25. Complete refrigeration cycle represented on a typical temperature–entropy diagram.

Using an ideal heat exchanger, the high-pressure gas stream could have been cooled down to some lower temperature, say T_5 with enthalpy h_5, in Fig. 23, and then expanded to the low pressure to T_6, with enthalpy h_6, in the two-phase region as shown. Refrigeration would then have been made available at constant temperature T_6 with an amount, per unit of gas, equal to the specific enthalpy change $h_5 - h_6$, which would be the same as $T_6 \, \Delta S$. Note that in this case the gas was not brought back to its initial temperature T_5 because of the need for an enthalpy balance in the heat exchanger. The enthalpy of the high-pressure stream was reduced from h_1 to h_5, that is from T_1 to T_5, while the enthalpy of the return stream was increased from h_5 to h_1, as required for enthalpy balance, but the temperature was increased from T_6 to T_1 because of the different specific heats of the high- and low-pressure streams.

To complete the refrigeration cycle, the return stream at low pressure must be recompressed to the high-pressure level. Referring to Fig. 25, an isothermal compression from P_1 to P_2 at T_1, followed by an isothermal expansion at T_2, using an ideal heat exchanger between T_1 and T_2, would constitute an ideal Carnot refrigerator, with an efficiency:

$$W/Q = (T_1/T_2) - 1 = abb^1a^1/cdd^1c^1$$

Both isothermal compression and isothermal expansion are very difficult to achieve (in practice, not at all) and single stage, *a* to *e*, or multistage, *a* to *f* and *f* to *g*, adiabatic compression is followed by adiabatic expansion, *c* to *h*, or as close to these ideals as possible as is more common.

2. *Isenthalpic expansion* (*Joule–Thomson*)

The molecules of an ideal gas are too far apart for either attractive or repulsive intermolecular forces to play any part, and the gas obeys ideal gas laws. However, at high density, when the gas molecules are close together, significant deviation from the ideal gas laws does occur.

General physical chemical evidence shows that two molecules weakly attract each other at distances of a few angstroms apart; at smaller distances, chemical bond may result or strong repulsion will result.

The strong repulsion at very small molecular separation is called the van der Waals repulsion. Energy is released, when a gas expands lessening the repulsive force, and the molecular kinetic energy increases and the gas warms up. The range of the repulsive force increases with increasing temperature; however, the effect relative to the increased kinetic energy decreases.

The weak attractive force at larger molecular separation is called the van der Waals attraction. Expansion absorbs energy doing work to overcome the attractive forces, and the molecular kinetic energy decreases and the gas cools. The range and effect of the attractive forces decreases with increasing temperature.

The highest temperature at which these two forces neutralize each other is called the inversion temperature. At any lower temperature there is some density condition at which these two forces balance, which enables an inversion curve to be generated. Figure 26 shows such a curve plotted for helium on a temperature–density graph. An isenthalpic expansion (i.e., one in which no work is done on any other entity, as an expansion across a valve) to the right and above the inversion curve will result in heating. Expansion to the left and below the inversion curve will result in cooling. Expansion from some state point on the right to some state point of the left may result in either heating or cooling. This is illustrated in Fig. 27. Expansion along the isenthalp A from P_1 to P_2 will result in heating. Expansion from P_2 to P_3 will result in cooling. Expansion along isenthalp A from P_1 to P_3 results in a net cooling.

Most gases have an inversion temperature above room temperature, and will therefore cool on simple isenthalpic expansion. Hydrogen and helium have inversion temperatures 210 and 55°K, respectively, and therefore must be cooled to below these temperatures before further cooling or isenthalpic expansion will result.

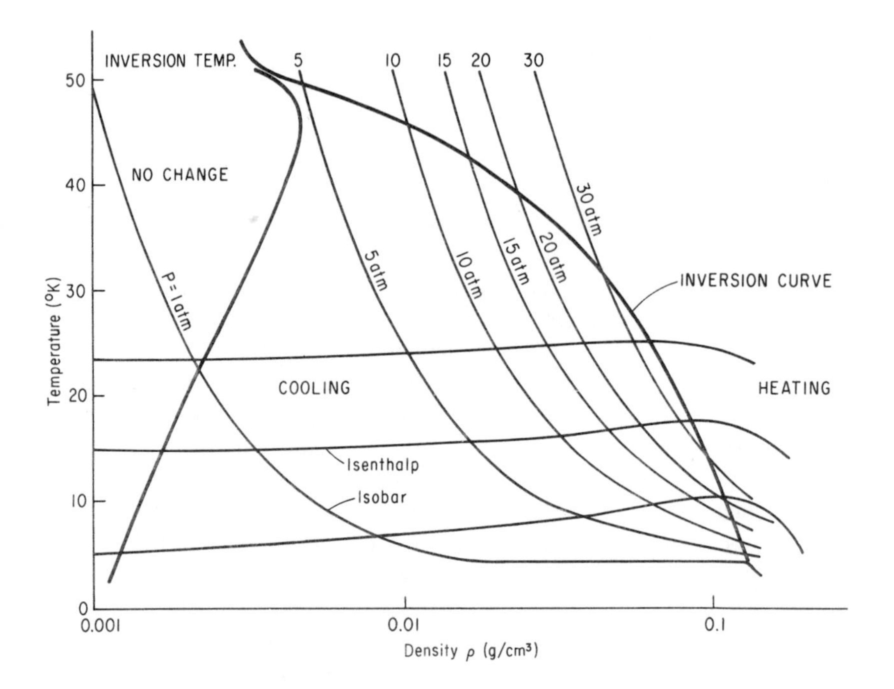

FIG. 26. Inversion curve for helium on a temperature–density diagram.

Using a countercurrent heat exchanger, the cooling effect can be transferred to lower temperatures and eventually to the liquefaction of the gas. The process is as follows: The high-pressure gas, cooled to below its inversion temperature, passes through the high-pressure side of a countercurrent heat exchanger to the expansion valve, where it is allowed to expand isenthalpically, see Fig. 28. This expansion causes some cooling of the gas.

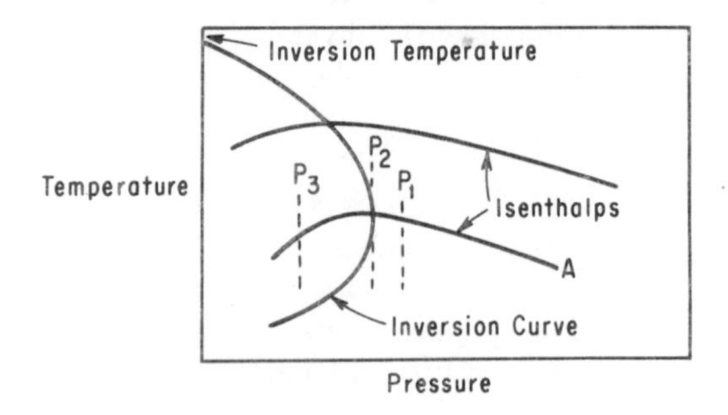

FIG. 27. Inversion curve for helium on a temperature–pressure diagram.

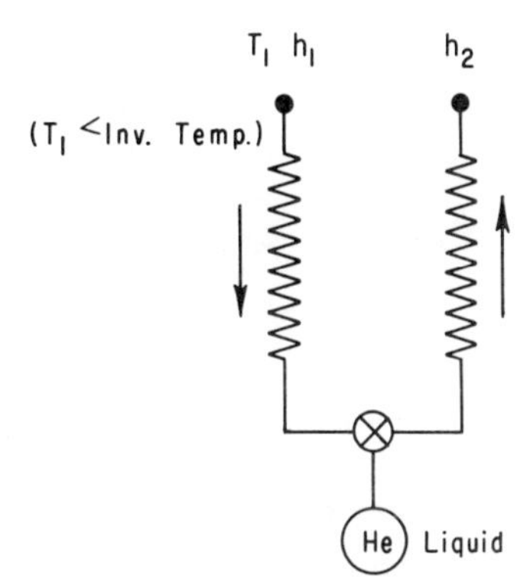

FIG. 28. Joule–Thomson heat exchanger.

The cooled gas passes countercurrent through the low-pressure side of the heat exchanger, giving up its cold to the high-pressure side. During the starting condition, therefore, the cooling at the expansion valve is regenerative because of the action of the heat exchanger, and the temperature drops progressively. The cooling of the high-pressure stream continues until, on expansion a fraction of the gas is able to liquefy, and a smaller mass of gas returns through the low-pressure side.

When the steady, liquefying, state is reached, assuming ideal conditions, the emerging gas regains the inlet temperature, and the total enthalpy is conserved. A fraction x of the gas is liquefied.

If h_1 is the enthalpy of a unit mass of the supply gas, $(1 - x)h_2$ the enthalpy of the emerging gas, and xh_L the enthalpy of the liquefied fraction, then for enthalpy balance, $h_1 = (1 - x)h_2 + xh_L$, and the fraction that liquefies may be expressed as

$$x = (h_2 - h_1)/(h_2 - h_L)$$

Figure 29a shows a typical temperature–entropy diagram of isenthalpic liquefaction, and Fig. 29b shows the typical temperature distribution in the ideal heat exchanger. The heat balance in the heat exchanger is

$$(h_1 - h_3) = (1 - x)(h_2 - h_4) \tag{11}$$

(with a "pinch" in the heat exchanger at the hot end).

If the liquid produced on isenthalpic expansion is removed and boiled off to provide refrigeration at the liquefaction temperature, the refrigeration is equal to the fraction of liquid x times the latent heat of evaporation LH; or referring to Fig. 29a,

$$xLH = x(h_4 - h_L) = h_4 - h_3$$

If, instead of removing the liquid from the system, as in a liquefier, the liquid produced is boiled off in the system, then all the gas which passes

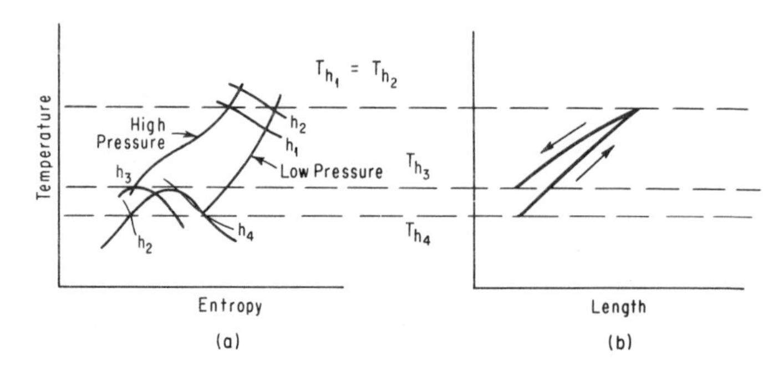

FIG. 29. (a) Temperature–entropy diagram corresponding to the (b) temperature–length distribution in a Joule–Thomson heat exchanger with a temperature pinch at the cold end.

through the high-pressure side of the heat exchanger is returned countercurrent through the low-pressure side. The heat balance in the exchanger is then

$$h_1 - h_3 = h_2 - h_4 \tag{12}$$

(with a heat exchanger "pinch" at either or both ends).

Comparing (11) with (12) it is obvious that for given values of h_1, h_2, and h_4 (that is, with a pinch at the hot end of the heat exchanger and a given refrigeration temperature), the value of h_3 is decreased by the amount $x(h_2 - h_4)$ and the refrigeration $(h_4 - h_3)$ is increased by this amount.

The amount $x(h_2 - h_4)$ was the extra sensible heat exchanged in the countercurrent heat exchanger. The refrigeration is then

$$h_4 - h_3 = h_2 - h_1$$

The maximum refrigeration is obtained when the temperature of the

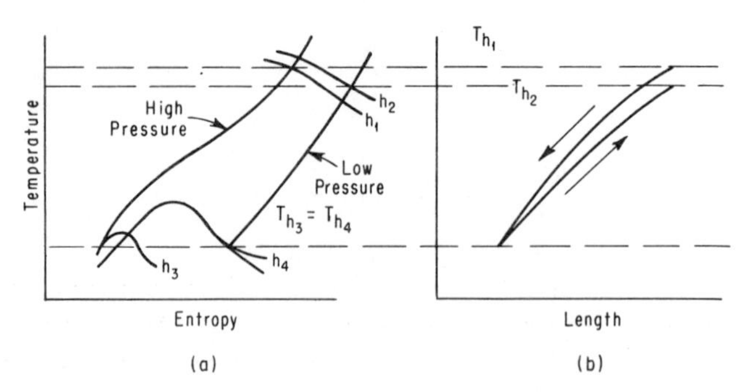

FIG. 30. (a) Temperature–entropy diagram corresponding to the (b) temperature–length distribution in a Joule–Thomson heat exchanger with a temperature pinch at the cold end.

returning stream is the same as that of the supply stream at the hot end of the heat exchanger. The exit temperature of the supply stream cannot get lower than that of the inlet temperature of the return stream, since that would be a violation of the second law of thermodynamics, and, illustrated in Fig. 30, refrigeration may have to be rejected from the system.

B. Practical Refrigeration Cycles

All practical refrigeration cycles for temperatures below 20°K use helium as the working fluid. The way in which refrigeration is affected by gas expansion depends on the cycle used. The gas may be expanded isentropically in an expansion engine, or isenthalpically through an expansion valve. In either case, work must be applied to recompress the gas if the refrigeration system is to be of the closed-loop continuous type. The same comment applies, of course, to a liquefier.

All refrigerators for below 20°K or helium liquefiers must use highly efficient heat exchangers. In conventional industrial practice, the design of extremely high-effectiveness exchangers is seldom justified, since an increased temperature differential usually results in decreasing the overall thermal efficiency by a relatively small percentage. In the refrigeration systems under consideration, heat-exchanger inefficiencies cause a direct reduction of the amount of usable refrigeration. To illustrate, using a 100%-efficient expander over a 10 to 1 ratio and a 97%-effective heat exchanger, 60% of the refrigeration produced by the expander would be available at about 30°K. If the heat-exchanger effectiveness decreased to just 94%, there would be only 20% refrigeration available at about 30°K; the rest would be taken up by heat-exchanger losses.

At first, one would be tempted to increase the performance and therefore the size of the heat exchangers. However, not only do pressure drops become limiting, but secondary losses associated with the larger exchangers also reduce overall performance. Typical items are end-to-end conduction, radiation losses, and larger supporting requirements. For these reasons, all low-temperature refrigerators utilize specially fabricated extended surface exchangers to increase performance without increasing size.

In the same way that heat-exchangers losses reduce the amount of refrigeration that could be obtained from a given circulation of gas, less than ideal expansion reduces the refrigeration. To illustrate this, using an example similar to above, if the expander had been only 75% efficient, the percentage of refrigeration that would have been available at 30°K using a 97%-effective heat exchanger would have dropped to 35% of the ideal amount, and if the heat-exchanger effectiveness had dropped to only 94%, the net refrigeration would have dropped to zero!

The losses occurring in engines are caused by such things as: (a) pressure drop across inlet and exhaust valves, (b) clearance volumes, (c) pressure drop in heat exchangers and piping, (d) irreversible mixing of gases at different temperatures or pressures, (e) heat exchange with engine components, and (f) nonwork-producing gas flows in the expander. The effects of these factors are collectively termed the "enthalpy efficiency" of the expander; that is, the change in enthalpy of the gas on expansion as compared with the enthalpy change which would take place in an ideal isentropic expansion. These efficiency factors which may be as high as 80 to 90% for large systems, fall off for small systems producing only a few watts in the temperature range of interest. Typical values for reciprocating devices are 50 to 80%, whereas for small turbine expanders, the efficiency is usually below 50%.

For the ideal production of refrigeration at a certain level, there is no benefit to be gained by cascading or staging the process. In practice, however, when there is a series of refrigeration requirements to be met, it is desirable to multistage. For example, the cooling of the Joule–Thomson stream would ideally require only the production of precooling refrigeration equal to the Joule–Thomson effect. Since the heat exchangers are not perfect, we must provide refrigeration to overcome the heat-exchanger losses. By setting up an expression for the work required to provide the temperature-dependent load and finding the optimum intermediate-stage temperature, the expression for the intermediate temperature takes the same form as that for a multistage compressor; i.e., for two-stage refrigeration,

$$T_1/T_0 = T_2/T_1 = (T_2/T_0)^{1/2}$$

and for three-stage refrigeration,

$$T_1/T_0 = T_2/T_1 = T_3/T_2 = (T_3/T_0)^{1/3}$$

Analysis for the minimization of the work required to support a device in which the principal load is in the form of thermal radiation results in the following equation for two-stage refrigeration:

$$\frac{T_1}{T_0} = \frac{T_2}{T_1} = \left[\frac{E_1 A_1}{4E_2 A_2}\left(\frac{T_2}{T_0 - T_2}\right)\right]^{1/5}$$

where E and A are the emissivity and area terms for the shields at the two refrigeration temperatures. In practice the optimum intermediate-stage temperature for radiation loss will not be the same as the optimum for conductive and heat-exchanger losses.

The production of refrigeration at a given temperature level requires a certain amount of work. To produce the refrigeration at a constant temperature level, the minimum amount of work required per unit of refrigeration is given by:

$$W/Q = (T_h/T_c) - 1$$

To produce 10 W of refrigeration at 4°K requires 750-W input using an ideal Carnot refrigeration. Now let us compare the work required in a real system, taking into account the departures from ideality (Hogan and Stuart, 1963). For a 200-psi Joule–Thomson (J–T) system, the optimum precooling temperature is 10°K. The input-work requirement for the J–T loop alone is 1000 W, and the minimum engine input requirement is 300 W, for a total of 1300 W.

Suppose we consider a 97%-effective heat exchanger. It is not possible to design a single-stage engine which will operate at 15°K using a 97%-effective heat exchanger and a high-side pressure of 200 psi, and even if we were to improve the heat-exchanger effectiveness, which is no mean task, the efficiency of the system would be extremely poor. (For example, a single-stage device with a $98\frac{1}{2}$%-effective heat exchanger would require 10 times the work of a two-stage device with 97%-effective exchangers.) The work requirement for a two-stage expansion system to precool the J–T loop to 10°K is about 2000 W, making a total of 3000 W, including the 1000 W for the J–T loop. If the expanders are only 75% efficient, the work increases to 5000 W. An allowance for the adiabatic compression instead of isothermal compression increases the work to 6500 W, while an allowance for the 70% efficiency of the compressor and the 80% efficiency of the motor driving it makes the power input requirement now 11,500 W for 10 W of refrigeration at 4°K.

Other losses must be overcome, such as radiation shielding and conductive loads. Therefore, the actual work input into most systems exceeds 2 kW per watt of refrigeration at 4°K. For very small machines, in which fixed losses predominate, even higher power requirements are not unusual.

Since an ideal Carnot device requires only 75 W/1 W at 4°K, it can be seen that these systems have thermal efficiencies of less than 5% of Carnot efficiency. This seemingly poor efficiency was obtained even though good efficiency was assumed in the heat exchangers, expansion engines, and compressors. Therefore, it should be apparent that even small departures from ideality can have serious consequences in the design of real systems. It should be emphasized that these efficiency terms are only illustrative. The important point is that without high efficiency the refrigerator will not operate at all.

1. *Cascade Joule–Thomson (J–T) liquefier (or refrigerator)*

The first practical helium liquefier utilized three J–T loops in cascade. Helium must be precooled to below its inversion temperature of 55°K

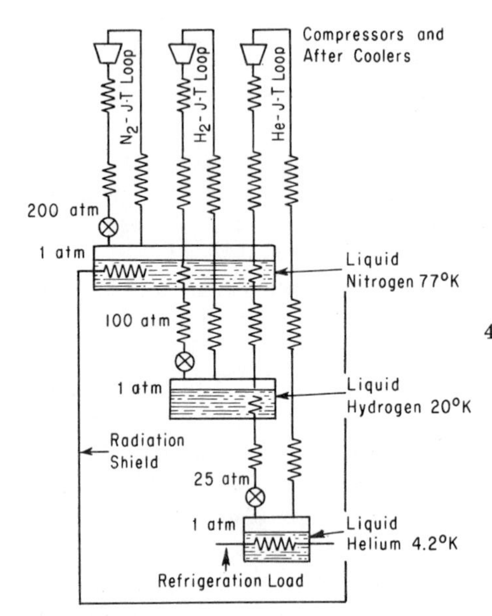

FIG. 31. Schematic of a cascade J–T 4.2°K refrigerator–liquefier.

before any cooling by isenthalpic expansion can be detected, and as a practical matter it must be precooled to the region of at least 25°K. This can be done by precooling with liquid hydrogen. Hydrogen also must be precooled to below its inversion temperature before its liquefaction can be effected, and as a practical matter, it is best if precooled to liquid-nitrogen temperature. Nitrogen is below its inversion temperature at room temperature. Figure 31 shows schematically such a system. The first loop, nitrogen, is expanded from 100 or 200 atm down to 1 atm, to provide refrigeration at about 77°K to two other gas streams and a radiation shield which intercepts about 90% of the radiation heat leak from room-temperature surfaces. The second loop, hydrogen, precooled by liquid nitrogen is expanded from ~100 atm to 1 atm, to provide refrigeration at about 20°K to the helium gas stream. The third loop, helium is precooled to liquid-nitrogen temperatures to overcome heat-exchanger losses, and also by liquid hydrogen to precool it to below its inversion temperature. The helium is then expanded from ~25 atm to 1 atm to provide a fraction of liquid helium at 4.2°K (or some other temperature corresponding to its expanded pressure) which can be boiled off to provide refrigeration at constant temperature.

The complexity of the room-temperature section, containing compressors, aftercoolers, and gas purifiers, is obvious. Three separate compressors, two with very high compression ratios, and three separate gas-purification systems (even with "dry" piston compressors) make this system less attractive than other systems developed later. The reliability of this system depends on the reliability of the compressor and the purification systems, in cooperation. It is clearly more sensitive than a single-gas system.

The apparent simplicity of the refrigerator section is appealing, since there are no moving parts. However, the complexity of heat exchangers and the limitation of stage temperatures (limited to the liquefaction temperatures of the precool fluids) makes this section also less attractive than others superficially more difficult. Again, the cooperation of three separate systems is needed for successful operation. Removing the hazards associated with hydrogen by the use of neon is not too attractive either: Neon, still very expensive, would liquefy at a higher temperature than hydrogen (27°K) and would make the helium loop correspondingly less efficient.

2. *Brayton cycle*†

In the cascaded J–T cycle discussed, nitrogen and hydrogen precooling circuits were used to remove energy from the helium circuit. The same energy removal can be effected by one or more mechanical expanders. The Brayton cycle as conceived by George Brayton was a heat engine, and the cycle is sometimes referred to as the reverse Brayton cycle when used for refrigeration.

A schematic of the Brayton cycle is shown in Fig. 32a, and the process path in Fig. 32b. A compressor delivers gas to an aftercooler, and then to the high-pressure inlet of a heat exchanger at T_1h_1. This stream is cooled in the heat exchanger to T_2h_2, enters the expander and is cooled to T_3h_3. If the expansion had been ideally isentropic, it would have cooled to T_4h_4. The efficiency of the expander η_e is the ratio of the actual enthalpy change to the maximum enthalpy change that could have occurred:

$$\eta_e = (h_2 - h_3)/(h_2 - h_4)$$

The refrigeration load is absorbed between T_3 and T_5. The return stream enters the heat exchanger at T_5 and is warmed up to T_6. In the heat exchanger there must be an enthalpy balance: $h_1 - h_2 = h_6 - h_5$ for balanced flow, as in this case, and the heat-exchanger efficiency, which is the ratio of the actual enthalpy exchange to the maximum that could have been exchanged is

$$\epsilon = (h_1 - h_2)/(h_1 - h_5) = (h_6 - h_5)/(h_1 - h_5)$$

The first law-energy balance on this system is

$$\dot{m}h_1 = \dot{m}h_6 + \dot{m}(h_2 - h_3) - \dot{m}(h_5 - h_3)$$

The available refrigeration is $\dot{m}(h_5 - h_3)$. This becomes zero when the heat-exchanger inefficiency loss equals the enthalpy change in the expander,

† See Muhlenhaupt and Strobridge (1968).

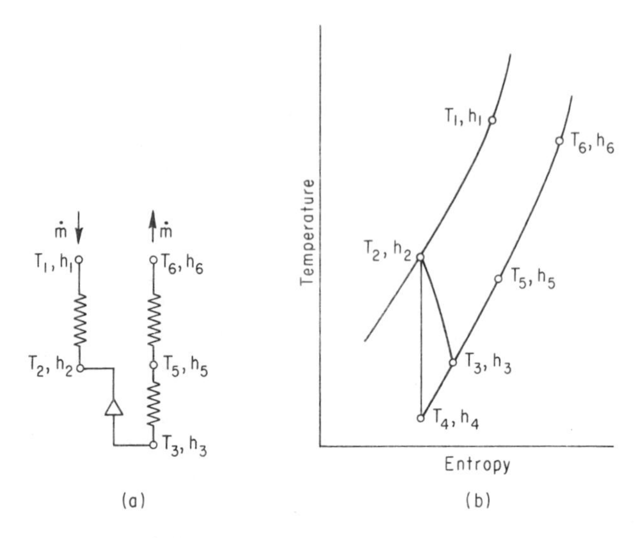

FIG. 32. Schematic of the (a) reverse Brayton-cycle heat exchanger and expander arrangement and (b) associated temperature–entropy diagram.

or $h_1 - h_6 = h_2 - h_3$. This cycle is seldom used on its own for refrigeration below 20°K. However, when precooled to 80°K by the use of liquid nitrogen, this system is quite effective.

3. *Claude cycle*

The Claude cycle is commonly used for the air liquefaction, and is in reality a J–T loop precooled by a Brayton-cycle refrigerator, even though the two typically share the same heat exchangers and compressors.

A schematic of the Claude cycle is shown in Fig. 33a, and the process path in Fig. 33b. The upper parts of these figures are identical to Fig. 32 for the Brayton cycle, but the refrigeration load that was made available by the expansion engine is now used to further cool a bypass stream to below its inversion temperature. Expansion through a J–T valve can provide liquefaction of some fraction of the working fluid.

First law-energy balances on the heat exchangers and the system would be

$$(\dot{m}_1 + \dot{m}_2 + \dot{m}_x)(h_1 - h_2) = (\dot{m}_1 + \dot{m}_2)(h_6 - h_5)$$

$$(\dot{m}_1 + \dot{m}_x)(h_2 - h_7) = (\dot{m}_1 + \dot{m}_2)(h_5 - h_3)$$

$$(\dot{m}_1 + \dot{m}_x)(h_7 - h_8) = \dot{m}_1(h_3 - h_9)$$

$$(\dot{m}_1 + \dot{m}_2 + \dot{m}_x)h_1 = (\dot{m}_1 + \dot{m}_2)h_6 + \dot{m}_2(h_2 - h_3) + \dot{m}_x h_L$$

$$(\dot{m}_1 + \dot{m}_2)(h_1 - h_6) + \dot{m}_x(h_1 - h_L) = \dot{m}_2(h_2 - h_3)$$

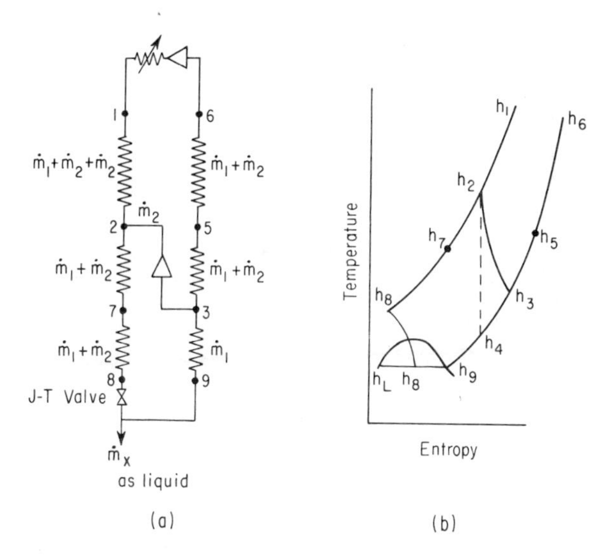

FIG. 33. Schematic of the (a) Claude-cycle heat exchanger and expander arrangement and (b) associated temperature–entropy diagrams.

The last statement says that the heat-exchanger thermal loss from the system, plus the enthalpy change of a fraction of fluid $\dot{m}_x$ from its initial condition to liquefaction is equal to the enthalpy change, or refrigeration produced, in the expansion engine.

If the system is used as a refrigerator, the energy balances in the heat exchangers remains the same, with the value of $\dot{m}_x = 0$. The system energy balance would be

$$(\dot{m}_1 + \dot{m}_2)h_1 = (\dot{m}_1 + \dot{m}_2)h_6 + \dot{m}_2(h_2 - h_3) - \dot{m}_1(h_9 - h_8)$$

and

$$\dot{m}_1[(h_1 - h_6) + (h_9 - h_8)] = \dot{m}_2[(h_2 - h_3) - (h_1 - h_6)]$$

As with the Brayton cycle, the Claude cycle is not often used for refrigeration at the liquid-helium temperature level. The requirements on both engine and heat-exchanger efficiency are too stringent.

This cycle used with liquid-nitrogen precooling, so that effectively the entering gas is at 80 instead of 300°K, is commonly used for helium liquefaction and liquid-helium-temperature refrigeration. Such a cycle is often referred to as the Kapitza cycle, after the scientist who used this cycle to liquefy helium.

4. *Collins cycle*

The Collins cycle (Collins, 1947) was the first cycle used in any commercially available equipment for the liquefaction of helium or for refrigera-

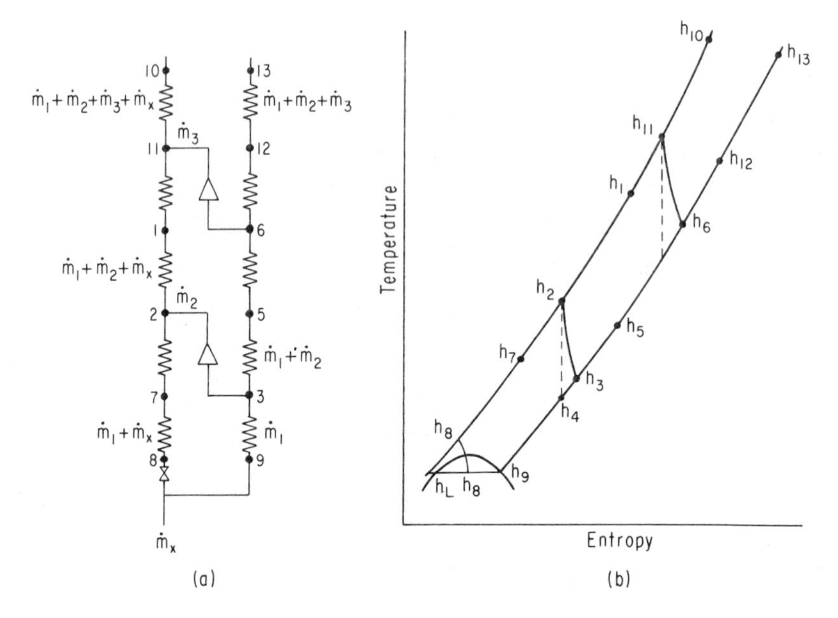

FIG. 34. Schematic of the (a) Collins-cycle heat exchanger and expander arrangement and (b) associated temperature–entropy diagram.

tion at temperatures below 20°K. A standard commercial helium liquefier that has been available for over 20 years is known as the Collins cryostat.

The name "cryostat" comes about because of its original design purpose. Collins wanted to design a cryostat in which experiments could be performed at any temperature from that of liquid helium up to room temperature. To do this he designed a wide neck Dewar and built his refrigerator into the upper portion of it. Only one or two were ever, in fact, used in this way, and many hundreds more were used as the standard helium liquefier in low-temperature laboratories all over the world.

Figures 34a and 34b show the schematic and process path of the Collins cycle. The lower parts of these figures are the same as that shown for the Claude cycle, with the temperature and enthalpy subscripts numbered in the same sequence.

In this arrangement a single stream of helium gas is compressed at room temperature, typically to ~15 atm, passes through an aftercooler, and enters the high-pressure side of the countercurrent heat exchanger. At a temperature of ~100°K, 15–20% of the main gas stream is tapped off and expanded in an expansion engine to the pressure of the return side of the heat exchanger.

This gas is cooled, in proportion to the work extracted from the engine, to a lower temperature (~50°K) and is returned to the low-pressure side of the countercurrent heat exchanger.

At a lower temperature, ~30°K, 50–75% of the remaining high-pressure stream is tapped off and expanded in a second expansion engine, again to the pressure of the return side of the heat exchanger. This portion of the gas is cooled to ~15°K and returned to the low-pressure side of the heat exchanger, thereby providing refrigeration to the remaining part of the high-pressure stream to below its inversion temperature. The remaining part of the high-pressure stream, after passing through a further length of heat exchanger, is expanded isenthalpically through an expansion valve to provide some fraction of liquid helium.

First-law energy balances on the heat exchangers and the system would be

$$(\dot{m}_1 + \dot{m}_2 + \dot{m}_3 + \dot{m}_x)(h_{10} - h_{11}) = (\dot{m}_1 + \dot{m}_2 + \dot{m}_3)(h_{13} - h_{12})$$

$$(\dot{m}_1 + \dot{m}_2 + \dot{m}_x)(h_{11} - h_1) = (\dot{m}_1 + \dot{m}_2 + \dot{m}_3)(h_{12} - h_6)$$

$$(\dot{m}_1 + \dot{m}_2 + \dot{m}_x)(h_1 - h_2) = (\dot{m}_1 + \dot{m}_2)(h_6 - h_5)$$

$$(\dot{m}_1 + \dot{m}_x)(h_2 - h_7) = (\dot{m}_1 + \dot{m}_2)(h_5 - h_3)$$

$$(\dot{m}_1 + \dot{m}_x)(h_7 - h_8) = \dot{m}_1(h_3 - h_9)$$

$$(\dot{m}_1 + \dot{m}_2 + \dot{m}_3 + \dot{m}_x)h_{10} = (\dot{m}_1 + \dot{m}_2 + \dot{m}_3)h_{13} + \dot{m}_3(h_{11} - h_5) + \dot{m}_2(h_2 - h_3) + \dot{m}_x h_L$$

$$(\dot{m}_1 + \dot{m}_2 + \dot{m}_3)(h_{10} - h_{13}) + \dot{m}_x(h_{10} - h_L) = \dot{m}_3(h_{11} - h_6) + \dot{m}_2(h_2 - h_3)$$

The last statement says that, as with the Claude cycle, the heat-exchanger thermal loss plus the enthalpy change of the liquefied fraction is equal to the total enthalpy change, or refrigeration produced, in the two expansion engines.

Also, if this system is used as a refrigerator, the heat-exchanger energy balances remain the same, with the value of $\dot{m}_x = 0$. The system energy balance would be

$$(\dot{m}_1 + \dot{m}_2 + \dot{m}_3)h_{10} = (\dot{m}_1 + \dot{m}_2 + \dot{m}_3)h_{13} + \dot{m}_3(h_{11} - h_6) + \dot{m}_2(h_2 - h_3) - \dot{m}_1(h_9 - h_8)$$

The Collins cycle as discussed works effectively for refrigeration below 20°K or for helium liquefaction; however, this cycle too is enhanced if liquid-nitrogen precooling is used. Typically, both liquefaction rate and refrigeration can be doubled on any given machine, by the use of liquid-nitrogen precooling.

It is interesting to compare the temperature profile in the countercurrent heat exchanger of a liquefier with that of the refrigerator.

Figure 35 shows such a profile in exaggerated form for a liquefier. Note that each section of heat exchanger has a pinch either at its hot end or its cold end. This comes about because of the unbalanced flows caused both by the engines and by the liquefied fraction that is withdrawn from the system. As will be discussed later, high efficiencies can be obtained with relatively small heat exchangers [as measured by the number of transfer units (NTUs) required] when significant imbalance in the $\dot{m}c_P$ enthalpy flows occurs.

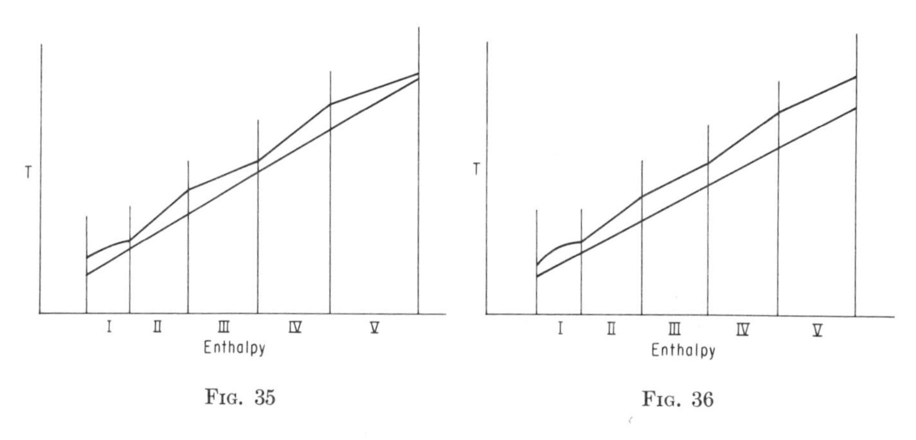

FIG. 35 FIG. 36

FIG. 35. Temperature distribution in the heat exchanger of a Collins-cycle liquefier.
FIG. 36. Temperature distribution in the heat exchanger of a Collins-cycle refrigerator.

Figure 36 shows a similar profile for a refrigerator. Note that the coldest heat exchanger has balance flow and a pinch at either the hot end, or the cold end, or at both ends. Two other sections of the heat exchanger also have balanced flow. This means that to achieve maximum effectiveness, or minimum heat-exchanger loss, very effective heat exchangers must be used—significantly larger for the same $\dot{m}c_P$ enthalpy flows than are required for the liquefier.

Because of this difference in the requirements of heat exchangers, it can be said that any effective Collins-cycle refrigerator will be an effective liquefier, but the converse is not necessarily so. For example, a Collins liquefier designed for 10-liter/h liquefaction rate may only give 10 W of refrigeration at 4.2°K while one designed for refrigeration may, with the same compressor input power, give 40 W of refrigeration at 4.2°K, but would still be only a 10-liter/h liquefier.

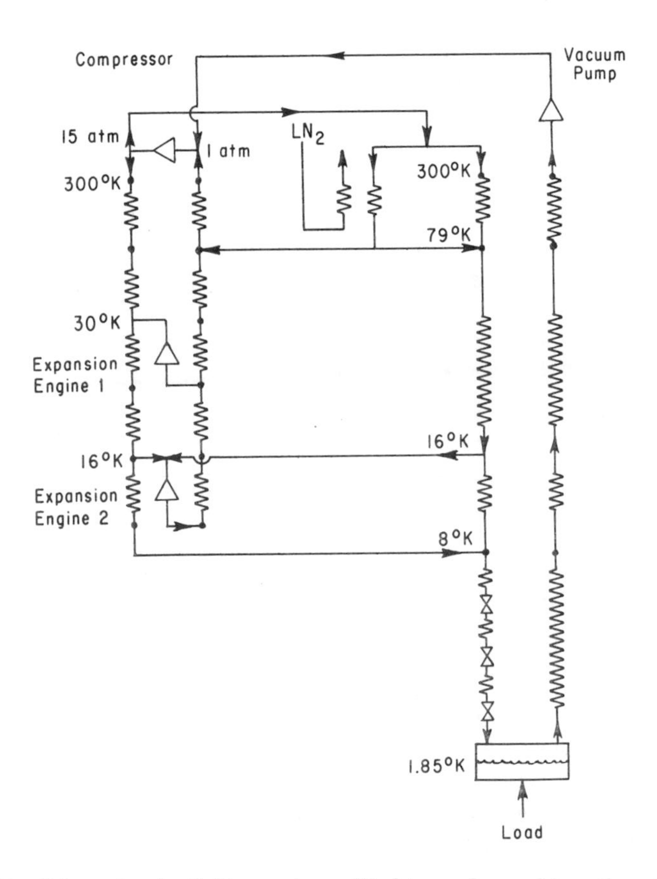

FIG. 37. Schematic of a Collins cycle modified to produce refrigeration at 1.85°K.

It will be noted that in the diagrams for the Claude and Collins cycles, the mass flow rates are numbered to build up from the lowest temperature, rather than decrease from an input mass flow as parts are taken successively by the engines. The reason for this is that in the design steps, one has to start with what is required, either as a liquefaction or a refrigeration rate, and then work back up to the next-higher-temperature section to see what is required there, and so on back up to the final compressor requirement.

There are many variations to the cycles discussed so far. Figure 37 shows the schematic of a Collins cycle used for refrigeration at 1.85°K (Collins *et al.*, 1967). A conventional engine circuit with two expanders operating between a high-side pressure of 15 atm and a low-side pressure of 1 atm, is associated with an ancilliary heat exchanger operating at 15 atm on the high side and 0.015 atm on the other side. The two systems are linked to

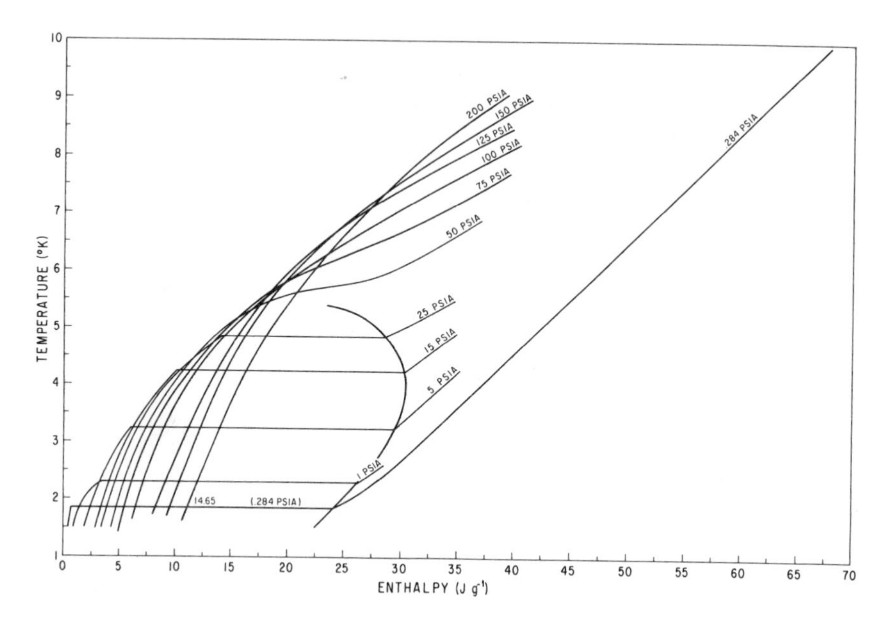

FIG. 38. Temperature–enthalpy diagram for helium.

provide in such a way as to create pinches in the heat exchangers at the most crucial points, thereby greatly increasing the effectiveness of the heat exchangers.

The final J–T stream has three J–T valves in series to enable maximum refrigeration to be transferred to the lower level. This point is illustrated in Fig. 38, which looks at the temperature–enthalpy plots for different pressures. Expansion from 200 to 0.284 psia would have limited refrigeration at 1.85°K to 13 J g^{-1}, regardless of the precooling conditions. By expanding from 200 to 75 psia at 6°K, and then a subsequent expansion to 0.284 psia, the maximum refrigeration is increased to 17.5 J g^{-1}. By a second intermediate expansion from 75 to 15 psia at 4.5°K, and then expansion to 0.284 psia, a total of 20 J g^{-1} can be realized—a significant improvement over the best for a single expansion from 200 psia.

Systems of this kind have been installed at the University of Chicago, for cooling a superconducting magnet focusing coil on an electron microscope. This unit will provide about 15 W of refrigeration at 1.85°K. Another system is installed at Stanford University to provide 300 W of refrigeration at 1.85°K to cool a superconducting linear accelerator.

Another variation of the Collins cycle is in the use of a third expansion engine in place of a J–T valve (Johnson *et al.*, 1967). In one installation at MIT a liquefier using a J–T valve gave 60-liters/hr liquefaction of helium.

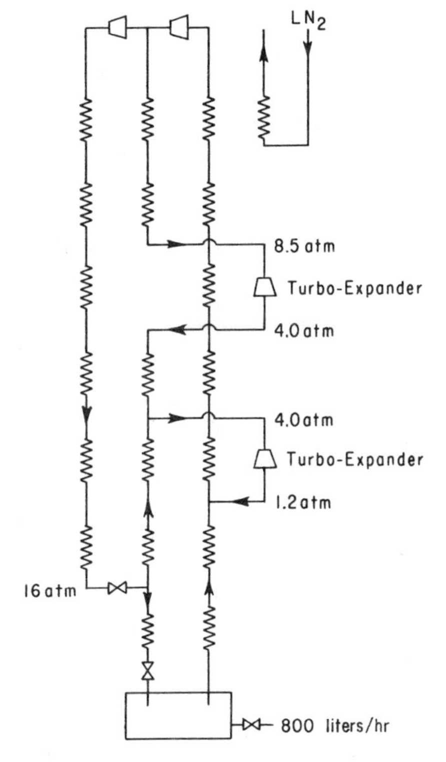

FIG. 39. Schematic flow diagram of a Sulzer 800 liters/hr helium liquefier.

The same installation using an expansion engine in place of the J–T valve, gave a liquefaction rate of 80 liters/hr.

High-speed turboexpanders, using oil-lubricated or gas-lubricated bearings, are often used for helium-liquefaction or helium-temperature refrigeration. Typically, these expanders operate over a 2:1 to 4:1 expansion ratio. Seldom over 10:1. The J–T process typically requires over a 10:1 expansion for best results. Figure 39 shows a schematic of one such application of turbo expanders used in a plant to produce 800 liters/hr of liquid helium.

5. *Stirling-type cycles*

The group of cycles discussed previously can be thought of as continuous flow devices. Even though the expansion engines used typically receive gulps of gas at high pressure to be discharged at low pressure, the heat capacity of the heat exchangers is such that little temperature variation occurs as a result of this fluctuation in flow, and the flow rate in the heat exchangers can be assumed to be constant. The process can be shown with little ambiguity on a temperature–entropy chart.

There is another group of refrigeration cycles which do not use (countercurrent) heat exchangers but which use regenerators, or periodic flow heat exchangers, wherein the flow is first in one direction through a heat-exchange path, and then in the opposite direction. A common form of regenerator is a stack of numerous perforated metal plates, or sheets of wire gauze, or small metallic spheres forming a matrix through which the gas can pass freely and which act as a series of heat reservoirs. The heat, or thermal energy, is stored in the matrix material of the flow path when the flow is from the hot to the cold end, and this energy is given up by the matrix material when the flow is opposite, from cold to hot end. The process of these types of machines cannot properly be shown on any T–S diagram because the temperature–entropy conditions are continuously changing. However, what does take place can be shown reasonably clearly by the use of pressure–volume (P–V) diagrams.

The first commercial refrigeration devices using regenerators were developed by the Philips Company of Eindhoven. They took the Stirling heat engine, on which they had been working for many years, and turned it upside-down, so to speak, and made a refrigerator out of it. They had considerable success and since their first introduction of such machines, other devices using regenerators have had practical application. It is for this reason, and for lack of a better name, that this family of devices are referred to under the generic term of Stirling-type cycles. This undoubtedly does great injustice to the many significant inventors and engineers who made great contributions, such as Erickson, Kirk, Solvay, Vuilleumier, Taconis, Bush, Gifford, and McMahon, to name just a few. Perhaps "regenerative refrigeration cycles" would have been a better classification, but the typical confusion between the terminology of the recuperator and a regenerator would have made the classification meaningless. Anyone who gets involved with cryogenics is going to come to grips with a Stirling cycle as something different from other classic cycles.

Consider, in Fig. 40a, the piston moving back and forth in the sealed cylinder. The pressure in the closed end will vary with the piston position, and the work put into the gas to compress it is returned to the piston on expansion. If the system were adiabatic and frictionless, once started it could keep going, with the energy supply and storage being provided by the flywheel on the driving crank shaft. Since there would be some energy loss, a small driving motor is added to the drive shaft to supply energy and overcome the losses. Figure 40b shows a thermal equivalent of this. Consider the displacer in a cylinder closed at both ends, except for access of the drive shaft, and with one end maintained cold and the other end maintained hot. The two ends are joined by a path which contains a regenerator. As the displacer is moved to the right, gas is displaced from

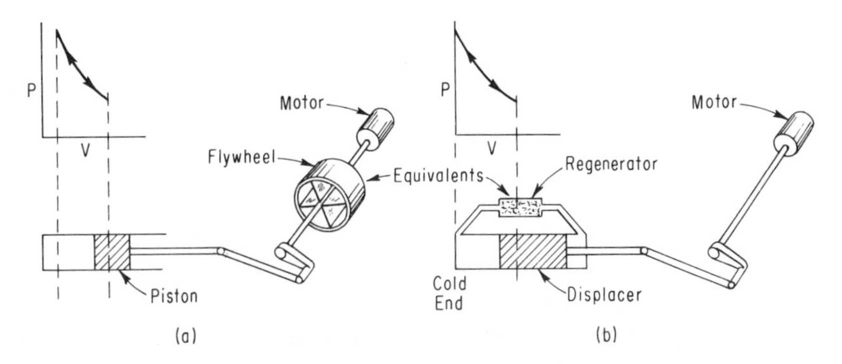

FIG. 40. Comparison of (a) mechanical and (b) thermal energy storage means.

the hot to the cold end, giving up its thermal energy to the regenerator matrix material. Since the volume of gas is fixed, and its temperature has been lowered, the gas pressure will drop. Moving the displacer to the left reverses the process. The thermal energy stored in the matrix of the regenerator is returned to gas as it is displaced from the cold to the hot end, and the pressure increases. The regenerator is a thermal flywheel in this case. As before, such a system would not be ideal, so again a small driving motor is shown on the crank shaft to overcome the losses in the system.

Figures 41a and 41b show the same systems as in Fig. 40 except that valves are included which allow gas to escape out when the pressure reaches a certain high value and which let more gas into the system when the pressure reaches a certain low level. Figure 41a is of course a conventional compressor. Net work is done by the piston on the gas, as shown by the

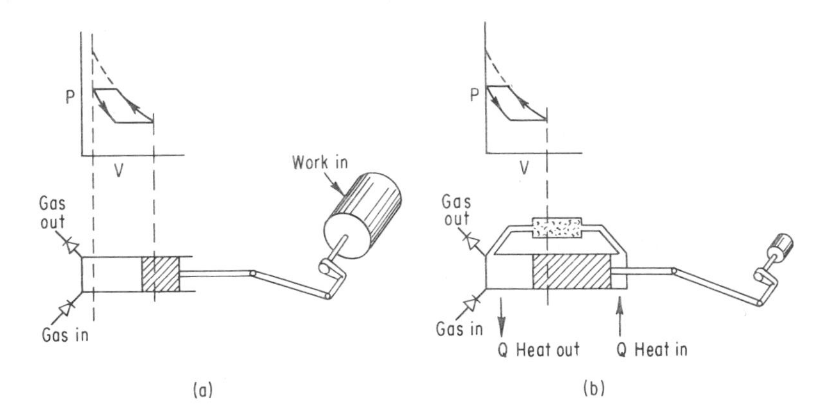

FIG. 41. Comparison of (a) mechanical and (b) thermal compressors.

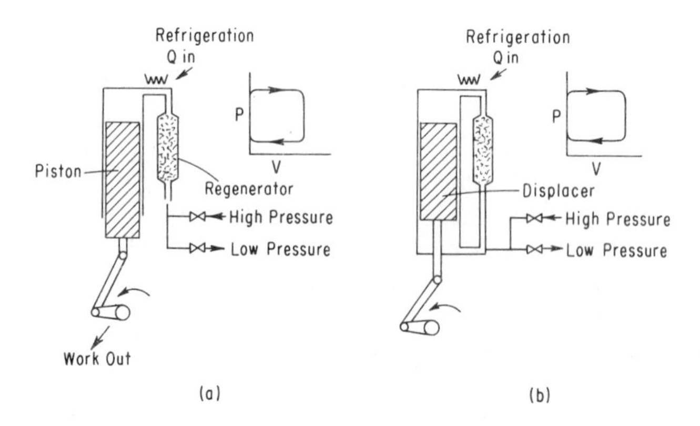

FIG. 42. Schematic of the Gifford–McMahon (a) work-cycle refrigerators and (b) no-work-cycle refrigerators. (After McMahon and Gifford, 1963.)

area of the P–V diagram. This work is supplied by the motor. In Fig. 41b, thermal energy is supplied to the system to effect the work of compression. Heat is put into the gas at the hot end and the same amount of heat is rejected at the cold end, to an extent proportional to the area of the P–V diagrams developed. Note the direction of the P–V diagram shown is counter clockwise, requiring a heat rejection at the cold end. A P–V diagram could be developed for the hot end which would be the same area but clockwise in development. Again, only a small motor is required to overcome the fraction- and pressure-drop losses in the system. The maximum compression ratio in the conventional compressor is limited only by the void volumes. In the thermal compressors the maximum compression ratio is limited to the ratio of the hot and cold absolute temperatures and further reduced by void volumes.

Figures 42a and 42b show two versions of the Gifford–McMahon (1963) refrigeration cycles. We can visualize the operation of the cycles by following discrete steps in the cycles. Consider, in Fig. 42a, the piston to be at the top of its stroke and the inlet valve connected to a high-pressure source, perhaps the discharge of a compressor. Assuming a steady-state operation has been achieved, as the high-pressure gas enters the cylinder region, it will be cooled down in the regenerator. Leaving the inlet valve open, move the piston to its bottom position, filling the space between the piston and cylinder with high-pressure cold gas. An exhaust valve is now opened, allowing the gas to expand to the low-pressure sink, which may be the suction side of the compressor. The gas cools on expansion. Now the piston is moved back to its top position, expelling the cooled gas through a first heat-exchange path in which the refrigeration load is absorbed, and then through the regenerator wherein it is warmed up substantially to its entry temperature. The heat that enters as low-temperature refrigeration is rejected as work out from the piston which has to be absorbed by a breaking mechanism. The refrigeration is equal to or proportional to the area of the *P–V* diagram, as is the work output. A well-designed cycle of this kind can achieve temperatures in a single stage as low as 25°K. Cycles of this kind are often referred to as Solvay cycles, although the lowest temperature reported by Solvay was 178°K and no description of the actual apparatus has been found. The language of Solvay's patent is vague, and a close scrutiny of the method of operation given in the patent might suggest that it would never work (Collins and Cannaday, 1958). However, the embodiments of Gifford and McMahon are explicit and they work.

The version shown in Fig. 42b is often referred to as the no-work Gifford–McMahon cycle, because the work output is zero and the work input is only that required to overcome friction and pressure drop losses. The refrigeration heat input is extracted from the system thermally. The gas leaving is hotter than the gas entering. The cycle may be understood from the following stepwise description. With the displacer at its top position and the cylinder filled with gas at the exhaust pressure, open the inlet valve connected to the high-pressure source. The gas in the lower room-temperature end, is compressed and is heated by the compression. With the inlet valve still open, move the displacer to its lowest position, filling the upper cold end with gas at high pressure after having been cooled by passage through the regenerator. The gas which enters the regenerator is a mixture of hot gas displaced from the hot-end volume and new gas coming from the high-pressure source. Its mixed temperature on entering the regenerator is higher than the new gas entering. When the displacer is at its bottom position, the inlet valve is closed and the exhaust valve is opened, allowing the gas to expand to the exhaust pressure, thereby creating further cooling at the cold end. The gas leaving the cold end passes through a heat exchanger in which it can absorb the refrigeration load so that the gas entering the regenerator is substantially at the same temperature as it left it. Passage through the regenerator warms the gas to a temperature substantially the same as it is when it enters it, that is a little higher than the initial temperature of the high-pressure gas source. In this way the heat input at the low temperature is extracted as a heat output at the discharge of the system.

Such systems can either be operated from a high-pressure gas bottle and discharged to atmosphere or other suitable sink, or can be connected to a compressor of the kind suggested in Figs. 41a and 41b. Such combinations require inlet and outlet valves at the refrigerator and at the compressor (if one is used). If compressors are used, it is obvious that the operating

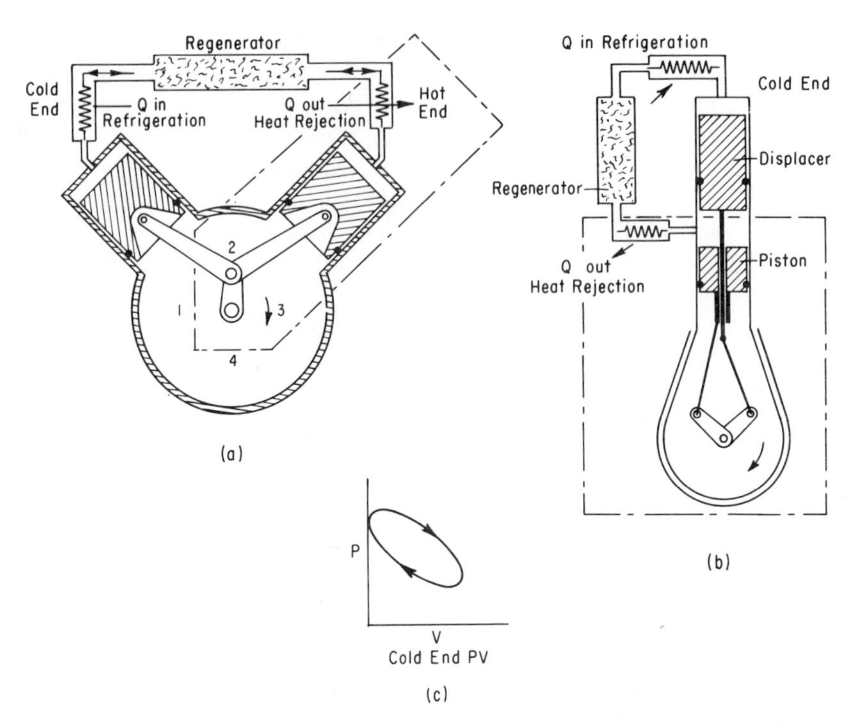

FIG. 43. Schematic of two commonly used Stirling refrigerators. (a) Classical version. (b) Philips version. (c) Associated PV diagram.

speeds of the refrigerator and the compressor can be selected separately judged on the optimum requirements of each.

The speeds of the refrigerator, or expander, and compressor can be synchronized to permit the elimination of both sets of valves, and therefore permit higher thermodynamic efficiencies. Figures 43a and 43b show two such versions that are in use today. In both, the motions of the pistons or displacers are sinusoidal and are phased approximately 90° apart. Their operation is similar to that described for the Gifford–McMahon cycles if one considers the dotted line in each figure. Outside the dotted line is the refrigerator portion shown in Fig. 42 and inside the dotted line is shown a typical single-stage compressor without inlet or exhaust valves.

The operation of the two versions shown in Fig. 43 can be understood by considering four idealized steps: (1) compression of the gas, mostly at the warm end, (2) transfer of compressed gas through the regenerator to the cold end, (3) expansion of the gas causing further cooling at the cold end, and (4) transfer of the expanded gas through a heat exchanger to absorb the refrigeration and then through the regenerator, to the warm end. If, in Fig. 43a, we consider the first step to be accomplished as the crank moves from position 1 to 2, and the second step as a crank motion from 2 to 3, and so on, we can visualize the pressure changes in the system and consequent temperature changes. Mostly the compression is at the warm end, and the heat of compression removed by passage through a heat exchanger, and mostly the expansion is at the cold end, with the refrigeration absorbed in a heat exchanger, and the warm and cold ends isolated by a regenerator.

The process of the version shown in Fig. 43b is identical to that in Fig. 43a and may be a little easier to visualize. Start with the displacer in the uppermost position and the piston in its bottom position.

Step 1 Move the piston up to compress the gas in the system.

Step 2 Move the displacer downwards, transferring compressed gas, first through a heat-rejection heat exchanger, and then through the regenerator to the cold end. During this step the piston moves upwards to maintain the pressure.

Step 3 The piston moves downwards to expand the gas in the system.

Step 4 The displacer is moved upwards, displacing the expanded cold gas, first through a heat exchanger to absorb the heat of expansion and then through the regenerator.

These steps are identical to the step for the Gifford–McMahon systems shown in Fig. 42, but less distinct since they are tied to a sinusoidal motion.

The cycles shown in Fig. 43 are often referred to as a Stirling refrigeration cycle. They are actually Kirk cycles. Stirling invented the heat engine. It was Kirk who reversed the cycle to provide refrigeration. Both of these cycles are like those of Fig. 42, but with a synchronized means of compression of the kind of Fig. 41a. If a compressor of the kind shown in Fig. 41b is used in a synchronized way we have a new cycle, first invented by Vuilleumier, and later in a different sequence of steps, by Taconus.

The Vuilleumier cycle, often referred to as the VM cycle because of the difficulty in pronouncing the name, is ideally an all thermal cycle (Chellis and Hogan, 1963). As shown in Fig. 44, this cycle requires two displacers. No drive mechanism is indicated here, but obviously the friction of the displacers and the pressure drop loss through the regenerators has to be overcome, and typically a drive motor is used. The motion relationship between the two displacers is sinusoidal phased 90° apart.

The steps of this can be considered, starting with the two displacers in remote relationship with the cylinder volume all at the intermediate temperature.

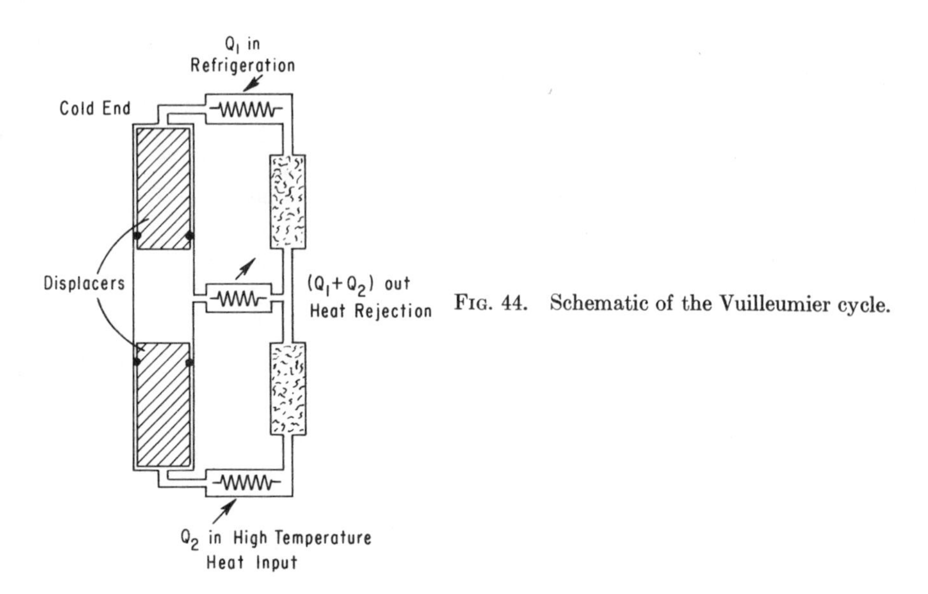

FIG. 44. Schematic of the Vuilleumier cycle.

Step 1 Move the lower displacer upward, displacing gas from an intermediate temperature to a higher temperature region, thereby increasing the pressure in the system.

Step 2 Move the upper displacer downward to transfer the high-pressure gas to the cold end of the system.

Step 3 Move the lower displacer back to its downward position, transferring gas from the hot end to the intermediate-temperature level, thereby lowering the pressure in the system, and of course, causing a temperature drop, by expansion of the gas in the cold end.

Step 4 Move the upper displacer back to its uppermost position, displacing the cold expanded gas, through the heat-absorption refrigerator and then through the regenerator.

Refrigeration devices based on the cycles shown in Figs. 42–44 are all capable of producing refrigeration efficiently at liquid-nitrogen temperature and of reaching terminal temperatures about 25°K. All can be extended by adding stages to the refrigeration cycle, as indicated in Fig. 45. Using two stages, useful refrigeration can be obtained at 15°K, and using three stages useful refrigeration can be obtained at 10°K. The lowest temperature reported for this type of device is 6.5°K. The lower limitation is imposed by the disappearance of heat capacity in most materials.

Since two- and three-stage refrigerators, similar to that shown in Fig. 45, can produce temperatures below the inversion temperature of helium, these

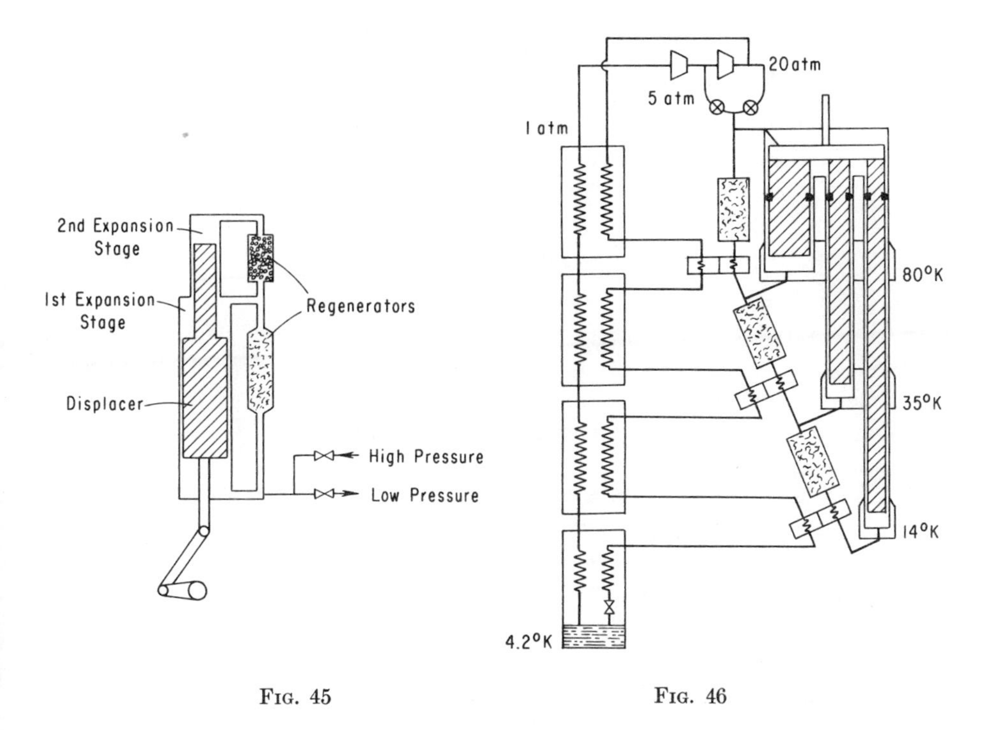

FIG. 45 FIG. 46

FIG. 45. Schematic of a two-stage Gifford–McMahon refrigerator.

FIG. 46. Schematic of a three-stage Gifford–McMahon refrigerator coupled to a J–T heat exchanger loop to provide refrigeration at 4.2°K. [After Gifford-McMahon, 1963.]

expansion engines can be used to precool a high-pressure helium gas stream to produce J–T cooling at liquid-helium temperatures. Such an arrangement is shown schematically in Fig. 46. This shows a three-stage Gifford–McMahon no-work cycle precooling a J–T heat-exchanger loop. The refrigerator expansion engine circuit would operate between the pressure limits of 20 and 5 atm. The J–T loop would typically operate between 15 and 20 atm on the high side and a pressure on the low side dependent on the temperature required, typically 1 atm for a temperature of 4.2°K.

Refrigerators of the kind shown in Fig. 46 have been used since 1961 for cooling maser amplifiers for the NASA deep-space program. The refrigerators on that particular network have accumulated over 500,000 hr of operation with a demonstrated average time between failures of over 13,000 hr (Little, 1967). Refrigerators of the type shown in Fig. 45 are used in over a hundred commercial and military satellite communication ground stations for cooling low-noise parametric amplifiers. These systems which are accumulating experience at about one million hr/yr have demonstrated the high-reliability potential of cryogenic equipment, with over 15,000-hr average time between failure (Hogan, 1968).

VI. Heat Exchangers

As important as the method of gas expansion to create refrigeration is the function of heat exchangers to conserve that refrigeration. The heat exchanger must act as a thermal isolator between the ambient temperature at which the gas is compressed and the lower temperatures at which it is expanded.

The primary function of a heat exchanger is to facilitate the removal of energy from the high-pressure inlet stream, to cool it to a desired temperature by transferring that energy to the low-pressure exit gas stream, and to heat it up to ambient temperature. The primary characteristic of a heat exchanger in which we are interested is the effectiveness with which it performs that function.

$$\epsilon = \text{actual energy transfer/maximum possible energy transfer}$$

There are an infinite variety of heat exchangers, but in cryogenic work they will mostly be of two kinds—recuperators and regenerators. Recuperators are heat exchangers with a dividing wall separating fluid streams across which heat is transferred away from one stream to be recuperated by the other. Most typically, the streams run countercurrent to each other, and also, mostly, the two streams are unbalanced in mass flow rate. Regenerators are periodic flow heat exchangers, in which a fluid stream will first flow in one direction, transferring heat to a colder matrix of solid material, as it cools down, and then returns in the opposite direction, regenerating heat from the solid matrix as it warms up. Typically, in cryogenics, regenerators have balanced mass flow rates.

A. Recuperators

Figure 47 shows a typical temperature–surface area plot for a countercurrent flow recuperator. The incoming hot stream is cooled from T_{h_1} to T_{h_2} while the outgoing cold stream is warmed from T_{c_1} to T_{c_2}. The dotted line indicates the possible temperature of the dividing wall. The tempera-

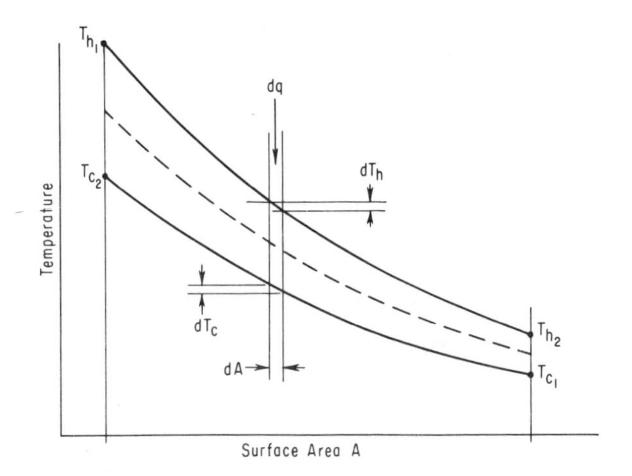

FIG. 47. Typical temperature–surface area plot for a recuperator with unbalanced heat capacity flow rates.

ture T_{h_2} cannot be less than T_{c_1}, because of the second law of thermodynamics, and it cannot, in fact, be equal to T_{c_1}, since a finite temperature difference must exist for a finite heat flow. Therefore, the heat exchanger cannot be 100% efficient.

The figure indicates that the heat-capacity flow rate C_h of the hot stream, that is, the mass flow rate times the specific heat $\dot{m}c_P$ of the hot stream, is less than the heat capacity rate C_c of the cold stream. The heat-transfer rate may be written as

$$\dot{q} = UA\ \Delta T_{av}$$

where U is the overall conductance for heat transfer, A is the surface area on which U is based, and T_{av} is some average temperature difference between the two fluid streams. Further

$$\dot{q} = C_h(T_{h_1} - T_{h_2}), \qquad \dot{q} = C_c(T_{c_2} - T_{c_1})$$

Also

$$\dot{q}_{max} = C_{min}(T_{h_1} - T_{c_1}) = C_{min}\ \Delta T_{max}$$

The effectiveness is then

$$\epsilon = \dot{q}/\dot{q}_{max} = UA\ \Delta T_{av}/C_{min}\ \Delta T_{max}$$

If we define UA/C_{min} as the number of heat transfer units in the heat exchanger, NTU, then

$$\epsilon = \text{NTU}\ \Delta T_{av}/\Delta T_{max}$$

To analyze the heat-transfer situation in the heat exchanger, in which the temperatures are changing, consider an incremental area as indicated in Fig. 47:

$$d\dot{q} = U(T_h - T_c)\, dA = -C_h\, dT_h = -C_c\, dT_c$$

and

$$dT_h - dT_c = d(T_h - T_c) = \left(\frac{1}{C_c} - \frac{1}{C_h}\right) d\dot{q} = -\left(1 - \frac{C_h}{C_c}\right)\frac{d\dot{q}}{C_h}$$

so that

$$\int_{T_{h_1}-T_{c_2}}^{T_{h_1}-T_{c_2}} \frac{d(T_h - T_c)}{(T_h - T_c)} = -\left(1 - \frac{C_h}{C_c}\right)\frac{U}{C_h}\int_0^A dA$$

Then

$$(T_{h_2} - T_{c_1})/(T_{h_1} - T_{c_2}) = \exp[-(UA/C_h)(1 - C_h/C_c)] \qquad (13)$$

In the figure, the hot stream was indicated as the minimum $\dot{m}c_P$ stream, so Eq. (13) can be rewritten as

$$\frac{\Delta T_{\min}}{\Delta T_{\max}} = \frac{T_{\min\,\text{out}} - T_{\max\,\text{in}}}{T_{\min\,\text{in}} - T_{\max\,\text{out}}} = \exp\left[-\frac{UA}{C_{\min}}\left(1 - \frac{C_{\min}}{C_{\max}}\right)\right]$$

$$\frac{\Delta T_{\min}}{\Delta T_{\max}} = \exp\left[-\text{NTU}\left(1 - \frac{C_{\min}}{C_{\max}}\right)\right] = \frac{1 - \epsilon}{1 - (C_{\min}/C_{\max})\epsilon}$$

Solving for effectiveness we find

$$\epsilon = \frac{1 - \exp[-\text{NTU}(1 - C_{\min}/C_{\max})]}{1 - (C_{\min}/C_{\max})\exp[-\text{NTU}(1 - C_{\min}/C_{\max})]}$$

Where the heat capacity flow rates are equal, $C_{\min} = C_{\max}$, we find

$$\epsilon = \text{NTU}/(1 + \text{NTU})$$

and for the case of boiling or condensing on one side of the heat exchanger $C_{\max} = \infty$, $C_{\min}/C_{\max} = 0$, and

$$\epsilon = 1 - \exp(-\text{NTU})$$

Figure 48 plots the relationship between NTU's and effectiveness for a range of $C_{\min}/C_{\max}$, in the range of interest in cryogenics.

From a knowledge of the mass flows required and the heat-exchanger loss that can be tolerated, a size of heat exchanger can be determined in terms of NTU's. This determines the product UA required, and this in turn is determined by the design concept of the heat exchanger and the allowable pressure drop.

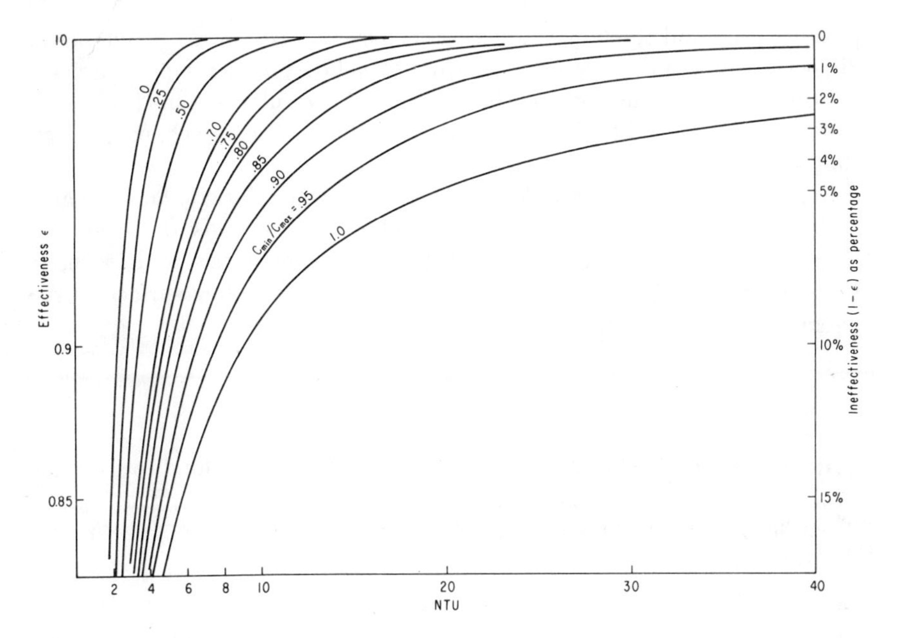

FIG. 48. Relationship between counterflow recuperator effectiveness and NTUs for various ratios of heat capacity rates.

The overall thermal conductance U is based on a temperature potential $(T_h - T_c)$ and a unit transfer area. The reciprocal of U is an overall thermal resistance which can be considered as having the following components in series:

(1) a hot-side film conduction component from the gas to the wall;
(2) a cold-side film conduction component from the wall to the gas; and
(3) a wall conduction component

$$\frac{1}{U} = \frac{1}{h_h A_h} + \frac{1}{h_c A_c} + \frac{1}{(k/t)A} \tag{14}$$

where h_h and h_c are the heat-transfer coefficients between the hot and cold gas streams and the heat-transfer surface, A_h and A_c are the surface areas to or from which heat is transferred, k is the thermal conductivity of the wall material and t the thickness of the wall, A is the prime surface area separating the two streams. Typically heat exchangers have extended surfaces, such as fins or pins that extend from the prime surface.

The last term in (14) is usually insignificant in the determination of U, so we are mostly concerned with the determination of the magnitude of pressure drop and the film conduction coefficients. To do this, a number of dimensionless parameters for similarity of velocity and temperature distribution are used that enable the experimental results obtained on a few systems to be applied to the almost infinite variety of heat-exchanger design that could be employed.

Similarity exists if two systems (1) are geometrically similar in shape, surface characteristics, etc., (2) have similar velocity fields at the boundaries, (3) have similar temperature fields at the boundaries, and (4) have the same ratio of the temperature gradient at the bounding surface to the average gradient in the fluid.

The dimensionless parameters follow.

1. *The Reynolds number*

Reynolds conceived that transmission of heat from a hot fluid to a surface was directly related to the fluid friction exerted on the surface. The similarity of flow of viscous fluids is given by similarity of the ratio of the inertia forces to the viscous forces:

$$\mathrm{R_e} = D_e G/\mu$$

where D_e is the equivalent diameter of a tube which would present the same heat transfer and pressure drop as the duct actually used. It is usually

4 × cross-sectional area of flow path/wetted perimeter of flow path

The total wetted perimeter is used in pressure-drop calculations, but only the perimeter through which heat is transferred is used in heat-transfer calculations. These two are most frequently the same. G is the mass flow rate per unit of cross-sectional area, and μ is the kinematic viscosity.

2. *The Prandtl number*

This is the ratio between the thermal diffusivity of the gas and the kinematic viscosity. It is a property of the gas:

$$\mathrm{P_r} = c_\mathrm{P}\mu/k$$

where c_P is the specific heat at constant pressure of the gas and k its thermal conductivity.

3. *The Stanton number*

This is the ratio of the heat transmission perpendicular to the surface to the enthalpy transport in the flow direction, for a unit temperature difference and area:

$$\mathrm{S_t} = h/Gc_\mathrm{P}$$

4. *The Nusselt number*

This directly indicates what multiple of mere thermal conduction is transferred with the aid of forced convection:

$$N_u = hD_e/k$$

where h is the film coefficient. These dimensionless parameters are interrelated as follows:

$$N_u = S_t P_r R_e$$

5. *Correlations*

The number of correlations between the dimensionless groupings R_e, P_r, and N_u is almost limitless. However, in most heat-exchanger designs where the gas flow is along the surface, as through a tube, the following correlations have been found acceptable:

$$N_u = 0.023(R_e)^{0.8}(P_r)^{0.4}, \qquad \Delta P = f(L/D_e)(G^2/\rho)$$

where L is the length of the flow path, ρ is the density of the gas, and

$$f = 64/R_e \qquad \text{for laminar flow,} \qquad R_e < 2000$$

$$f = 0.316/R_e^{0.25} \qquad \text{for turbulent flow,} \qquad R_e > 2000$$

Also, in most heat-exchanger designs where the gas flow is across the surface, as across tubes or through perforated disks, the following correlations have been found successful:

$$N_u = 0.33(R_e)^{0.6}(P_r)^{0.3}$$

and f in the pressure drop equation is

$$f = 30/R_e^{0.9} \quad \text{for} \quad R_e < 100 \qquad \text{and} \qquad f = 5/R_e^{0.39} \quad \text{for} \quad R_e > 100$$

Using the approach and formulas given most countercurrent heat exchangers or recuperators can be designed or evaluated.

B. Regenerators

The regenerator is a periodic flow heat exchanger. The same space is occupied alternately by the hot fluid and the cold fluid while the energy to be transferred is stored in and released from the regenerator packing material.

The temperature of gases and the temperature of the regenerator matrix material are functions of both position and time. After a period of operation,

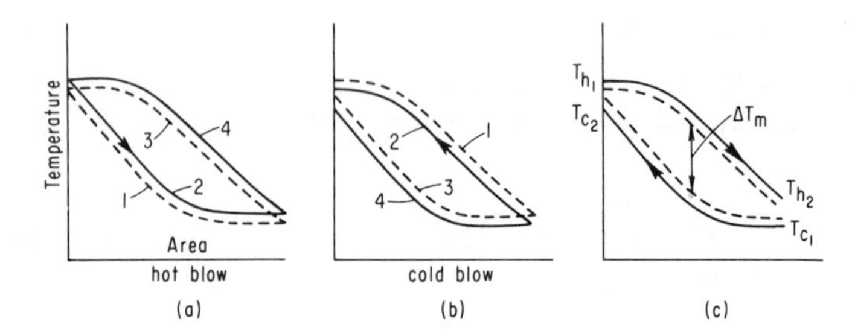

FIG. 49. Temperature distribution variations in a regenerator. Mean Temp Swing per cycle.

a steady-state operation is achieved in which the temperature distribution in the regenerator repeats itself every cycle. Figure 49 tries to illustrate this. In Fig. 49a imagine line 1 to represent the temperature profile of the matrix material just at the end of a blow down of cold gas. Initially the incoming hot gas will give up enthalpy to the cold matrix with a temperature difference established by the heat-transfer coefficient in the same way as for countercurrent heat exchangers. So, initially the hot gas will acquire a temperature profile as indicated by line 2. As the flow of hot gas combines, the temperature of the matrix increases, until a final condition exists which may be a temperature profile for the matrix as line 3 and for the gas as line 4. When the cold gas flows into the regenerator, the reverse procedure occurs. The initial conditions might be as shown by lines 1 and 2 in Fig. 49b for the matrix and gas, and the final conditions by lines 3 and 4. Figure 49c shows the final condition of a complete cycle of the regenerator. The energy stored and released in the regenerator per unit volume is ρ_m, the density of the matrix, times c_m, the specific heat of the matrix, times ΔT_m, the temperature swing of the matrix.

Some specific dimensionless parameters are used in the evaluation of a regenerator. The reduced length λ relates to the heat-transfer characteristics:

$$\lambda = (h/c_P G_o) A_V L$$

where h is the heat-transfer coefficient, c_P the specific heat of the gas, G_o the mass rate of flow per unit area of open frontal area in the direction of flow, A_V the heat-transfer surface area per unit volume, and L the length of the regenerator.

The reduced time τ relates to the length of time gas is being passed through the regenerator:

$$\tau = ht/c_m \rho_m r_e$$

where t is the actual time of gas flow in one direction, c_m the heat capacity of the matrix, ρ_m the density of the matrix material, and r_e the equivalent radius or volume of packing per unit area of matrix.

The ratio of these two quantities is called the utilization factor $\mathcal{U}$ which is the ratio of the heat capacity of the gas per blow to the matrix heat capacity:

$$\mathcal{U} = \tau/\lambda = m_g c_P / m_m c_m$$

where m_g is the total mass of gas per blow, c_P is the specific heat of the gas, m_m is the total mass of the matrix material, and c_m its specific heat.

The heat-transfer coefficient can be evaluated for a number of regenerators from: $N_u = 0.63 R_e^{0.55}$ for wire-screen-type matrices and $N_u = 0.21 R_e^{0.69}$ for sphere-packed matrices.

In the calculation of Reynolds number for regenerator matrices, the following parameters are of interest:

- f void fraction or fraction open volume in a matrix;
- A_V surface area of matrix per unit volume;
- r_b volume of packing/transfer surface area $= (1 - f)/A_V$;
- p porosity, or fraction open in the frontal plane;
- r_h hydraulic radius $= p/A = r_b p/(1 - f)$;
- D_e characteristic dimension for pressure drop and heat exchange;
- D_e $= 4r_h$;
- G_o mass flow rate per unit of open frontal area; and

$$R_e = \frac{D_e G_o}{\mu} = 4r_b \frac{p}{(1 - f)} \frac{G_o}{\mu}$$

The efficiency of a regenerator with a short period or a large heat capacity approaches the efficiency of a countercurrent recuperator with the same heat-transfer area. This efficiency is $\epsilon = \lambda/(\lambda + 2)$.

As the regenerators deviate from this ideal, the efficiency decreases and a common expression is used:

$$\epsilon = [\lambda/(\lambda + 2)](1 - \tfrac{1}{9}U^2)$$

Figure 50 plots the inefficiency $(1 - \epsilon)$ for symmetrical period flow regenerators as a function of reduced length and reduced time, for the range of values of interest in cryogenics.

The foregoing approach and formulas for regenerator design are at best only a guide to regenerator design. It was assumed in the presentation that all the gas which enters one end of the heat exchanger leaves the other end. This is very often not the case in real refrigerators using regenerators. Also, heat capacities were assumed to be constant, and while that is reasonable for gases, it is not so for matrix materials when low temperatures are

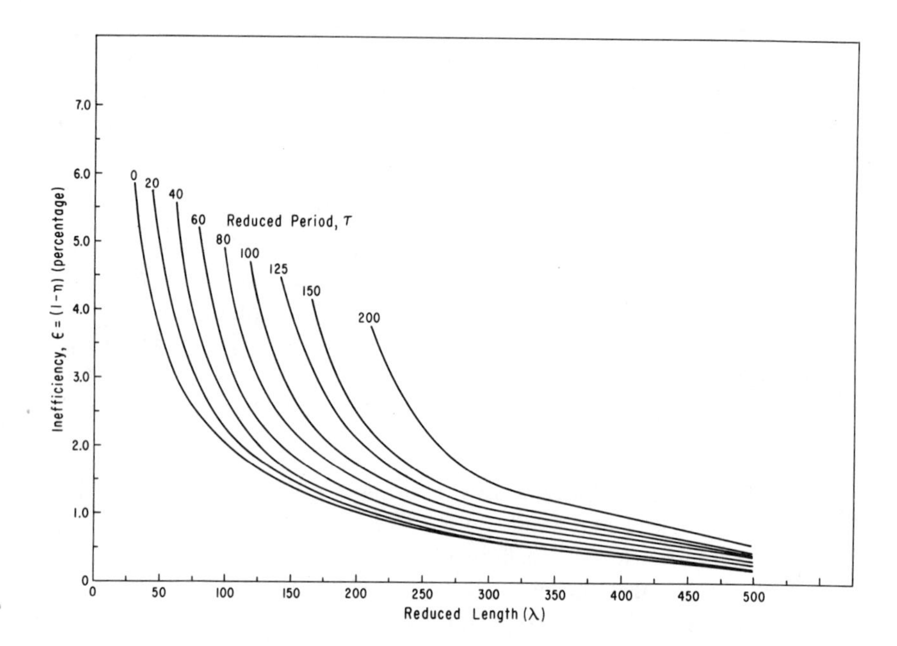

FIG. 50. Plot of the inefficiency for a balanced flow regenerator as a function of reduced length and reduced time. [After Kays and London, 1964.]

experienced and the matrix temperature swings considerably. However, the approach outlined will enable an appreciation of and an ability to take a first cut at regenerator design.

VI. Expanders

We have seen from the foregoing that when a gas expands and performs work, it cools. The expansion may be of several kinds; e.g., simply pushing on a piston; pushing itself, i.e., other molecules of its mass, through an opening to a lower pressure; converting its internal energy to kinetic energy through a properly shaped orifice, or merely by overcoming, by expansion, the attractive forces acting between molecules.

At temperatures where the gas acts much like an ideal gas, obeying Boyle's law, the extent of cooling can be readily calculated. However, at lower temperatures where the gas behavior deviates from the ideal gas laws it is necessary to use empirical relationships. Most empirical relationships lend themselves to computer calculation, but tabular or graphical presentations of the properties of gases are adequate for most calculations of thermodynamic cycles. Fortunately for those interested in refrigeration below 20°K, the tabular and graphical forms of data for normal helium exist in a form suitable for approximate cycle analysis or design. Many such graphics are included in this chapter.

A. Joule–Thomson Expansion

With few exceptions, helium liquefaction or refrigeration at liquid-helium temperatures is effected by the J–T expansion of helium. This process depends on the nonideality of helium at low temperatures. It cannot be determined by simple formulas, but can be by reference to available tables and graphs.

If we refer back to Fig. 33, the refrigeration or liquefaction that can be made available by J–T expansion is dependent on the temperature and pressure of the supply stream entering the heat exchanger at point 7. Assuming the supply stream to be well below the inversion temperature and assuming the heat exchanger to be reasonably efficient, the specific enthalpy of the return stream at point 3 has to be greater than that of the supply at point 7. Therefore, if the mass flow rate of supply and return streams are equal, then the increased enthalpy has to be made up by a heat input; that is refrigeration. If the total enthalpies of the supply and return streams are equal, an imbalance of mass flow must occur, which is due to the fraction of the supply stream that is liquefied.

Figure 51 shows a plot of the ratio of mass rate of liquefaction $\dot{m}_x$ at 1 atm to the mass rate of circulated helium $\dot{m}_1$, versus supply pressure P_1 and supply temperature T_1 and using an ideal heat exchanger.

If the heat exchanger were 95% efficient, the values $\dot{m}_x/\dot{m}_1$ would be approximately 0.05 lower over most of the range of interest, and if the heat exchanger were only 90% efficient, the value of the ratio would be approximately 0.10 lower. This puts considerable emphasis on heat-exchanger efficiency when the ratio of $\dot{m}_x/\dot{m}_1$ is low.

Figure 52 shows a plot of the difference of specific enthalpy between the supply and the 1-atm return stream at the warm end of the heat exchanger versus supply pressure and temperature. The solid lines are for a heat exchanger of 100% efficiency and the dotted lines for a heat exchanger of 90% efficiency. The difference in enthalpy is made up by a heat input, Q load, at a constant temperature of 4.2°K. In the case of liquefaction, the pinch in the heat exchanger is almost always at the warm end. However, this is not always the case for balanced flow refrigeration circuits. When using a single J–T expansion, the refrigeration is limited by a pinch in the

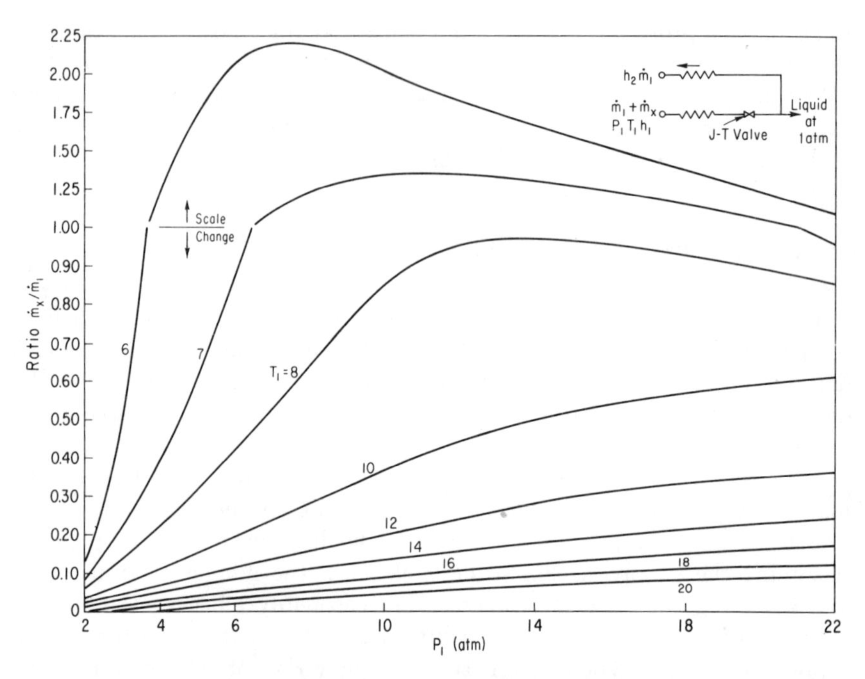

FIG. 51. Ratio of the liquid fraction of the gaseous fraction of helium on J–T expansion to 1 atm as a function of inlet temperature and pressure to the J–T heat exchanger.

heat exchanger at the warm end for those cases represented by a position under the sloping dotted line in Fig. 52. On the dotted line the heat exchanger could be pinched at both ends. Above the dotted line the heat exchanger would be pinched at the cold end and the refrigeration available would be limited to the value of the dotted line corresponding to the supply pressure. This limitation can be overcome by using additional J–T expansion, but even then there is a limit when the exit conditions from the heat exchanger are 1 atm and 4.2°K. For example, expansion from 11 atm will yield increasing refrigeration as the supply temperature is reduced to 10°K. Further reduction of supply temperature will still yield only the same refrigeration, 16 J g^{-1}, from a single expansion and an ideal heat exchanger. Increased refrigeration, up to 20.9 J g^{-1}, could be obtained by multiple J–T expansion as the supply temperature is dropped to 8°K. However, further reduction to 7°K would not increase the refrigeration available because of the limiting pinch at the cold end of the heat exchanger. In normal practice, it is seldom necessary to use more than one J–T expansion for 4.2°K refrigeration.

For the case of refrigeration, the minimum refrigeration that must be

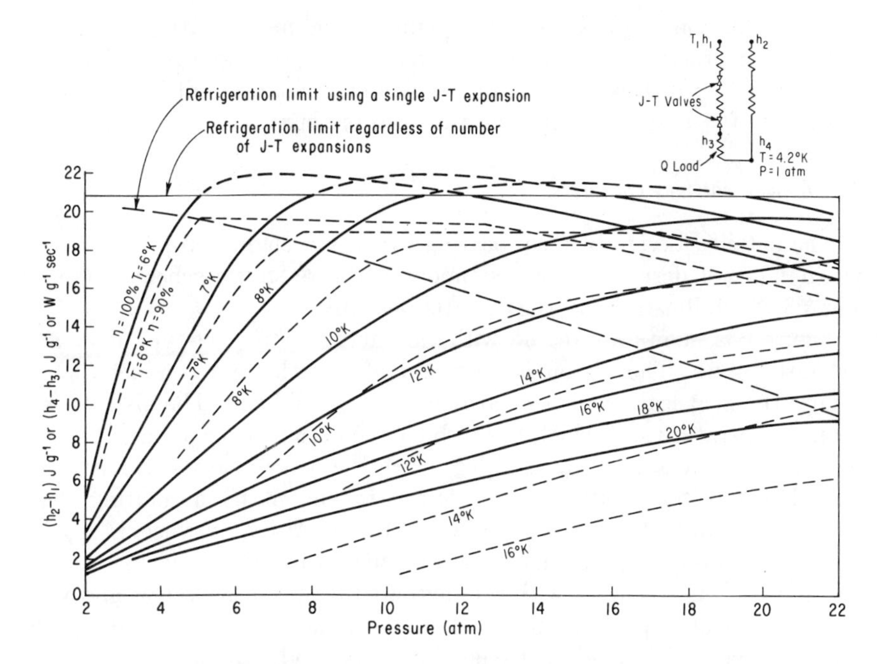

FIG. 52. Refrigeration at 4.2°K available from J–T expansion of helium gas to 1 atm as a function of inlet temperature and pressure to the J–T heat exchanger.

supplied to the helium stream to cool it below the inversion temperature is only that required at 4.2°K. This might best be done using a cycle shown in Fig. 33 if very efficient heat exchangers are used. More practical heat exchangers and less than ideal expansion engines would suggest that the cycle shown in Fig. 34 is better. To liquefy 1 g of helium requires a minimum refrigeration of 1570 J. This is best done with two or more stages of cooling, for example, precooling with liquid nitrogen, and two further levels of cooling by using expansion engines.

B. EXPANSION ENGINES

For systems using countercurrent heat exchangers, i.e., recuperators, there are, in general, two kinds of expansion engine: reciprocating and rotating engines, i.e., turbines. Both are work extracting, and for the adiabatic case, the work extracted equals the refrigeration produced.

The ideal isentropic work extracted per unit mass flow is

$$-W/\dot{m} = h_2 - h_1 = c_P(T_2 - T_1) = c_P T_1[(T_2/T_1) - 1]$$
$$= [k/(k-1)]RT_1[(P_2/P_1)^{(k-1)/k} - 1]$$

This equation applies to both reciprocating and rotating expansion engines. In practice the change of enthalpy is often taken from graphics, such as appear in this chapter, or from tables that take into account the variation of c_P and k with temperature and pressure.

1. *Reciprocating expanders*

The detailed design of reciprocating expansion engines are as varied as the number of designers. Two common types are shown schematically in Fig. 53 which illustrate all the essential features.

Figure 53a shows an engine with cam-actuated poppet-type inlet and exhaust valves. The piston is long and of low thermal conductivity to reduce the heat leak from the warm end to the cold end. The piston has a room-temperature seal which has to be a good seal. Leakage past this seal would represent a large thermal loss since the leakage gas first has to be cooled to low temperature. The work is taken out by a rotating shaft connected to any of a number of kinds of brake. Also a fly wheel is usually connected to the rotating shaft to ensure the return of the piston.

Figure 53b is an engine with cam-actuated poppet-type valves and with the pressure on top of the piston so that the actuating rod is always in tension. The seal is made between the piston and cylinder by designing

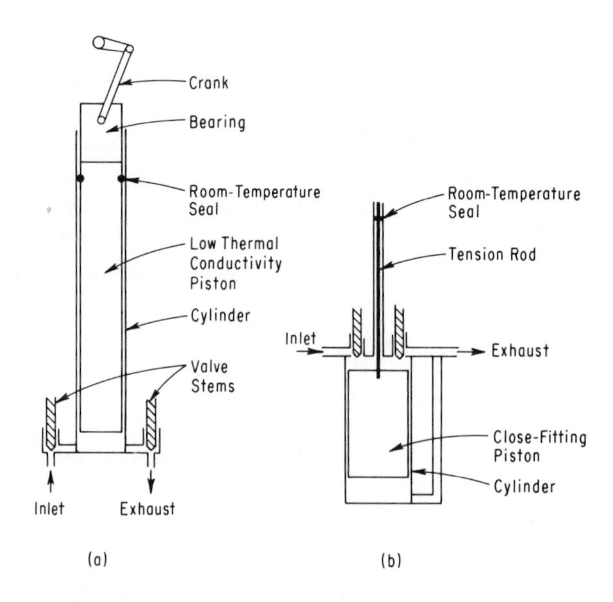

FIG. 53. Schematic of reciprocating expansion engines. (a) with room-temperature seal and (b) with cold seal.

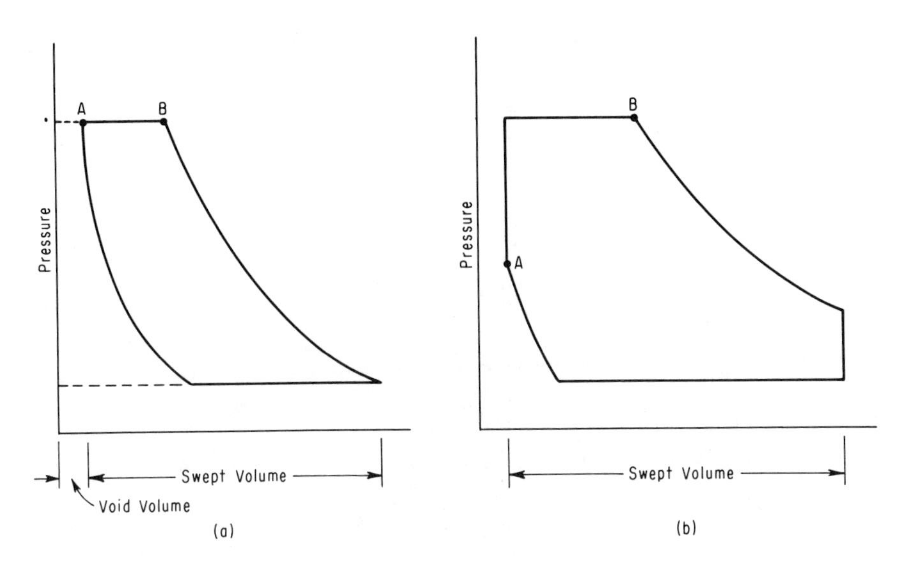

FIG. 54. Pressure–volume or indicator diagrams of piston and cylinder reciprocating expansion engines. (a) Ideal cycle; (b) under expansion and recompression.

very small clearances. The use of a slender tension rod allows the piston to be self-centering in the cylinder. The underside of the piston is connected to the low-pressure discharge. Leakage through the clearance between the piston and cylinder does not represent a thermal loss as leakage past the seal in the other kind of engine, but does represent an isenthalpic expansion bypassing the engine.

The indicator, or P–V, diagrams of these types of engines are shown in Fig. 54. Figure 54a shows the ideal cycle—dotted lines for a zero clearance volume engine and heavy lines for a real engine with clearance volume. Both can be ideally isentropic. Figure 54b shows the more practical situation with less than complete expansion and less than complete recompression. The adiabatic efficiency is based on the work extracted from the mass of gas charging the cylinder, that is, the mass of gas at condition B less the mass of gas at condition A. Figure 54a shows 100% efficiency.

Other factors affecting the efficiency are pressure drop at the inlet and exhaust valves, and heat leak into the gas during expansion due to conduction and radiation heat leaks from warm parts of the system. Figure 55 illustrates these losses on a T–S diagram (Barron, 1966).

Reciprocating expanders are most typically used as precoolers for a J–T stream to provide liquefaction of helium in the size range from 1 liter/hr up to a few hundred liters per hour, or a few watts of refrigeration at 4.2°K up to a kilowatt or more at 4.2°K. Their efficiencies range from about 50%

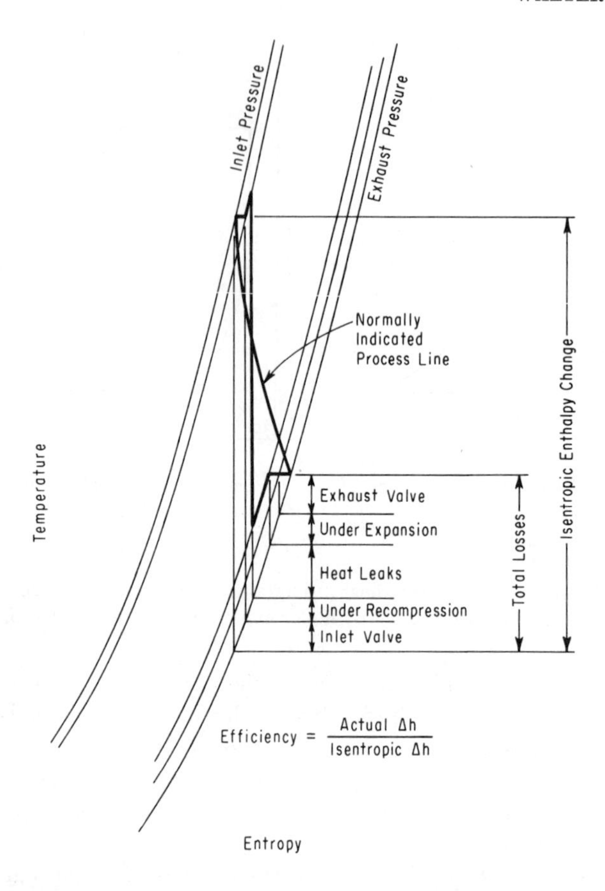

Fig. 55. Temperature–entropy diagram illustrating typical losses associated with reciprocating expansion engines.

at the lowest capacities to as high as 95% at the upper range, with more typical values of 80%. They typically operate at low speeds, 100–600 strokes/min, depending on the design. They have the advantage that they can be operated at good efficiency over a wide range of pressure and speed conditions.

2. *Turbine expanders*

The concept of using turbo-expanders for gas liquefaction and refrigeration is nearly as old as the use of reciprocating expansion engines. Lord Rayleigh suggested its use in 1898. However, successful application did not come until the 1930s with the use of a turbo-expander at the Linde works in Germany.

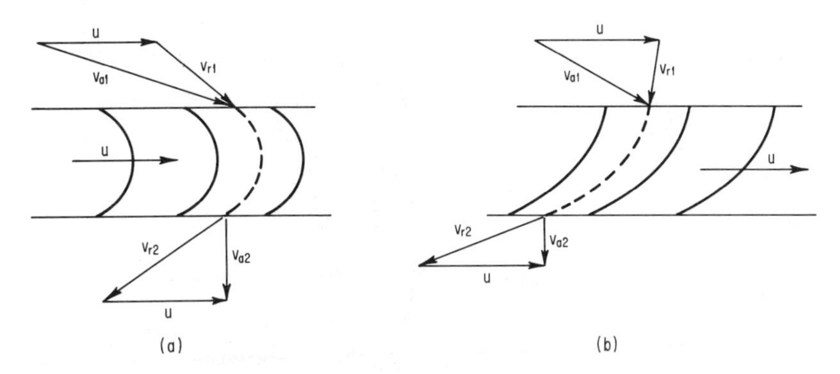

FIG. 56. Velocity diagram for (a) impulse and (b) reaction turbine expansion engines with wheel velocity.

A general expression for work done by a unit weight of gas passing through the rotor of a turbine is as follows:

$$W = (1/2g)[(v_{a1}^2 - v_{a2}^2) + (v_{r2}^2 - v_{r1}^2) + \omega^2(r_1^2 - r_2^2)]$$

where v_{a1} and v_{a2} are the absolute velocities entering and leaving the rotor; v_{r1} and v_{r2} are the velocities of the fluid relative to the rotor velocity at the entrance and exit; ω is the angular velocity of the rotor; g is the gravitational constant; and r_1 and r_2 are the outer and inner radii of the rotor.

The first two terms of the right-hand side of the above equation apply to the work done on the blades of an axial flow turbine. The third term represents the work done by a unit weight of gas moving inward against the centrifugal force from r_1 to r_2 in an inward radial flow turbine.

Turbines may be classified by fluid action as impulse or reaction type, and by flow pattern by axial or radially inward. In a pure impulse turbine all gas expansion takes place in the stationary nozzles to give maximum absolute velocity to the fluid for the maximum enthalpy drop. There is no pressure drop across the blades in an ideal impulse turbine. In a pure reaction turbine the enthalpy drop is taken half in the stationary nozzles and half in the moving blades. There will be a pressure drop across the blades and an acceleration of the gas in the moving blades due to their function as secondary nozzles. Figure 56 shows the velocity diagram for an impulse and a reaction turbine.

In actual practice neither pure impulse nor pure reaction is achieved, all are mixed impulse and reaction. Larger turbines are frequently axial flow, but for cryogenic work the flow pattern is more often mixed flow, being part radial inward and part axial.

The losses of a turbo-expander are similar to those of a reciprocating

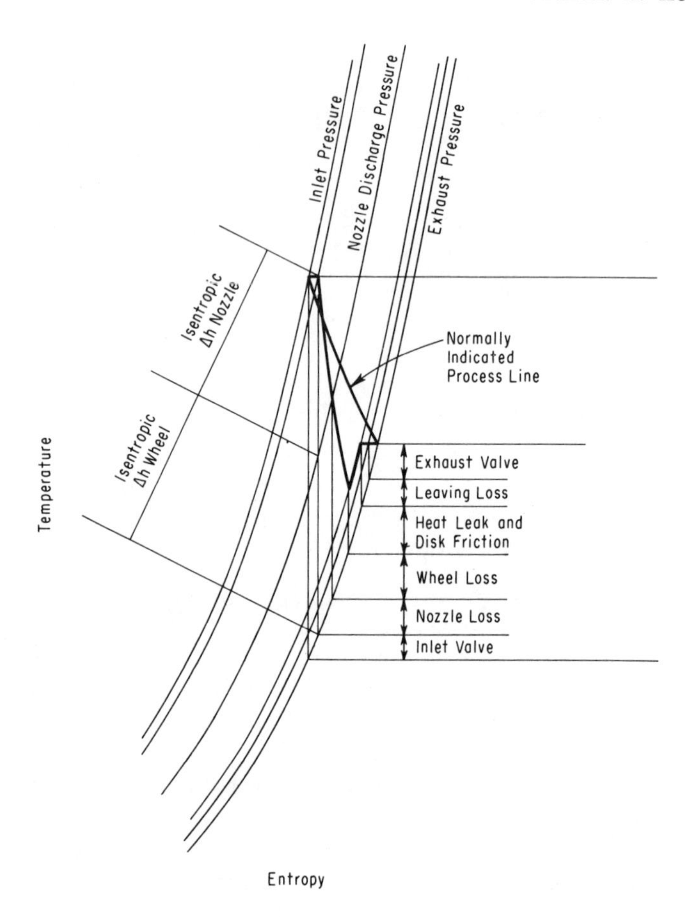

FIG. 57. Temperature–entropy diagram illustrating typical losses associated with turbine-type expansion engines.

expander. There are the inlet and exhaust valve or port pressure drops; the heat leak loss, and the gas leakage loss through the bearings. Additionally there are losses associated with disk friction, which is frictional energy dissipation between the rotor and the gas in the space between the rotor and the housing; impeller losses when the gas velocity and the rotor velocity are not correct for the nozzle and the blade angles; and finally the leaving loss due to the kinetic energy of the outlet stream. Figure 57 illustrates these losses on a T–S diagram.

Large capacity turbo-expanders with refrigeration capacities from a kilowatt and upwards have been in use for many years. They use oil lubricated bearings; have rotary speeds up to 100,000 rpm and efficiencies in the 70 to 80% region. Smaller capacity turbo-expanders with refrigeration capacities in the 100- to 1000-W range have recently come into commercial application. These units use gas lubricated bearings with rotary speeds up to 250,000 rpm. Typically their efficiencies are in the 50 to 70% range. Smaller capacity turbo-expanders are still in the development stage.

References

Barron, R. (1966). "Cryogenic Systems." McGraw-Hill, New York.
Chellis, F. F., and Hogan, W. H. (1963). *In* Proc. 1963 Cryog. Eng. Conf.
Collins, S. C. (1947). *Rev. Sci. Instr.* **18,** 157.
Collins, S. C., and Cannaday, R. L. (1958). "Expansion Machines for Low Temperature Processes," p. 43. Oxford Univ. Press, London and New York.
Collins, S. C., Stuart, R. W., and Streeter, M. H. (1967). *Rev. Sci. Instr.* **38,** 1654.
Dalton, J. (1802). Mem. Manchester Lit. Phil. Soc. v.1802, p. 595.
Daunt, J. G. (1956). *In* "Encyclopedia of Physics" (S. Flügge, ed.), Vol. 14, p. 1. Springer-Verlag, Berlin and New York.
Gay-Lussac, J. L. (1802). *Ann. Chim. Phys.* **xliii,** 137.
Hogan, W. H. (1968). "Proc. 2*nd* Int. Conf. Cryog. Eng. I Liffe Sci. & Tech. Pub.
Hogan, W. H., and Stuart, K. W. (1963). ASME Paper No. 63-WA-292. ASME, New York.
Joule, J. P. (1851). Mem. Manchester Lit. Phil. Soc. **2Jix,** 107.
Johnson, V. J., (1960). Wadd Tech. Rep. No. 60-56, Phase I, pt. I.
Johnson, R. *et al.* (1970). Cryog. Eng. Conf., Boulder, Colo., 1970.
Kays, W. M., and London, A. L. (1964). "Compact Heat Exchangers." McGraw-Hill, New York.
Little, Arthur D., Inc. (1967). "Low Temperature Refrigeration for Microwave Systems," p. 253. Boston Technical Publishers, Cambridge, Massachusetts.
Loeb, Leonard B. (1938). "Kinetic Theory of Gases." McGraw-Hill, New York.
McMahon, H. O., and Gifford, W. E. (1963). *Advan. Cryog. Eng.* **8,** 190.
Muhlenhaupt, B. C., and Strobridge, T. R. (1968). Nat. Bur. Stand. Tech. Note No. 366. Nat. Bur. Stand., Washington, D. C.
Scott, Russell B. (1959). "Cryogenic Engineering." Van Nostrand, Princeton, New Jersey.
Stewart, R. B. and Johnson, V. J., (1961). Wadd Tech. Rep. No. 60-56, Phase II.
Van Wylen, G. J., and Sonntag, R. E. (1965). "Fundamentals of Classical Thermodynamics." Wiley, New York.

Index

D

T

U

V